Molecular Chaperones and Cell Signalling

This book reviews current understanding of the biological roles of extracellular molecular chaperones. It provides an overview of the structure and function of molecular chaperones, their role in the cellular response to stress, and their disposition within the cell. It also questions the basic paradigm of molecular chaperone biology – that these proteins are, first and foremost, protein-folding molecules. The current paradigms of protein secretion are reviewed, and the evolving concept of proteins (such as molecular chaperones) as multi-functional molecules, or "moonlighting proteins," is discussed. The role of exogenous molecular chaperones as cell regulators is examined, and the physiological and pathophysiological roles that molecular chaperones play are described. In the final section, the potential therapeutic use of molecular chaperones is described, and in the final chapter, the crystal ball is brought out and the question – what does the future hold for the extracellular biology of molecular chaperones – is asked.

Brian Henderson is Professor of Cell Biology at the Eastman Dental Institute, University College London, and Head of the Cellular Microbiology Research Group. His major research interests are concerned with bacterial interactions with the host and how such interactions control inflammation and associated tissue destruction. It is through these studies that he identified that molecular chaperones are bacterial virulence factors and started his interest in the direct immunomodulatory actions of cell stress proteins.

A. Graham Pockley is Professor of Immunobiology at the University of Sheffield Medical School and is Head of the Immunobiology Research Unit. Professor Pockley has long-standing interests in the immunobiology of transplant rejection, and his unit is currently focussed on research relating to the biology and immunotherapeutic potential of heat shock proteins, particularly their involvement in the rejection of organ transplants and the development and progression of cardiovascular disease.

Molecular Chaperones and Cell Signalling

Edited by

Brian Henderson
University College London

A. Graham Pockley
University of Sheffield

CAMBRIDGE
UNIVERSITY PRESS

CAMBRIDGE UNIVERSITY PRESS
Cambridge, New York, Melbourne, Madrid, Cape Town,
Singapore, São Paulo, Delhi, Tokyo, Mexico City

Cambridge University Press
32 Avenue of the Americas, New York, NY 10013-2473, USA

www.cambridge.org
Information on this title: www.cambridge.org/9780521177474

First published 2005
First paperback edition 2011

A catalog record for this publication is available from the British Library

Library of Congress Cataloging in Publication data
Molecular chaperones and cell signalling / edited by Brian Henderson, A. Graham Pockley.
  p. ; cm.
Includes bibliographical references and index.
ISBN 0-521-83654-9 (hardback)
1. Molecular chaperones.
[DNLM: 1. Molecular chaperones–physiology. QU 55 M716245 2005]
1. Henderson, Brian, PhD.   11. Pockley, A. Graham (Alan Graham), PhD. 1960–   111. Title.
QP552.M64M64   2005
572'.645 – dc22   2005000664

ISBN 978-0-521-83654-8 Hardback
ISBN 978-0-521-17747-4 Paperback

Contents

Contributors

A. Asea
Center for Molecular Stress Response
Boston Medical Center and Boston
  University School of Medicine
Boston
Massachusetts 02118
U.S.A.

P. P. Banerjee
Center for Immunotherapy of Cancer
  and Infectious Diseases
University of Connecticut School of
  Medicine
263 Farmington Avenue
Farmington
Connecticut 06030
U.S.A.

L. A. Bergmeier
Peter Gorer Department of
  Immunobiology
Guy's, King's and St. Thomas Hospital
  Medical School
London SE1 9RT
United Kingdom

T. Bowes
Department of Biochemistry
McMaster University
Hamilton
Ontario L8N 3Z5
Canada

S. K. Calderwood
Department of Radiation Oncology
Beth Israel Deaconess Medical Center
  and Harvard Medical School
21-27 Burlington Avenue
Boston
Massachusetts 02215
U.S.A.

G. Chimini
Centre d'Immunology de Marseille-
  Luminy
INSERM/CNRS et Université de la
  Mediterranée
Parc Scientifique de Luminy
Marseille 13288
France

A. R. M. Coates
Department of Medical Microbiology
St. George's Hospital Medical School
Tooting
London SW17 0RE
United Kingdom

I. R. Cohen
Department of Immunology
The Weizmann Institute of Science
Rehovot 76100
Israel

V. M. Corrigall
Academic Department of
  Rheumatology
GKT School of Medicine
King's College London
London SE1 9RT
United Kingdom

A. K. De
Immunobiology & Stress Response
  Laboratories
University of Rochester Medical Center
Department of Surgery
601 Elmwood Avenue
Rochester
New York 14642
U.S.A.

R. J. Ellis
Department of Biological Sciences
University of Warwick
Coventry CV4 7AL
United Kingdom

J. Frostegård
Unit of Rheumatology
Department of Medicine and Center
  for Molecular Medicine
Karolinska Hospital
171 76 Stockholm
Sweden

S. Furman
Department of Clinical Biochemistry
  and Interdepartmental Core Facility
Sackler School of Medicine
Tel Aviv University
Tel Aviv 69978
Israel

V. L. Gabai
Department of Biochemistry
Boston University School of Medicine
Silvio Conte Building
715 Albany Street

Boston
Massachussets 02118
U.S.A.

I. Gozes
Department of Clinical Biochemistry
  and Interdepartmental Core Facility
Sackler School of Medicine
Tel Aviv University
Tel Aviv 69978
Israel

R. S. Gupta
Department of Biochemistry
McMaster University
Hamilton
Ontario L8N 3Z5
Canada

B. Henderson
Division of Microbial Diseases
Eastman Dental Institute
University College London
256 Gray's Inn Road
London WC1X 8LD
United Kingdom

L. E. Hightower
Department of Molecular and Cell
  Biology
The University of Connecticut
91 North Eagle Road
Storrs
Connecticut 06269
U.S.A.

C. J. Jeffery
University of Illinois at Chicago
Department of Biological Sciences
MBRB 4252 M/C 567
900 South Ashland Avenue
Chicago
Illinois 60607

D. S. Latchman
The Master

Birkbeck College
University of London
Malet Street
London WC1E 7HX
United Kingdom

K. Laudanski
Immunobiology & Stress Response
  Laboratories
University of Rochester Medical Center
Department of Surgery
601 Elmwood Avenue
Rochester
New York 14642
U.S.A.

T. Lehner
Peter Gorer Department of
  Immunobiology
Guy's, King's and St. Thomas
  Hospital Medical School
London SE1 9RT
United Kingdom

Z. Li
Center for Immunotherapy of
  Cancer and Infectious Diseases
University of Connecticut School
  of Medicine
263 Farmington Avenue
Farmington
Connecticut 06030
U.S.A.

C. L. Miller-Graziano
Immunobiology & Stress Response
  Laboratories
University of Rochester Medical
  Center
Department of Surgery
601 Elmwood Avenue
Rochester
New York 14642
U.S.A.

L. Mizzen
Stressgen Biotechnologies Corporation
#350-4243 Glanford Avenue
Victoria
British Columbia V8Z 4B9
Canada

J. Neefe
Stressgen Biotechnologies Inc.
#201-409 2nd Avenue
Collegeville
Pennsylvania 19426
U.S.A.

G. S. Panayi
Academic Department of
  Rheumatology
Guy's, King's and St. Thomas' School
  of Medicine
King's College London
London SE1 9RT
United Kingdom

A. G. Pockley
Immunobiology Research Unit
Division of Clinical Sciences (North)
(University of Sheffield)
Northern General Hospital
Herries Road
Sheffield S5 7AU
United Kingdom

F. J. Quintana
Department of Immunology
The Weizmann Institute of Science
Rehovot 76100
Israel

A. Rubartelli
Department of Oncogenesis
Cell Transport Unit
National Cancer Research Institute
Genova
Italy

S. Sadacharan
Department of Biochemistry
McMaster University
Hamilton
Ontario L8N 3Z5
Canada

A. Shamaei-Tousi
Division of Microbial Diseases
Eastman Dental Institute
University College London
256 Gray's Inn Road
London WC1X 8LD
United Kingdom

M. Y. Sherman
Department of Biochemistry
Boston University School of Medicine
Silvio Conte Building
715 Albany Street
Boston
Massachussets 02118
U.S.A.

B. Singh
Department of Biochemistry
McMaster University
Hamilton
Ontario L8N 3Z5
Canada

I. Spivak-Pohis
Department of Clinical Biochemistry
   and Interdepartmental Core Facility
Sackler School of Medicine
Tel Aviv University
Tel Aviv 69978
Israel

A. Stephanou
Medical Molecular Biology Unit
Institute of Child Health
University College London
30 Guilford Street
London WC1N 1EH
United Kingdom

P. Tormay
Department of Medical Microbiology
St. George's Hospital Medical School
Tooting
London SW17 0RE
United Kingdom

R. M. Vabulas
Max-Planck-Institut für Biochemie
Am Klopferspitz 18
82152 Martinsried
Germany

I. Vulih
Department of Clinical Biochemistry
   and Interdepartmental Core Facility
Sackler School of Medicine
Tel Aviv University
Tel Aviv 69978
Israel

H. Wagner
Institut für Med. Mikrobiologie,
   Immunologie u. Hygiene
Technische Universität München
Troger Str. 9
81675 München
Germany

Y. Wang
Peter Gorer Department of
   Immunobiology
Guy's, King's and St. Thomas'
   Hospital Medical School
London SE1 9RT
United Kingdom

T. Whittall
Peter Gorer Department of
   Immunobiology
Guy's, King's and St. Thomas'
   Hospital Medical School
London SE1 9RT
United Kingdom

Preface

The last four decades of the 20th century saw the discovery of the heat shock or cell stress response and the identification of the proteins produced by cells in response to adverse environmental conditions. In 1987, the term 'molecular chaperone' was coined to describe several unrelated protein families which had the ability to assist the correct folding and assembly/disassembly of other proteins. The past 20 years have seen the elucidation of the structural mechanisms of protein chaperoning by several key molecules including chaperonin (Hsp) 60 and Hsp70 and the realisation that not all molecular chaperones are cell stress proteins and vice versa. The genesis of molecular chaperones was contemporaneous with the identification of these highly conserved proteins as paradoxical immunodominant antigens that appeared to be important in microbial infection and autoimmunity. Indeed, the administration of molecular chaperones such as Hsp60 and Hsp70 was found to inhibit experimental autoimmune disease. By the 1990s, it was realised that correct protein folding was the key to cellular homeostasis and the paradigm that developed was that molecular chaperones were intracellular proteins whose function was to assist in protein folding. The paradoxical immunogenicity and immunomodulatory effects of molecular chaperones remained unexplained.

Another strand of the molecular chaperone story began to develop in the late 1980s and early 1990s with reports of the appearance of certain molecular chaperones on the surface of cells. Later, reports began to appear that molecular chaperones when applied exogenously to cells in culture had effects similar to those of pro-inflammatory cytokines. Molecular chaperones, such as early pregnancy factor (chaperonin 10), Hsp27, and BiP can also have anti-inflammatory/immunosuppressive actions. In addition to having direct effects on cells, there is mounting evidence that certain molecular chaperones can bind peptides and present these to antigen-presenting cells to modulate T lymphocyte responses.

This rapidly expanding body of literature is suggesting a new paradigm in which molecular chaperones are moonlighting proteins – that is, proteins with more than one function that are able to interact with many cell types to produce a range of effects. Support for this paradigm comes from studies revealing the presence of a number of molecular chaperones in the body fluids of man and animals.

Molecular Chaperones and Cell Signalling provides the reader with an overview of our current understanding of the biological roles of extracellular molecular chaperones. The book is divided into a number of sections. The first provides an overview of the structure and function of molecular chaperones, their role in the cellular response to stress, and their disposition within the cell. It also questions the basic paradigm of molecular chaperone biology – that these proteins are, first and foremost, protein-folding molecules. A key concern of those working on molecular chaperones as extracellular signals is the mechanism of secretion from cells. Section 2 reviews the current paradigms of protein secretion from cells and discusses the evolving concept of proteins (such as molecular chaperones) as multi-functional molecules for which the term "moonlighting proteins" has been introduced. In Section 3, attention turns to the role of exogenous molecular chaperones as cell regulators. Section 4 describes the physiological and pathophysiological roles that molecular chaperones play, and in Section 5, the potential therapeutic use of molecular chaperones is described. The final chapter of this volume brings out the crystal ball and asks, 'What does the future hold for the extracellular biology of molecular chaperones?'.

This book will be of interest to biologists, biochemists, cell biologists, clinicians, immunologists, pharmacologists, and pathologists at both the graduate and postgraduate levels.

Molecular Chaperones and the Cell Stress Response

1

Chaperone Function: The Orthodox View

R. John Ellis

1.1. Introduction

The term molecular chaperone came into general use after the appearance of an article in *Nature* that suggested it was an appropriate phrase to describe a newly defined intracellular function – the ability of several unrelated protein families to assist the correct folding and assembly/disassembly of other proteins [1]. The identification of the chaperonin family of molecular chaperones in the following year [2] triggered a tidal wave of research in several laboratories aimed at unravelling how the GroEL/GroES chaperones, and later the DnaK/DnaJ chaperones, from *Escherichia coli* facilitate the folding of newly synthesised polypeptide chains and the refolding of denatured proteins. This wave continues to surge, with the result that much detailed information is available about the structure and function of those families of chaperone that assist protein folding [3].

It is now well established that a subset of proteins requires this chaperone function, not because chaperones provide steric information required for correct folding but because chaperones inhibit side reactions that would otherwise cause some of the chains to form non-functional aggregates. The number of different protein families described as chaperones is now more than 25 – some, but not all, of which are also stress proteins – and there is no slackening in the rate of discovery of new ones. The success of this wave of research has changed the paradigm of protein folding from the earlier view that it is a *spontaneous* self-assembly process to the current view that it is an *assisted* self-assembly process [4].

Is another paradigm shift in the offing? Other chapters in this volume discuss the evidence that some molecular chaperones may have extracellular roles as cell-cell signalling molecules in addition to their intracellular roles in protein folding. This view has not found general acceptance, partly because it is novel and partly because of the paucity of high-quality evidence compared with that available in support of the protein folding paradigm. The purpose of this introductory chapter is to summarise the conventional view of chaperone function to provide a context for the cell-cell signalling hypothesis discussed in later chapters.

1.2. Origins

The current association of the term molecular chaperone with protein folding in the cytoplasm overlooks the fact that this term was used first to describe the properties of a nuclear protein in assisting the assembly of nucleosome cores from folded histone proteins and DNA [5]. This acidic protein is abundant in the soluble phase of the nuclei of eggs and oocytes of the amphibian *Xenopus laevis* and is thus called nucleoplasmin. These nuclei are unusual in that they contain large amounts of histone proteins stored in preparation for the rapid assembly of nucleosomes associated with the rapid replication of DNA triggered by fertilisation. Nucleosomes fail to conform to the principle of protein self-assembly established by the pioneering work of Anfinsen for refolding proteins and by Caspar and Klug for virus assembly. This principle is an important corollary of the Central Dogma of molecular biology and states that all the steric information necessary for a protein chain to reach its functional conformation is present in the amino acid sequence of the primary translation product. This principle also applies to the assembly of macromolecular complexes from more than one subunit.

In nucleosomes, histones are bound to DNA by electrostatic interactions; disruption of these requires high salt concentrations, but exposing mixtures of isolated DNA and histones to the salt concentrations found inside the nucleus results in the formation of insoluble aggregates rather than nucleosomes. Nucleoplasmin solves this aggregation problem by transiently binding its acidic groups to positively charged groups on the histones, thus lowering their overall surface charge and allowing the intrinsic self-assembly properties of the histones to predominate over the incorrect interactions favoured by the high density of opposite charges. Control experiments show that nucleoplasmin does not provide steric information essential for histones to bind correctly to DNA, nor is it a component of assembled nucleosomes. It is these two latter features that laid the foundation for our current general concept of the function of chaperones [6].

The term molecular chaperone was later extended to include an abundant chloroplast protein called the rubisco large subunit binding protein, which functions to keep newly synthesised rubisco large subunits from aggregating until they assemble into the rubisco holoenzyme [7]. These subunits are notoriously prone to aggregation, not because of electrostatic interactions but because they expose highly hydrophobic surfaces to the aqueous environment. For a while the term was restricted to the two proteins that assist the assembly of amphibian nucleosomes and chloroplast rubisco. Its modern usage started when the author suggested that the term could be usefully extended to describe the function of a larger range of proteins that were postulated to assist folding and assembly/disassembly reactions in a wide range of cellular processes [1].

1.3. The general concept of chaperone function

The suggestions made in the first comprehensive description of the chaperone function have so far stood the test of time [8]. Molecular chaperones are defined as being a large and diverse group of proteins that share the property of assisting the non-covalent assembly/disassembly of other macromolecular structures but which are not permanent components of these structures when these are performing their normal biological functions. Assembly is used here in a broad sense and includes several universal intracellular processes: the folding of nascent polypeptide chains both during their synthesis and after release from ribosomes, the unfolding and refolding of polypeptides during their transfer across membranes, and the association of polypeptides with one another and with other macromolecules to form oligomeric complexes.

Molecular chaperones are also involved in macromolecular *dis*assembly processes, such as the partial unfolding and dissociation of subunits when some proteins carry out their normal functions, and the re-solubilisation and/or degradation of proteins partially denatured and/or aggregated by mutation or by exposure to environmental stresses, such as high temperatures and oxidative conditions. Some, but not all, chaperones are also stress or heat shock proteins as the requirement for chaperone function increases under stress conditions that cause proteins to unfold and aggregate. Conversely, some, but not all, stress proteins are molecular chaperones.

It is important to note that this definition is functional, not structural, and it contains no constraints on the mechanisms by which different chaperones may act; this is the reason for the use of the imprecise term 'assist'. Thus, molecular chaperones are defined neither by a common mechanism nor by sequence similarity. Only two criteria need be satisfied to designate a macromolecule a molecular chaperone. Firstly, it must in some sense assist the non-covalent assembly/disassembly of other macromolecular structures, the mechanism being irrelevant, and secondly, it must not be a component of these structures when they are performing their normal biological functions. In all cases studied so far, chaperones bind non-covalently to regions of macromolecules that are inaccessible when these structures are correctly assembled and functioning but that are accessible at other times.

The term non-covalent is used in this definition to exclude those proteins that catalyse co- or post-translational covalent modifications. These are often important for protein assembly, but are distinct from the proteins being considered here. Protein disulphide isomerase may appear to be an exception, but it is not. It is both a covalent modification enzyme and a molecular chaperone, but these activities lie in different parts of the molecule [9] and can be functionally separated by mutation. Other examples include peptidyl-prolyl isomerase, which possesses both enzymatic and chaperone activities in different regions of the molecule, and the α-crystallins, which in the lens of the eye combine two essential functions in the same molecule, contributing to the transparency and refractive index required for vision as well as to the chaperone function, which combats the loss of transparency as the protein chains aggregate with increasing age. The proteasome particle has a chaperone-like activity involved in unfolding proteins prior to their proteolysis. Thus, in principle, there is no reason why molecular chaperones should not possess additional functions, and the possibility that many possess cell-cell signalling functions is the central postulate argued in the other chapters of this volume.

The number of distinct chaperone families continues to rise, and examples occur in all types of cell and in most intracellular compartments. The families are defined on the basis that members within each family have high sequence similarity, whereas members in different families do not. Table 1.1 presents an incomplete list of proteins described as chaperones; however, it must be emphasised that in many cases this description rests on *in vitro* data only and needs confirmation by *in vivo* methods. There is evidence that some chaperones cooperate with each other in defined reaction sequences, but this, along with many other aspects of 'chaperonology', is beyond the scope of this chapter.

1.4. Common misconceptions

As with any new field, misconceptions abound. A common error is to use the term 'chaperonin' synonomously with the term 'chaperone', but it should be noted that the chaperonins are just one particular family of chaperone – i.e., the family that contains GroEL, Hsp60, and tailless complex polypeptide-1 (TCP-1) (see Chapters 5 and 6 for more details on chaperonins). The occasional use of the non-sense term 'molecular chaperonin' in some respectable journals suggests that some people use these terms casually without reference to either their meaning or their history. It should be obvious that the word 'molecular' is used to qualify 'chaperone' because in common usage 'chaperone' refers to a

Table 1.1. Proteins described as molecular chaperones

Family	Proposed roles
Non-steric chaperones	
Nucleoplasmins/nucleophosmins	Nucleosome and ribosome assembly/disassembly
Chaperonins	Folding of newly synthesised and denatured polypeptides
Hsp27/28	Prevention of stress-induced aggregation by adsorbing unfolded chains
Hsp40	Protein folding and transport, oligomer disassembly
Hsp47	Pro-collagen folding in the endoplasmic reticulum (ER)
Hsp70	Protein folding and transport, oligomer disassembly
Hsp90	Cell cycle, hormone activation, signal transduction
Hsp100	Dissolution of insoluble protein aggregates
Calnexin/calreticulin	Folding of glycoproteins in ER
SecB protein	Protein transport in bacteria
Lim protein	Folding of bacterial lipase
Syc protein	Secretion of toxic YOP proteins by bacteria
Protein disulphide isomerase	Prevention of misfolding in ER
ExbB proteins (may be structural rather than chaperones)	Folding of TonB protein in bacteria
Ubiquinated ribosomal proteins	Ribosome assembly in yeast
NAC complex	Folding of nascent proteins
Signal recognition particle	Arrest of translation and targeting to ER membrane
Trigger factor	Folding of nascent polypeptides in bacteria
Prefoldin	Cooperation with chaperonins in folding of newly synthesised polypeptides in *Archaea* and the eukaryotic cytosol
Tim9/Tim10 complex	Prevention of aggregation of hydrophobic proteins during import across mitochondrial intermembrane space
23S Ribosomal RNA	Folding of nascent polypeptides
PrsA protein	Secretion of proteins by *Bacillus subtilis*
Clusterin	Extracellular animal chaperone
Phosphatidylethanolamine	Folding of lactose permease
RNA binding proteins	Folding of RNA
P45	Protection against denaturation in halophilic *Archaea*
Steric chaperones	
PapD proteins	Assembly of bacterial pili
Propeptides (Class I)	Folding of some proteases

person. The term 'molecular chaperonin' is therefore as non-sensical as the term 'molecular immunoglobulin'.

Another common misconception is that molecular chaperones are necessarily promiscuous – i.e., that each assists the assembly of many different types of polypeptide chain. This is true for the Hsp70, Hsp40 and GroE chaperonin families but is not true for Hsp90, PapD, Hsp47, Lim, Syc, ExbB, PrtM/PrsA and prosequences, which are specific for their substrates. Similarly, it is not a universal property of chaperones that they hydrolyse ATP; Hsp100, Hsp90, Hsp70 and the chaperonins hydrolyse ATP, whereas trigger factor, Hsp40, prefoldin, calnexin, protein disulfide isomerase and papD do not. It is not even necessary that the chaperone function resides in molecules separate from their substrates. Thus, some pro-sequences are required for the correct folding of the remainder of the molecule but are then removed [10]. Another example of such intramolecular chaperones is the terminal ubiquitin residues of three ribosomal proteins in yeast; these residues promote the assembly of these proteins into the ribosome but are then removed, thus fulfilling the criteria suggested earlier for the chaperone function [11].

The term 'chemical chaperone' has been proposed to describe small molecules such as glycerol, dimethylsulfoxide and trimethylamine N-oxide that act as protein stabilising agents [12]. This terminology is unfortunate because proteins are also chemicals. However, its usage persists.

Finally, experience suggests that the distinction between molecular chaperones and stress proteins cannot be restated too often. The often-made interpretation, that because a protein accumulates after stress it must be a molecular chaperone, is incorrect, as is the belief that all molecular chaperones are stress proteins. For example, many heat shock proteins are ubiquitin-conjugating enzymes, while the cytosolic chaperonin of eukaryotic cells is not a stress protein.

1.5. Why do molecular chaperones exist?

Given that most denatured proteins that have been examined can refold into their functional conformations on removal of the denaturing agent *in vitro*, the question arises as to why molecular chaperones exist at all. Current evidence suggests that, with two possible exceptions [10], chaperones do not provide steric information for proteins to assemble correctly; rather they either prevent or reverse aggregation processes that would otherwise reduce the yield of functional molecules. Aggregation results because some proteins fold and unfold via intermediate states that expose some interactive surfaces (either charged or hydrophobic) to the environment. In aqueous environments hydrophobic surfaces stick together, while charged surfaces bind to ones bearing the opposite charge, a problem acute in the nucleus where negatively charged nucleic acids are bound to positively charged proteins. Thus, the existence of molecular chaperones does not cast doubt on the validity of the self-assembly principle. Rather, chaperones are required because, to operate efficiently under intracellular conditions, self-assembly needs assistance to avoid unproductive side reactions. This is why the term molecular chaperone is not an example of academic whimsy, but a precise description, because the role of the human chaperone is to improve the efficiency of 'assembly' processes between people without providing the steric information for these processes.

Protein aggregation has long been observed to occur during the *in vitro* refolding of many pure denatured proteins in dilute buffer solutions, but only recently has it been appreciated that the high degree of macromolecular crowding that characterises the intracellular environment makes the aggregation problem much more severe *in vivo*. Although the total concentrations of macromolecules inside cells are in the range 200–400 mg/ml, the properties of the isolated macromolecules are commonly studied *in vitro* at much lower concentrations in uncrowded buffers. The large thermodynamic effects of the high total concentrations of macromolecules inside cells are not generally appreciated and include increasing the association constants of protein aggregation reactions by one to two orders of magnitude [13]. Aggregation is a specific process involving identical or very similar chains and is driven by the interaction of both hydrophobic side chains and main-chain atoms in segments of unstructured backbone that are transiently exposed on the surface of partly folded chains; it is thus a high-order process that increases in rate as the concentration of similar chains or the temperature is raised. Refolding experiments suggest that large multi-domain proteins suffer from aggregation to a greater degree than small single-domain proteins because they fold more slowly via partly folded intermediate states. Thus proteins differ greatly in their propensity to aggregate, and it is likely that chaperones have evolved to combat this tendency of a particular subset of proteins.

These considerations can be reduced to a simple unifying principle. *All cells need a chaperone function to both prevent and reverse incorrect interactions that may occur when potentially interactive surfaces are exposed to the intracellular environment. Such surfaces occur on nascent and newly synthesised unfolded polypeptide chains, on mature proteins unfolded by stress or degradative mechanisms, and on folded proteins in near-native conformation.*

Thus, it is as valid to talk about the chaperone function as it is to talk about the transport function or the defence function of other proteins. We can view the chaperone function as one of the universal mechanisms that enable the crowded state of the cellular interior to be compatible with life.

The best understood chaperones are those involved in the folding of newly synthesised polypeptide chains in *E. coli*. The next section summarises what is known about their mechanisms of action.

1.6. Chaperones involved in *de novo* protein folding

The folding of newly synthesised proteins inside cells differs from the refolding of denatured proteins *in vitro* in two respects [14]. Firstly, protein chains fold inside cells in highly crowded macromolecular environments that favour aggregation. Secondly, protein chains are made vectorially inside cells at a rate slower than the rate of folding. It takes about 20 seconds for a cell of *E. coli* to synthesise a chain of 400 residues at 37 °C; however, *in vitro*, many denatured proteins will refold completely well within this time. Thus there is the possibility that the elongating nascent chain will either misfold because it is incomplete or aggregate with identical elongating chains on the same polysome. It is important to realise that misfolding is conceptually distinct from aggregation. Misfolding can be defined as the chain reaching a partly folded conformation from which it is unable to reach the final functional conformation on a biologically relevant time scale. Misfolded chains may or may not bind to one another to form non-functional aggregates that may be as small as a dimer or large enough to be insoluble. Thus all aggregates are, by definition, misfolded, but to what extent misfolded, but unaggregated, chains occur in cells is unclear.

Molecular chaperones assist the folding of both nascent chains bound to ribosomes and newly synthesised chains released from ribosomes – i.e., in both co-translational and post-translational modes. The chaperones working in these co-translational and post-translational modes are distinct and can be usefully termed small and large chaperones, respectively, because this is a case where size is important for function [15]. Small chaperones are less than 200 kDa in size and include trigger factor, nascent chain-associated complex, prefoldin, the Hsp70 and Hsp40 families and their associated co-chaperones. Co-chaperones are defined as proteins that bind to chaperones to modulate their activity; they may or may not also be chaperones in their own right. Large chaperones are more than 800 kDa in size and include the thermosome in *Archaea*, GroE proteins in *Eubacteria* and the eukaryotic organelles evolutionarily derived from them, and the TCP-1 or TRiC complexes and associated co-chaperones in the cytosol of *Eukarya*. The large chaperones are evolutionarily related and are collectively referred to as the chaperonins. There are no large chaperones in the endoplasmic reticulum lumen of eukaryotic cells, but small chaperones such as BiP (an Hsp70 homologue), calnexin, calreticulin, and protein disulphide isomerase that assist the folding of chains transported into the lumen after synthesis in the cytosol are present. Table 1.2 lists some of the chaperones that assist protein folding.

Table 1.2. Chaperones that assist protein folding

Family	Other names		Functions
	Eukaryotes	Prokaryotes	
Hsp100	Hsp104, 78	ClpA/B/X	Disassembly of oligomers and aggregates
Hsp90	Hsp82, Hsp83, Grp94	HtpG	Regulate assembly of steroid receptors and signal transduction proteins
Hsp70	Hsc70, Ssal-4, Ssb1-2, BiP, Grp75	DnaK, Hsc66, Absent from many *Archaea*	Prevent aggregation of unfolded protein chains
Chaperonins	Hsp60, TRiC, CCT, TCP-1, rubisco subunit binding protein	GroEL, GroES	Sequester partly folded chains inside central cage to allow completion of folding in absence of other folding chains
Hsp40	Ydj1, Sis1, Sec63p, auxilin, zuotin, Hdj2	DnaJ	Stimulate ATPase activity of Hsp70
Prefoldin	GimC	Absent from *Bacteria*, present in *Archaea*	Prevent aggregation of unfolded protein chains
Trigger factor	Absent from *Eukarya*	Present	Bind to nascent chains as they emerge from ribosome
Calnexin, calreticulin	Present	Absent from prokaryotes	Bind to partly folded glycoproteins; located in ER membrane and lumen, respectively
Nascent-chain associated complex (NAC)	Present	Absent from prokaryotes	Bind to nascent chains as they emerge from ribosome
PapD	Absent from *Eukarya*	Present in some	Prevent aggregation of subunits of pili

1.6.1. Small chaperones

Small chaperones bind transiently to small hydrophobic regions (typically seven or eight residues long) on both nascent and completed, newly synthesised chains and thus prevent aggregation both during and after chain elongation by shielding these regions from one another (Figure 1.1) [3]. Trigger factor (48 kDa) is the

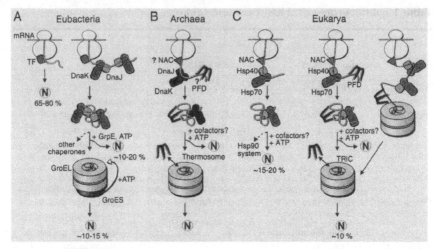

Figure 1.1. Models for the chaperone-assisted folding of newly synthesised polypeptides in the cytosol. (A) *Eubacteria*. TF, trigger factor; N, native protein. Most nascent chains probably interact with TF, and most small proteins (about 65–80% of total chain types) may fold rapidly upon synthesis without further chaperone assistance. Longer chains (10–20% of total chain types) interact subsequently with DnaK and DnaJ and fold after one or several cycles of ATP-dependent binding and release. About 10–15% of total chains fold within the chaperonin GroEL/GroES system. GroEL does not bind to nascent chains and is thus likely to receive its substrates after their release from DnaK. (B) *Archaea*. PFD, prefoldin; NAC, nascent-chain associated complex. Only some archaeal species contain DnaK/DnaJ. The existence of a ribosome-bound NAC homologue and the binding of PFD to nascent chains have not been shown. (C) *Eukarya*. Like TF, NAC probably interacts with many nascent chains. The majority of smaller chains may fold without further chaperone assistance. About 15–20% of chains reach their native states after assistance by Hsp70 and Hsp40, and a specific fraction of these are then transferred to Hsp90. About 10% of chains are passed to the TriC system in a reaction involving PFD. Reprinted from [3] with permission.

first chaperone to bind to nascent chains in prokaryotes because it is associated with the ribosomal large subunit that contains the tunnel from which the chains emerge [16]. A cell of *E. coli* contains about 20,000 copies of this chaperone, sufficient to bind to all nascent chains. Trigger factor shows peptidyl-prolyl isomerase activity and contains a hydrophobic groove which transiently binds to regions of the nascent chain enriched in aromatic residues. It binds to nascent chains as short as 57 residues and dissociates in an ATP-independent manner after the chain is released from the ribosome; this binding does not require prolyl residues in the nascent chain. The isomerase activity may provide a means of keeping nascent chains containing prolyl residues in a flexible state. The eukaryotic cytosol lacks trigger factor; however, its function may be replaced by that of a heterodimeric complex of 33- and 22-kDa subunits, termed the nascent chain–associated complex. Like trigger factor, this complex binds transiently to short nascent chains; however, unlike trigger factor, it does not possess peptidyl-prolyl isomerase activity [17].

Cells lacking trigger factor show no phenotype because its function can be replaced by that of the other major small chaperone, Hsp70 [18, 19]. The Hsp70 family has many 70-kDa proteins distributed between the cytoplasm of *Eubacteria* and some, but not all, *Archaea*, the cytosol of *Eukarya*, and eukaryotic organelles such as the endoplasmic reticulum, mitochondria and chloroplasts. Some, but not all, of these members are also stress proteins. Unlike trigger factor, most of the Hsp70 members do not bind to ribosomes but do bind to short regions of hydrophobic residues exposed on nascent and newly synthesised chains. Such regions occur statistically about every 40 residues and are recognised by a peptide-binding cleft in Hsp70; this recognition involves not just the hydrophobic side chains, but also main-chain atoms in the extended polypeptide backbone of the nascent chain [20].

Most is known about the Hsp70 member in *E. coli*, termed DnaK. Like all Hsp70 chaperones, DnaK contains an ATPase site and occupation of this site by ATP promotes rapid, but reversible, peptide binding. The importance of this ATPase site in cell-cell signalling is described in Chapter 10. ATP hydrolysis then tightens the binding through conformational changes in DnaK. The cycling of ATP between these states is regulated by a 41-kDa co-chaperone of the Hsp40 family, termed DnaJ in *E. coli*, and GrpE, a nucleotide exchange factor that is a co-chaperone, but not a chaperone. DnaJ binds to DnaK through its J domain and increases the rate of ATP hydrolysis, thus facilitating peptide binding. DnaJ, like all the Hsp40 proteins, acts as a chaperone in its own right because it also binds to hydrophobic peptides. Thus DnaK and DnaJ cooperate in binding each other to nascent chains; all Hsp70 chaperones are thought to cooperate with Hsp40 chaperones. The role of GrpE is to stimulate release of ADP from DnaK, allowing the latter to bind another molecule of ATP and so release the peptide. In the eukaryotic cytosol, the role of GrpE is fulfilled by an unrelated co-chaperone called Bag-1 [21]. Some *Archaea* lack Hsp70 proteins; however, their role in protein folding may be replaced by that of an unrelated 90-kDa dimer chaperone called prefoldin.

There is enough DnaK in each *E. coli* cell for one molecule to bind to each nascent chain. DnaK binds to longer chains than trigger factor and so probably binds after trigger factor. When the gene for trigger factor is deleted, the fraction of nascent and newly synthesised chains binding to DnaK increases from about 15% to about 40%. However, removal of the genes for both trigger factor and DnaK in the same cell causes the aggregation of many newly synthesised chains

[18, 19]. The redundancy of important control systems is as good a design principle for cells as it is for passenger planes.

Small chaperones function essentially by reducing the time that potentially interactive surfaces on neighbouring chains are exposed by cycling on and off these chains until they have folded; they do not appear to change the conformation of the chains. Such a simple mechanism can be thought of as analogous to tossing a hot potato from hand to hand until it has cooled enough to be held. However, the other major class of chaperones involved in protein folding function by a much more sophisticated mechanism enabled by their large size.

1.6.2. Large chaperones – the chaperonins

Most is known about GroEL and GroES, the chaperonin and co-chaperonin found in *E. coli*; however, the general principles of their mechanism (Figure 1.2) are thought to apply also to the thermosome found in *Archaea* and to the TCP-1 complex (also called the TRiC or CCT complex) found in the eukaryotic cytosol. GroEL (800 kDa) consists of two heptameric rings of identical 57-kDa ATPase subunits stacked back to back, containing a cage in each ring [22]. The term cage is used because the walls surrounding each central cavity contain gaps, perhaps to allow entry and exit of nucleotides and water. Each subunit contains three domains. The equatorial domain contains the nucleotide binding site and is connected by a flexible intermediate domain with the apical domain. The latter presents several hydrophobic side chains at the top of the ring orientated towards the cavity of the cage, an arrangement that permits either a partly folded polypeptide chain or a molecule of GroES to bind but prevents binding to another GroEL oligomer.

GroES is a single heptameric ring of 10-kDa subunits that cycles on and off either end of the GroEL in a manner regulated by the ATPase activity of GroEL. At any one time, GroES is bound to only one end of GroEL, leaving the other end free to bind a partly folded polypeptide chain after its release from the ribosome. GroEL does not bind to nascent chains, whereas TCP-1 may do so. The two rings of GroEL are coupled by negative allostery so that only one ring at a time binds nucleotide, but within each ring the binding of nucleotide is cooperative. When either ADP or ATP is bound to one GroEL ring, the GroES sits on top of this ring – now called the *cis* ring. The binding of GroES triggers a large rotation and upward movement of the apical domains, resulting in an enlarged cage and a change in its internal surface properties from hydrophobic to hydrophilic. This enlarged cage can accommodate a single partly folded

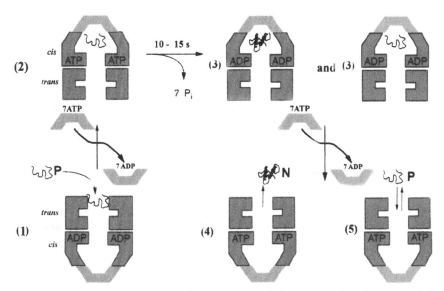

Figure 1.2. Mechanism of action of the GroEL/GroES system in *E. coli*. P, unfolded polypeptide chain; N, native folded chain. Dark shaded illustrates a section through GroEL, light shaded illustrates a section through GroES. For details, see text.

compact polypeptide chain up to about 60 kDa in size, perhaps depending on shape.

The reaction cycle starts with a GroEL-GroES complex containing ADP bound to the *cis* ring (Figure 1.2, step 1). The hydrophobic residues on the apical domains on the other ring, now called *trans*, bind to hydrophobic residues exposed on a partly folded polypeptide chain, presumably after release of small chaperones from this chain. GroES and ATP then bind to this ring, thereby converting it into a new *cis* ring and causing the release of GroES and ADP from the old *cis* ring (Figure 1.2, step 2). This binding of GroES to the *trans* ring displaces the bound polypeptide into the cavity of the cage because some of the hydrophobic residues of the apical domains that bind the polypeptide are the same residues that bind GroES. The displaced chain lying free in the cavity of the cage now has 10–15 seconds to continue folding, a time set by the slow but cooperative ATPase activity of the seven subunits in each ring (Figure 1.2, step 3). The chain thus continues its folding sheltered in a hydrophilic environment containing no other folding chain. Many denatured polypeptide chains will fold completely within 15 seconds in the classic Anfinsen renaturing experiment carried out in a test tube instead of inside GroEL. It is for this reason that I call this mechanism the 'Anfinsen cage model' [23].

The binding of ATP and GroES to the new *trans* ring then triggers the release of GroES and ADP from the *cis* ring containing the polypeptide chain, thereby allowing the latter to diffuse out of the cage into the cytoplasm. If this chain has internalised its hydrophobic residues, it remains free in the cytoplasm (Figure 1.2, step 4). However, any chain that still exposes hydrophobic residues rebinds to the same ring for another round of encapsulation (Figure 1.2, step 5). Rebinding to the same ring rather than the ring of another GroEL oligomer is favoured by the crowding effect created by the high concentration of macromolecules in the cytoplasm and reduces the risk that partly folded chains will meet one another in the cytoplasm in a potentially disastrous encounter [24].

This model was proposed to explain the results of many ingenious *in vitro* experiments; however, recent genetic studies confirm the importance of the Anfinsen cage mechanism in intact cells. Mutants in which the mechanism is prevented by blockage of the entrance to one of the rings of each GroEL oligomer are viable, but the cells form colonies only 10% the size of the wild-type colonies [23, 25]. That these mutants are viable at all suggests that the ring whose entrance is not blocked is acting rather like the small chaperones – reducing aggregation by binding and releasing from the hydrophobic regions on newly synthesised chains.

An unexpected added advantage of the Anfinsen cage mechanism is that, for proteins in a certain size range, encapsulation in the cage increases the rate of folding compared to the rate observed in free solution under conditions where aggregation is not a problem. Thus the rate of folding of bacterial rubisco (50 kDa) is increased four-fold by encapsulation, whereas that of rhodanese (33 kDa) is not affected [26]. This effect can be explained in terms of a type of macromolecular crowding called confinement, in which the proximity of the walls of the confining cage stabilises compact conformations more than extended ones and so enhances the rate of interactions leading to compaction of the folding chain [27]. A possible further advantage of the Anfinsen cage mechanism is that the rotation of the apical domains may cause the bound polypeptide to unfold to some extent and thus destabilises conformations that may have misfolded [28]; this interesting suggestion awaits experimental support.

Current estimates suggest that the fraction of newly synthesised polypeptide chains that bind *in vivo* to either Hsp70 proteins or the chaperonins is in the range 10–20% [3]. Whether the majority of newly synthesised chains bind to other, as yet undiscovered, chaperones or fold unassisted because their sequences have evolved to avoid aggregation is unknown. Nor is it understood what determines that only a defined minority of polypeptides bind to GroEL in the intact cell.

1.7. Other functions of molecular chaperones

The ubiquity and diversity of molecular chaperones is consistent with the possibility that some members have functions additional to the prevention of aggregation. Two examples are well established: the first concerns chaperones that regulate the properties of proteins that have largely folded but are not in their functional native states, whereas the second illustrates a chaperone family that redissolves insoluble protein aggregates.

The Hsp90 family assists in the regulation of signal transduction pathways, the best studied of which are the steroid response pathways mediated by specific receptor proteins [29, 30]. Hsp90 does not act generally in the folding of nascent protein chains as does Hsp70; instead, most of its known substrates are signal transduction proteins folded in a near-native state, ready for interacting with other molecules that trigger their signalling function. Steroid receptors have domains for binding their steroidal ligands, for dimerisation, and for binding regulatory proteins that determine the transcriptional activity of specific sets of genes. In the absence of steroid hormones these receptors are bound to several types of chaperone and co-chaperone, including Hsp70 as well as Hsp90. However, they remain monomeric, do not bind to DNA and are thus functionally inactive as transcription factors. This type of chaperone function involves changes in conformations of the substrate protein and therefore falls within the general area of chaperone-assisted protein folding, but with the aim of regulating function instead of preventing aggregation.

The second example concerns the reversal of protein aggregation. The success of molecular chaperones at preventing aggregation is not perfect, and there is evidence that it declines as organisms age [31, 32]. Protein damage accumulates with time and eventually overloads the repairing ability of chaperones, with Alzheimer's disease being the most dramatic and depressing consequence. The universal heat shock response can be regarded as a mechanism to increase the concentrations of chaperones when cells are subject to environments that cause protein denaturation and subsequent aggregation. The Hsp100 family has the unique ability to redissolve insoluble protein aggregates and acts to rescue or remove those proteins that evade the aggregation-prevention screen provided by other chaperones. The Hsp100 family is a functionally diverse group of oligomeric ATPases that are, in turn, a subfamily of the AAA+ superfamily. All AAA+ proteins form single-ring-shaped complexes reminiscent of the chaperonins. As with the latter, it is thought that the combination of several substrate-binding sites arranged around the ring is crucial to their function [33].

The Hsp100 family member in yeast is called Hsp104; removal of the Hsp104 function by mutation is not lethal, but it does prevent the cell from rescuing the activity of test enzymes inactivated by brief exposure of the cells to 44 °C. Such heat shock causes proteins to aggregate in both the cytoplasm and nucleus of yeast cells; when such cells are placed at 25 °C after the heat shock these aggregates disappear, but this function is lost in cells lacking the Hsp104 function [34]. *In vitro* studies show that, unlike many other chaperones, Hsp104 is unable to prevent the aggregation of denatured proteins but, when mixed with Hsp70 and Hsp40 from yeast, can mediate the recovery of enzymic activity from insoluble aggregates.

1.8. The cell–cell signalling hypothesis

The term cell-cell signalling is conventionally used to encompass all those processes by which cells secrete specific compounds that influence the behaviour of other cells. Given the universality and abundance of molecular chaperones it is plausible to speculate that some of them have such signalling roles, in addition to their known roles in protein folding, and establishing such roles would be an important advance. In my view the current evidence is suggestive rather than conclusive.

Much of the evidence that chaperones occur on cell surfaces and in the extracellular medium relies on the uses of antibodies (see Chapter 2). This evidence needs strengthening by physical characterisation of what the antibodies are detecting – are they detecting proteolytic fragments, subunits, or the oligomeric forms of these chaperones? In the case of the large chaperones such characterisation should be straightforward. It is also not clear in what form exogenous pure chaperones added to cells in culture bind to other cells, given that in some cases (e.g., the mycobacterial chaperonins) it is reported that effects on cell behaviour resist boiling or treatment with proteases (described in Chapter 6). A separate issue that needs further clarification is how these materials appear in the extracellular medium – are they secreted by specific mechanisms by live cells or are they simply released by dead or dying cells? (Partial answers to this question are to be found in Chapters 3, 5 and 12). The ten genes for the mitochondrial chaperonin 60 (Cpn60) of human cells all contain mitochondrial targeting sequences, but there is no established mechanism known for the transport of mitochondrial matrix proteins across the plasma membrane (see Chapter 3). Establishing the existence of such a mechanism would be an important contribution to cell biology.

The minimum hypothesis that is consistent with current evidence is a special form of cell-cell signalling in which cells respond to signals in the form of chaperone fragments released from cells damaged by infection or other stresses. Such damaged cells could include pathogenic bacteria as well as cells of the host. This type of signalling would broadcast the news that some cells in the organism had experienced stresses, including invasion by pathogenic bacteria and viruses, so that ameliorating responses could be triggered. There is convincing evidence that the Cpn60 from *Enterobacter aerogenes* cells found in the saliva of antlion larvae that prey on other insects acts to paralyse the prey, establishing an extracellular role for this particular chaperone [35]. It was shown that the paralytic principle is the intact oligomer of Cpn60 and that it is inactivated by trypsin, but whether it is actively secreted into the saliva or is released from dying bacterial cells was not determined. The Cpn60 from *E. coli* is not toxic to insects but can be converted into one simply by changing the isoleucine at position 100 to valine or changing aspartate at position 338 to glutamate. These remarkable findings are consistent with the possibility that at least some chaperones do have extracellular roles; however, more incisive studies are required to establish this as a widespread phenomenon.

REFERENCES

1. Ellis R J. Proteins as molecular chaperones. Nature 1987, 328: 378–379.
2. Hemmingsen S M, Woolford C, van der Vies S M, Tilly K, Dennis D T, Georgopoulos G C, Hendrix R W and Ellis R J. Homologous plant and bacterial proteins chaperone oligomeric protein assembly. Nature 1988, 333: 330–334.
3. Hartl F U and Hayer-Hartl M. Molecular chaperones in the cytosol: from nascent chain to folded protein. Science 2002, 295: 1852–1858.
4. Ellis R J and Hartl F U. Protein folding and chaperones. *Nature Encyclopedia of the Human Genome*. Macmillan Publishers Ltd 2003, pp 806–810.
5. Laskey R A, Honda B M and Finch J T. Nucleosomes are assembled by an acidic protein that binds histones and transfers them to DNA. Nature 1978, 275: 416–420.
6. Ellis R J. The general concept of molecular chaperones. In Ellis, R. J., Laskey, R. A. and Lorimer, G. H. (Eds.) *Molecular Chaperones*. Chapman and Hall for The Royal Society, London 1993, pp 1–5.
7. Musgrove J E and Ellis R J. The rubisco large subunit binding protein. Phil Trans R Soc Lond B 1986, 313: 419–428.
8. Ellis R J and Hemmingsen S M. Molecular chaperones: proteins essential for the biogenesis of some macromolecular structures. Trends Biochem Sci 1989, 14: 339–342.
9. Puig A and Gilbert H F. Protein disulfide isomerase exhibits chaperone and antichaperone activity in the oxidative folding of lysozyme. J Biol Chem 1994, 269: 7764–7771.
10. Ellis R J. Steric chaperones. Trends Biochem Sci 1998, 23: 43–45.
11. Finley D, Bartel B and Varshavsky A. The tails of ubiquitin precursors are ribosomal proteins whose fusion to ubiquitin facilitates ribosome biogenesis. Nature 1989, 338: 394–401.
12. Welch W J. Role of quality control pathways in human diseases involving protein misfolding. Semin Cell Devel Biol 2004, 15: 31–38.

13. Ellis R J. Macromolecular crowding: obvious but underappreciated. Trends Biochem Sci 2001, 26: 597–603.
14. Hartl F U. Molecular chaperones in cellular protein folding. Nature 1996, 381: 571–579.
15. Ellis R J and Hartl F U. Principles of protein folding in the cellular environment. Cur Opin Struct Biol 1999, 9: 102–110.
16. Schlieker C, Bukau B and Mogk A. Prevention and reversion of protein aggregation by molecular chaperones in the *E. coli* cytosol; implications for their applicability in biotechnology. J Biotech 2002, 96: 13–21.
17. Beatrix B, Sakai H and Wiedmann M. The α and β subunits of the nascent polypeptide complex have distinct functions. J Biol Chem 2000, 275: 37838–37845.
18. Deuerling E, Schulze-Specking A, Tomoyasu A, Mogk A and Bukau B. Trigger factor and DnaK cooperate in folding of newly synthesized proteins. Nature 1999, 400: 693–696.
19. Teter S A, Houry W A, Ang D A, Tradler T, Rockabrand D, Fischer G, Blum P, Georgopoulos C and Hartl F U. Polypeptide flux through bacterial hsp70: DnaK cooperates with trigger factor in chaperoning nascent chains. Cell 1999, 97: 755–765.
20. Bukau B and Horwich A L. The Hsp70 and Hsp60 chaperone machines. Cell 1998, 92: 351–366.
21. Sondermann H, Scheufler C, Schneider C, Hohfeld J, Hartl F U and Moarefi I. Structure of a Bag/hsc70 complex: convergent functional evolution of hsp70 nucleotide exchange factors. Science 2001, 291: 1553–1557.
22. Xu Z, Horwich A L and Sigler P B. The crystal structure of the asymmetric GroEL-GroES- (ADP)$_7$ chaperonin complex. Nature 1997, 388: 741–750.
23. Ellis R J. Protein folding: importance of the Anfinsen cage. Cur Biol 2003, 13: R881–R883.
24. Martin J and Hartl F U. The effect of macromolecular crowding on chaperonin-mediated protein folding. Proc Natl Acad Sci USA 1997, 94: 1107–1112.
25. Farr G W, Fenton W A, Rospert S and Horwich A L. Folding with and without encapsulation by cis and trans-only GroEL-GroES complexes. EMBO J 2001, 22: 3220–3230.
26. Brinker A, Pfeifer G, Kerner M J, Naylor D J, Hartl F U and Hayer-Hartl M. Dual function of protein confinement in chaperonin-assisted protein folding. Cell 2001, 107: 223–233.
27. Takagi F, Koga N and Takada S. How protein thermodynamics and folding are altered by the chaperonin cage: molecular simulations. Proc Natl Acad Sci USA 2003, 100: 11367–11372.
28. Thirumalai D and Lorimer G H. Chaperonin-mediated protein folding. Ann Rev Biophys Biomol Struct 2001, 30: 245–269.
29. Young J C, Moarefi I and Hartl F U. Hsp90: a specialized but essential protein-folding tool. J Cell Biol 2001, 154: 267–273.
30. Smith D F. Chaperones in progesterone receptor complexes. Semin Cell Devel Biol 2000, 11: 45–52.
31. Csermley P. Chaperone overload is a possible contributor to 'civilization diseases'. Trends Genetics 2001, 17: 701–704.
32. Soti C and Csermley P. Molecular chaperones and the aging process. Biogerontology 2000, 1: 225–233.

33. Glover J R and Tkach J M. Crowbars and rachets: hsp100 chaperones as tools in reversing aggregation. Biochem Cell Biol 2001, 79: 557–568.
34. Parsell D A, Kowal A S, Singer M A and Lindquist S. Protein disaggregation by heat shock protein hsp104. Nature 1994, 372: 475–477.
35. Yoshida N, Oeda K, Watanabe E, Mikami T, Fukita Y, Nishimura K, Komai K and Matsuda K. Protein function. Chaperonin turned insect toxin. Nature 2001, 411: 44.

2

Intracellular Disposition of Mitochondrial Molecular Chaperones: Hsp60, mHsp70, Cpn10 and TRAP-1

Radhey S. Gupta, Timothy Bowes, Skanda Sadacharan and Bhag Singh

2.1. Introduction

This chapter reviews work on the intracellular disposition of a number of molecular chaperones that are generally believed to be localised and function mainly within the mitochondria of eukaryotic cells. However, in recent years, compelling evidence has accumulated from many lines of investigation indicating that several of these mitochondrial (m-) chaperones are also localised and perform important functions at a variety of other sites/compartments within cells (see [1, 2]). The four chaperone proteins that are the subjects of this chapter include the following: (i) the 60-kDa heat shock chaperonin protein (Hsp60, also known as chaperonin 60, Cpn60), which is a major protein in both stressed and unstressed cells and plays an essential role in the proper folding and assembly into oligomeric complexes of other proteins [3–6]; (ii) the 10-kDa heat shock chaperonin (Hsp10 or Cpn10), which is a co-chaperone for Hsp60 in the protein folding process [7]; (iii) the mitochondrial homologue of the major 70-kDa heat shock protein (mHsp70), which plays a central role in the import of various proteins into mitochondria and their proper folding [4, 6]; and (iv) the mitochondrial Hsp90 protein, which was originally identified in mammalian cells as the tumour necrosis factor receptor-associated protein-1 (TRAP-1) [8, 9] and is commonly referred to by this latter name.

All of these proteins are encoded by nuclear genes, and, after translation of their transcripts in the cytosol, their protein products are then imported into mitochondria. The import of protein into mitochondria is generally a highly efficient process which occurs very rapidly and generally to completion [10, 11]. Further, once imported into mitochondria, the proteins are not known to exit under normal physiological conditions. Hence, their presence at extra-mitochondrial sites raises important questions regarding the possible mechanisms by which they have reached these locations [1, 2]. The presence of these proteins at other sub-cellular locations also greatly broadens the range of functions with which they are likely to be involved within the cell. This chapter provides a brief review of the cellular distributions of these chaperone proteins and the significance of their distributions on their cellular functions. Other aspects of these proteins are covered in various reviews [3–5, 11, 12] and elsewhere in this volume. The reader is referred to Chapter 3 for a discussion of non-classical pathways of protein export in eukaryotic cells and to Chapter 12 for a review of molecular chaperone release from cells and of molecular chaperones in the circulation.

2.2. Sub-cellular localisation of the mitochondrial molecular chaperones

2.2.1. Hsp60/chaperonin 60

Hsp60 or chaperonin 60 (Cpn60), which is the eukaryotic homologue of the bacterial GroEL protein, constitutes one of the major and most characterised molecular chaperone proteins in both stressed and unstressed cells [4, 13]. In eukaryotic organisms, this protein is primarily found in organelles such as mitochondria and chloroplasts, which have originated from bacteria belonging to the proteobacteria and cyanobacteria groups, respectively [14, 15]. Of these, the mitochondrial Hsp60 has been extensively studied. It is encoded by nuclear DNA and synthesised as a larger precursor containing an N-terminal mitochondrial targeting sequence (MTS), which is cleaved during import of the precursor protein into the matrix compartment [16].

The mature form of the Hsp60 found in mitochondria and various other compartments lacks the MTS sequence [17–19]. Hsp60 was initially discovered in mammalian cells as a protein (P_1) that was specifically altered in Chinese hamster ovary (CHO) cell mutants resistant to the microtubule inhibitor podophyllotoxin [20–22]. Cellular sub-fractionation and immunofluorescence studies indicated that this protein was primarily localised in the mitochondrial matrix compartment [23, 24]. The matrix localisation of Hsp60 (P_1) was totally unexpected in view of the earlier genetic and biochemical studies that strongly indicated that this protein interacted with tubulin, which is not found in mitochondria [25]. Although earlier cell fractionation and immunofluorescence studies suggested that Hsp60 was exclusively a mitochondrial protein, subsequent work reviewed below provides strong evidence that it is also present outside of mitochondria, including on the cell surface (see [1, 2]).

One of the earliest observations pointing to the presence of Hsp60 on the cell surface came from studies on murine and human T cells that recognised the mycobacterial Hsp60 (GroEL). These cells were also found to be stimulated by a protein present on the surface of stressed macrophages and certain tumour cells [26, 27], and this stimulation was blocked by both polyclonal and monoclonal antibodies to Hsp60 [28], indicating that such cells were expressing Hsp60 on their cell surface.

More definitive evidence for the presence of Hsp60 on the surface of cells was provided by its immunoprecipitation from surface-iodinated or surface-biotinylated proteins by polyclonal and monoclonal antibodies specific for Hsp60 [28, 29]. Hsp60 has also been identified on the cell surface by chemical cross-linking of live cells, where it was found associated with the plasma membrane resident $p21^{ras}$ protein [17], suggesting its possible involvement in signal transduction events. Hsp60 in the plasma membrane is also found to be concomitantly enhanced in CHO cell mutants exhibiting an increase in the A system of amino acid transport, suggesting its possible association with the corresponding amino acid transporter [18].

In another study, the plasma membrane–associated Hsp60 was found to be specifically phosphorylated upon activation of Type I protein kinase A [19]. Interestingly, this study also found that histone 2B formed a complex with plasma membrane–associated Hsp60. Phosphorylation of both Hsp60 and histone 2B by Type I protein kinase A disrupted their association, leading to expulsion of histone 2B, but not Hsp60, from the membrane. Hsp60 in the plasma membrane has also been shown to bind to the high-density lipoprotein [30] and has been indicated to play a role in the peptide presentation process [31, 32]. The presence of both Hsp60 and tubulin in the plasma membrane also provides a plausible explanation for the puzzling observation that led to the discovery of Hsp60 in mammalian cells [21, 22] – that mutational changes in this protein cause resistance to anti-mitotic drugs that bind to tubulin. In our earlier studies, in which Hsp60 (P_1) was identified as a tubulin-associated protein [20, 33], we suggested that tubulin in the plasma membrane is associated with Hsp60, such that mutational changes in Hsp60 can alter drug binding to tubulin [2, 34]. All of the aforementioned studies strongly suggest that Hsp60 in the plasma membrane functions as a membrane chaperone, which enables other soluble proteins to exhibit membrane association.

A number of studies have reported an increased cell surface expression of Hsp60 under stressed or apoptotic conditions. In aortic endothelial cells exposed to cytokines or high temperature, increased expression of Hsp60 has been detected on the cell surface by fluorescence imaging [35], and this has been shown to make such cells susceptible to complement-dependent lysis by Hsp60-specific antibodies [36]. Increased expression of Hsp60 and Hsp70 on the cell surface has also been observed in T cells undergoing apoptosis [37]. In a recent study, Hsp60 was found to interact with Bax in the cytoplasm of cardiomyocytes [38].

However, during hypoxia, Hsp60 is re-localised to the plasma membrane, allowing Bax to translocate to mitochondria to induce apoptosis. Although certain types of cells or conditions might enhance or induce cell-surface Hsp60 expression, it is important to recognise that surface expression of Hsp60 is a common characteristic of eukaryotic cells and is not limited to stressed or apoptotic cells [1, 2].

Extensive work has been carried out on the sub-cellular localisation of Hsp60 in different cultured cell lines as well as tissues by means of immunogold labelling (or immuno-electron microscopy (Immuno-EM)), employing monoclonal and polyclonal antibodies. Although immuno-EM labelling of cultured mammalian cells (CHO, BSC-1 kidney cells, PC12 neuronal, Daudi Burkitt's lymphoma and human diploid fibroblasts) has demonstrated the majority of Hsp60 labelling to be found within mitochondria [39], 15–20% of the reactivity is consistently observed at discrete extra-mitochondrial sites, including unidentified cytoplasmic vesicles and granules, sites on endoplasmic reticulum, and at the cell surface (Figure 2.1A) [29, 39]. Using backscattered electron imaging of intact cells, the cell surface copy number of Hsp60 has been estimated to be approximately 200–2000 molecules per cell, in CHO and CEM-SS human T lymphocyte cell lines, which appears to represent about 1–10% of the total cellular Hsp60 [29].

The sub-cellular distribution of Hsp60 has also been examined in different mammalian tissues using a high-resolution immuno-EM technique [40–42]. In some tissues, such as heart, kidney (proximal and distal tubules), skeletal muscle, adrenal gland and spleen, reactivity to Hsp60 antibody was primarily restricted to mitochondria [41]. However, in a number of other tissues, strong and specific labelling due to Hsp60 antibody has been observed in a number of other compartments in addition to mitochondria. In pancreatic β-cells, strong reactivity with Hsp60 antibodies is also seen in mature insulin secretory granules (ISGs) (Figure 2.1B) [40]. In this instance, the Hsp60 antibodies specifically labelled the central core of the mature ISGs, but no labelling was seen in immature secretory granules [40]. In rat liver, specific labelling with Hsp60 antibodies has been observed in mitochondria and peroxisomes [39, 41, 43]. The Hsp60 reactivity in peroxisomes is primarily associated with the urate oxidase crystalline core, which is a distinguishing characteristic of rat liver peroxisomes. In pancreatic acinar cells and pituitary, strong labelling with Hsp60 antibodies has been observed in zymogen granules (ZGs) (Figure 2.1C) and growth hormone granules (GHGs), respectively [41]. The labelling of these compartments with Hsp60 antibodies is completely abolished upon pre-adsorption of the antibodies with recombinant Hsp60, thereby providing evidence that it is specific for Hsp60 [41].

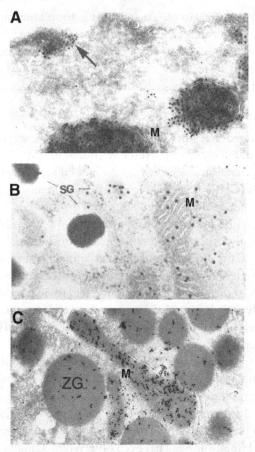

Figure 2.1. High-resolution immuno-EM visualisation of Hsp60 distribution in different cells and tissue using Hsp60-specific antibodies. (A) Immunogold labelling of cryosections of CHO cells; labelling is clearly seen in both mitochondria (M) and on the cell surface. (B) In mouse pancreatic beta cells, labelling is seen both within mitochondria and in the dense core of mature insulin secretory granules (SG). (C) In pancreatic acinar cell sections, strong labelling is observed both within mitochondria and zymogen granules (ZG). In all of the preceding cases, very little labelling is seen in the ER and Golgi compartments. From Brudzynski et al. [40]; Soltys and Gupta [39]; Cechetto et al. [41].

It is of much interest that, although strong labelling due to Hsp60 antibodies is observed in ZGs, GHGs and ISGs, there is negligible (i.e., close to background) labelling seen in immature ISGs, as well as in the endoplasmic reticulum (ER) and Golgi compartments [41]. These results are in marked contrast to those obtained with antibodies to other proteins such as insulin and amylase, which are targeted to the above compartments and in which strong labelling of the entire

ER–Golgi pathway is observed [44, 45]. The absence of significant Hsp60 labelling in the cytoplasm, as well as the ER–Golgi compartments, using different monoclonal and polyclonal antibodies, suggests that Hsp60 could be reaching these granules via a novel mechanism that is different from that which uses the classical ER–Golgi pathway [44, 46]. The presence of Hsp60 in secretory granules suggests that certain cell types should secrete Hsp60. Indeed, the secretion of an Hsp60-like protein has been reported for cultured neuroglial cells and a neuroblastoma cell line [47]. Velez-Granell and colleagues, employing a polyclonal antibody against the bacterial GroEL (from *Chromatium vinosum*), have observed considerable Hsp60 reactivity along the ER–Golgi secretory pathway [42]. However, this antibody also exhibited very high background labelling [42], hence the significance of these results is not clear.

The possible physiological function of Hsp60 in these compartments is presently not known. Hsp60 is associated with the central core of the mature ISGs, but it is not present in immature ISGs. The main difference between these two types of granules is that, during transition from immature ISGs to mature ISGs, pro-insulin is enzymatically cleaved to form insulin, which then, by a poorly understood process, is extensively condensed to form the highly compacted core of the mature granules [48]. The central core of these mature ISGs thus represents a highly organised, supra-molecular structure, the main function of which appears to be to maintain insulin at a very high concentration in a functional form that is ready to be secreted. The Hsp60 in other types of granules such as ZGs and GHGs, may be playing an analogous role. In a similar manner, in peroxisomes, Hsp60 is associated with the urate oxidase crystalline cores, which also constitute a higher order supra-molecular structure which likely requires a chaperone for its assembly. In accordance with its established functions in the formation of oligomeric protein complexes, and in protein secretion in bacteria [4, 49], we have suggested that Hsp60 in different types of granules (ZG, ISGs, GHGs) and peroxisomes also plays a chaperone role in the condensation of proteins within these compartments, and in maintaining the highly compacted proteins in functional forms required for their biological actions [39–41].

2.2.2. Hsp10 (Cpn10)

Hsp10, or Cpn10, is the eukaryotic homologue of the bacterial GroES protein, which serves as a co-chaperone for Hsp60 (GroEL) in the protein folding and assembly processes [3, 12]. Similar to Hsp60, this protein is present in eukaryotic cells in organelles such as mitochondria and chloroplasts [7, 15, 45]. Unlike most other mitochondrial matrix proteins, Hsp10 does not contain an N-terminal cleavable MTS; rather its N-terminal sequence has the ability to form an amphipathic alpha helix which possibly enables it to cross the mitochondrial membrane [50]. Surprisingly, Hsp10 has also been shown to be identical to a protein previously identified as early pregnancy factor (EPF), which appears in maternal serum within 24 hours after fertilisation [51, 52]. The evidence that EPF from human platelets is identical to the Hsp10 protein is provided by several observations: (i) The amino acid sequences of three different fragments covering more than 70% of the EPF show complete identity with the human Hsp10 protein [52]. (ii) Purified rat Hsp10 is found to be as active in the EPF bioassay as the platelet-derived EPF, and this activity can be neutralised by a monoclonal antibody to EPF. In contrast to the mammalian Hsp10, bacterial GroES is not active in the bioassay, providing evidence of specificity [53]. (iii) In the presence of ATP, EPF (similar to Hsp10) forms a stable complex with Hsp60 which co-elutes from a gel filtration column. Further, immobilised Hsp60, in the presence of ATP, removes all EPF activity from pregnancy serum, provid-ing evidence of a specific interaction between these proteins [52, 53]. Fletcher and colleagues have indicated that EPF in mouse cells may be encoded by an intronless gene, the pattern of expression of which is similar to that of EPF activity [54].

The sub-cellular distribution of Hsp10 in rat tissues has been examined in detail using the high-resolution immuno-EM technique employing polyclonal antibodies raised against different regions of human Hsp10 [45]. In all rat tis-sues examined including liver, heart, pancreas, kidney, anterior pituitary, sali-vary gland, thyroid and adrenal gland, antibodies to Hsp10 strongly labelled mitochondria. However, in a number of tissues, in addition to mitochondria, strong and specific labelling with the Hsp10 antibodies is also observed in several extra-mitochondrial compartments. These sites included ZGs in pan-creatic acinar cells (Figure 2.2A), GHGs in anterior pituitary (Figure 2.2B) and pancreatic polypeptide granules in islet cells. These granules likely pro-vide the pathway for the secretion of this protein into the blood stream, in which it can serve the function of EPF. The Hsp10 labelling in these com-partments is at least comparable to (if not higher than) that seen in mitochon-dria and it has been shown to be specific by different means [45]. In con-trast to these secretory granules, the labelling in cytoplasm, nucleus and ER is generally very weak and in most cases at, or near, background levels [45]. These observations are very similar to those attained using Hsp60 antibodies (Figure 2.1) and indicate that the Hsp10 is reaching these compartments by a novel pathway [45].

In addition to these granules, specific reactivity of the Hsp10 antibodies has also been observed within mature erythrocytes (Figure 2.2C) [45]. This observation is surprising because erythrocytes are believed to be devoid of

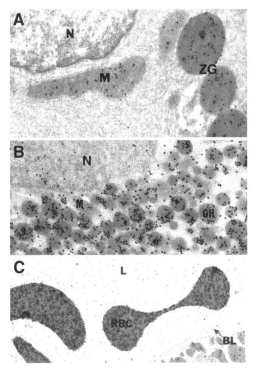

Figure 2.2. Immuno-EM localisation of Hsp10 (Cpn10) in different rat tissues using specific antibodies. (A) Labelling of mitochondria and zymogen granules (ZG) in pancreatic acinar cells. (B) Labelling of GHGs in anterior pituitary sections. (C) Immunogold labelling of red blood cells (RBC). BL, basal lamina; L, lumen. From Sadacharan et al. [45].

mitochondria and various other organelles [55, 56]. However, it is possible that, although mature erythrocytes extrude all mitochondria, they retain specific mitochondrial proteins that may be required for particular functions. Our studies indicate that, similar to Hsp10, cross-reactive proteins to Hsp60 antibodies are also present in erythrocytes (unpublished results). The possible roles that Hsp60 and Hsp10 may play in erythrocytes is presently unclear; however, based on their established chaperone function, it is possible that these proteins are involved in either the assembly or functioning of haemoglobin, which is their primary constituent [55, 56].

2.2.3. mHsp70/DnaK chaperone

Mitochondrial Hsp70 (mHsp70) is the mitochondria-targeted member of the highly conserved Hsp70/DnaK family of proteins. This protein has been independently identified as mortalin, as the 74-kDa peptide binding protein (PBP74) [57], as the 75-kDa glucose regulating protein (Grp75) [58], and as mHsp70 [59, 60]. Similar to Hsp60, it contains an N-terminal MTS that is responsible for its targeting to the mitochondrial matrix [59, 60]. Within mitochondria, it functions as a monomeric ATPase that binds to exposed hydrophobic amino acid residues in proteins to prevent their aggregation or misfolding [4, 12]. Additionally, mHsp70 plays a central role in the mitochondrial import of proteins by binding to, and pulling in, unfolded polypeptide chains entering through the translocase of the inner (outer) membrane translocon [61].

As with the other DnaK homologues, mHsp70 works in conjunction with its co-chaperones, mDnaJ (hTid-1/Hsp40 homologue) and GrpE, which modulate its ATP exchange and ATPase activity [62]. In an earlier study, in which mHsp70 was identified as the PBP74, a protein involved in the processing of antigens, it was shown to be localised in a number of sites including the endocytic vesicles of B cells containing internalised antigen, the ER, and the plasma membrane [63]. mHsp70 has also been shown to interact with exogenously added fibroblast growth factor type-1 as well as with interleukin receptor type-1 and thought to play a role in their internalisation [58, 64].

Studies on mHsp70 localisation by high-resolution immuno-EM demonstrate that, in addition to its expected localisation within mitochondria, this protein is also present at discrete sub-cellular locations including the plasma membrane, endocytic vesicles, and unidentified cytoplasmic granules in both CHO and BSC-1 cells (Figure 2.3A) [60]. Our recent studies show that, similar to Hsp60 and Hsp10, antibodies to mHsp70 also show strong reactivity towards a number of different types of granules including ZGs (Figure 2.3B). Further, as in the case of Hsp60 and Hsp10, very little reactivity can be observed in the cytoplasm and the ER–Golgi compartments, indicating that these proteins are reaching these granules via some novel, yet to be discovered, pathway (see Chapter 3).

Wadhwa and Kaul's groups have independently identified mHsp70 as 'mortalin', a protein that is implicated in conferring the senescent phenotype on cultured mammalian cells [65, 66]. Immunofluorescence studies with antibodies to mortalin in normal fibroblasts generally show a pancytosolic labelling, whereas its localisation varies from a fibrous peri-nuclear to granular staining of the juxtanuclear cap in different immortal cell lines [66]. Based on the distribution pattern of this protein in normal and transformed cells, a number of different complementation groups have been identified [65]. Although these findings are of much interest for understanding the cellular function of mHsp70 in cellular senescence, due to the limited resolution of confocal microscopy it is difficult to determine whether the observed differences in mHsp70 distribution are simply due to altered mitochondrial distributions

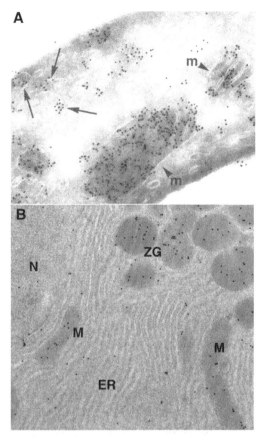

Figure 2.3. Immuno-EM localisation of mHsp70 in (A) cryosections of BSC-1 cells, and (B) in tissue sections of pancreatic acinar cells. In panel (B), the labelling is specific to mitochondria and ZGs and only background labelling is seen in the ER and the nucleus. From Singh et al. [60] and unpublished results.

(or morphology) in different cells, or whether mHsp70 is present at certain discreet extra-mitochondrial sites/compartments in any of these cells [65].

2.2.4. Hsp90 or TRAP-1 protein

Due to their endosymbiotic origin from bacteria, mitochondria are known to contain distinct homologues of various heat shock chaperone proteins. The mitochondrial homologues of Hsp60, Hsp70, Hsp10, as well DnaJ (hTid-1/Hsp40) and GrpE have all been well characterised [67, 68]. In contrast, until recently, no homologue of the major 90-kDa heat shock protein (Hsp90) was identified in mitochondria. However, sequence homology studies have led to the realisation that a protein, TRAP-1, corresponds to the mitochondrial homologue of the Hsp90 protein [8, 9]. TRAP-1 shows greater sequence homology to the Hsp90 homologue from Gram negative bacteria (HtpG), from which mitochondria have originated, than to either the cytosolic or ER-resident forms of human Hsp90 [8, 69]. Additionally, both TRAP-1 and HtpG sequences lack a charged region that is present in all eukaryotic nucleocytosolic Hsp90s [69, 70].

The evidence that TRAP-1 is primarily a mitochondrial protein is provided by several lines of investigations: (i) In immunofluorescence studies, different monoclonal and polyclonal antibodies to TRAP-1 protein are all found to specifically stain mitochondria [8, 69]. (ii) Sub-cellular fractionation of rat liver mitochondria indicates that TRAP-1 is primarily present in mitochondria. Within mitochondria, TRAP-1 reactivity is primarily observed in the matrix and the outer membrane fractions [69]. (iii) TRAP-1 in both human and *Drosophila* is synthesised as a longer precursor protein containing an N-terminal targeting sequence bearing various characteristics of a typical mitochondrial matrix targeting sequence [8]. Similar to other mitochondrial proteins, this sequence is not present in the mature protein and is likely cleaved off during mitochondrial import. (iv) Immunogold labelling of different tissue sections shows strong and specific labelling of mitochondria in all cases [69].

In a number of tissues such as liver and spleen, labelling with TRAP-1 antibody was exclusively seen in mitochondria. However, in several other tissues, labelling at additional discrete locations has also been observed (discussed later). Although TRAP-1 is clearly a mitochondrial protein, and also an Hsp90 homologue, its function within mitochondria is presently not known. Similar to other Hsp90 homologues, TRAP-1 binds ATP and shows ATPase activity which can be blocked by the Hsp90 inhibitors, geldanamycin and radicicol [8]. However, its lack of binding to the Hsp90 co-chaperones p23 and Hop (p60) and its inability to substitute for Hsp90 in other functions [8] indicate that its cellular function has diverged from other well-characterised Hsp90 homologues.

Although the function of TRAP-1 (mHsp90) within mitochondria is not known, it interacts with a variety of proteins involved in diverse functions outside of mitochondria. TRAP-1 was first identified on the basis of its interaction with the intracellular domain of the type I tumour necrosis factor receptor (TNFR-1) in the yeast two-hybrid system [9]. TNFR-1 is a cell surface receptor involved in a variety of cellular events, including cytotoxicity, fibroblast proliferation and antiviral responses (see subsequent chapters that deal with pro-inflammatory cytokine induction by various chaperones). In other studies,

TRAP-1 has been shown to interact with the retinoblastoma protein [71] as well as two proteins involved in hereditary multiple exostoses, EXT1 and EXT2 [72]. The retinoblastoma protein is a nuclear protein involved in cell cycle progression and differentiation [73]. The functions of EXT1 and EXT2 however, are not definitively known, although indirect evidence indicates that these proteins, found primarily in the ER, are involved in glycosaminoglycan synthesis and act as tumour suppressors [72].

In a recent study the *Dictyostelium* homologue of TRAP-1 has been found to display differential sub-cellular localisation in response to nutrient starvation [74]. This protein is normally localised to mitochondria and the cortical membrane region of the cells. However, after 6 hours in media devoid of nutrients, TRAP-1 was found only within mitochondria. TRAP-1 translocates back to the mitochondria from the cortical membrane during starvation, and this translocation could be induced by an as yet uncharacterised secreted factor(s) present in the media of starved cells [74]. In all cases, TRAP-1 was determined to be in its mature form by Western blotting. These results are interesting because they suggest that TRAP-1 is able to cross back into mitochondria without the need for its MTS. In another interesting study, exogenous tumour necrosis factor (TNF) was found to be delivered to mitochondria [75]. Although it remains unclear how exogenously added TNF is directed to mitochondria, TNF binding to TRAP-1 and the subsequent translocation of the complex to the mitochondria is a distinct possibility.

In accordance with its interaction with many extra-mitochondrial proteins, immuno-EM studies on TRAP-1 show that, in addition to mitochondria, strong and significant reactivity to TRAP-1 antibodies is seen in a number of tissues at discrete extra-mitochondrial sites [69]. In pancreatic acinar cells, TRAP-1 is present in both ZGs (Figure 2.4A) as well as glucagon granules [69]. Lower, but significant, reactivity of TRAP-1 has also been seen in the nuclei of pancreatic acinar cells [69]. In endothelial cells lining blood vessels, strong reactivity to TRAP-1 antibody has been observed on the apical cell surface (Figure 2.4B). The labelling is concentrated in foci, which could represent sites where TNFR-1 is localised, or areas in which vesicles containing this protein are either fusing or pinching off from the cell membrane. Significant reactivity of TRAP-1 has also been observed in cardiac sarcomeres, the functional significance of which is presently unknown [69].

2.3. Conclusions and future directions

The presence of various mitochondrial heat shock proteins (viz. Hsp60, Hsp10, mHsp70 and TRAP-1) outside of the mitochondrion, and their involvement in

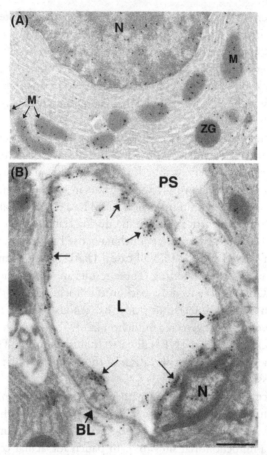

Figure 2.4. Immuno-EM localisation of TRAP-1 (mHsp90) in rat tissue sections using specific antibodies. (A) Localisation of TRAP-1 in mitochondria, ZGs and nucleus in pancreatic acinar cells. (B) Presence of TRAP-1 on the surface of endothelial cells lining a blood vessel. BL, basal lamina; PS, peri-capillary space. From Cechetto and Gupta [69].

important cellular functions, is widely recognised. However, the question as to how these proteins arrive at these locations is generally ignored and has not received due attention. The primary reason for this is that most mitochondrial proteins are encoded by nuclear DNA and translated in the cytosol before being targeted to mitochondria. Hence, it is generally assumed that these proteins can easily move from the cytosol to various other cellular destinations directly, and their presence at extra-mitochondrial location presents no conceptual problem [10, 76]. However, the available facts do not support this simplistic assumption. In this context, it is important to recognise that most of these proteins are synthesised as larger precursor molecules containing an N-terminal leader sequence that targets the proteins to mitochondria. Most of these proteins are also encoded by single copy genes and there is no evidence for the occurrence of alternate processing at either the transcription or translation levels.

The targeting sequences of these proteins are specifically cleaved by a mito-chondrial matrix resident protease and these sequences are not present in the mature protein found within mitochondria [10, 76]. Extensive evidence now indicates that various proteins discussed in this review (viz. Hsp60, mHsp70, TRAP-1), as well as other mitochondrial proteins that have been identified at extra-mitochondrial locations, all comprise mature forms of the protein lacking the targeting sequence. Further, for Hsp60, mHsp70 and several other mitochon-drial proteins, it has been demonstrated that the conversion of the precursor into the mature form requires entry of the precursor (or at least its targeting sequence) into mitochondria. In cells treated with the potassium ionophores such as nonactin or valinomycin, which inhibit mitochondrial import of the precursor proteins, no conversion of the precursor forms into the mature pro-teins is observed [39, 60]. Further, under normal conditions the amounts of the precursor proteins that are present in cells are too low to be detected, indicating that the mitochondrial targeting is highly efficient and leading to no significant accumulation of the precursor proteins in the cytosol, or their mis-targeting to other compartments.

In addition to the proteins discussed in this chapter, most other mitochondrial proteins whose sub-cellular distributions in different cells and tissues have been examined in detail by means of high-resolution techniques are found to be present, besides mitochondria, at other specific sites in cells. Examples of such proteins include mitochondrial aspartate aminotransferase, which also functions as a fatty acid transporter on the cell surface [77, 78], and the p32 protein, which is involved in nuclear functions and also acts as a cell surface receptor for complement C1q [79, 80]. Interestingly, cytochrome C, the exit of which from mitochondria is believed to play a central role in apoptosis [81], is also present in ZGs and GHGs in normal rat tissues [82]. Additionally, our recent work indicates that a number of proteins that are encoded by mitochondrial DNA and that are transcribed and translated within mitochondria are also present outside of mitochondria at specific locations (unpublished results). The presence of these proteins at extra-mitochondrial locations challenges the widely held notion that mitochondria constitute a dead-end compartment in cells, from which proteins cannot exit under normal physiological conditions [1, 2, 10, 83, 84].

The preceding observations have led us to postulate the existence of spe-cific mechanisms for protein transport from mitochondria to other cellular compartments [1, 2]. The existence of specific transport mechanisms from mitochondria should not be surprising in view of their origin from Gram negative bacteria, which possess an ever-growing number of mechanisms for protein export/secretion across the cell membrane [2]. Although such mechanisms remain to be characterised in eukaryotic cells, the presence of various mitochondrial chaperones (and other proteins) at extra-mitochondrial sites indicates that their cellular functions are not restricted to mitochondria, and many of these proteins likely play important roles in diverse cellular processes (see [1, 2]).

Acknowledgements

This work has been supported by a research grant from the Canadian Institute of Health Research to R.S.G.

REFERENCES

1. Soltys B J and Gupta R S. Mitochondrial-matrix proteins at unexpected locations: are they exported? Trends Biochem Sci 1999, 24: 174–177.
2. Soltys B J and Gupta R S. Mitochondrial proteins at unexpected cellular locations: export of proteins from mitochondria from an evolutionary perspective. Int Rev Cytol 2000, 194: 133–196.
3. Bukau B and Horwich A L. The Hsp70 and Hsp60 chaperone machines. Cell 1998, 92: 351–366.
4. Craig E A, Gambill B D and Nelson R J. Heat shock proteins: molecular chaperones of protein biogenesis. Microbiol Rev 1993, 57: 402–414.
5. Ellis R J and Hartl F U. Protein folding in the cell: competing models of chaperonin function. FASEB J 1996, 10: 20–26.
6. Hartl F U, Martin J and Neupert W. Protein folding in the cell: the role of molecular chaperones Hsp70 and Hsp60. Ann Rev Biophys Biomol Struct 1992, 21: 293–322.
7. Lubben T H, Gatenby A A, Donaldson G K, Lorimer G H and Viitanen P V. Identification of a groES-like chaperonin in mitochondria that facilitates protein folding. Proc Natl Acad Sci USA 1990, 87: 7683–7687.
8. Felts S J, Owen B A, Nguyen P, Trepel J, Donner D B and Toft D O. The hsp90-related protein TRAP1 is a mitochondrial protein with distinct functional properties. J Biol Chem 2000, 275: 3305–3312.
9. Song H Y, Dunbar J D, Zhang Y X, Guo D and Donner D B. Identification of a protein with homology to hsp90 that binds the type 1 tumor necrosis factor receptor. J Biol Chem 1995, 270: 3574–3581.
10. Herrmann J M and Neupert W. Protein transport into mitochondria. Cur Opin Microbiol 2000, 3: 210–214.
11. Pfanner N and Neupert W. The mitochondrial protein import apparatus. Ann Rev Biochem 1990, 59: 331–353.
12. Hendrick J P and Hartl F U. Molecular chaperone functions of heat-shock proteins. Ann Rev Biochem 1993, 62: 349–384.

13. Hartl F U. Molecular chaperones in cellular protein folding. Nature 1996, 381: 571–579.
14. Gray M W. Evolution of organellar genomes. Cur Opin Genetics Develop 1999, 9: 678–687.
15. Gupta R S. Evolution of the chaperonin families (Hsp60, Hsp10 and Tcp-1) of proteins and the origin of eukaryotic cells. Mol Microbiol 1995, 15: 1–11.
16. Singh B, Patel H V, Ridley R G, Freeman K B and Gupta R S. Mitochondrial import of the human chaperonin (HSP60) protein. Biochem Biophys Res Commun 1990, 169: 391–396.
17. Ikawa S and Weinberg R A. An interaction between p21ras and heat shock protein hsp60, a chaperonin. Proc Natl Acad Sci USA 1992, 89: 2012–2016.
18. Jones M, Gupta R S and Englesberg E. Enhancement in amount of P1 (hsp60) in mutants of Chinese hamster ovary (CHO-K1) cells exhibiting increases in the A system of amino acid transport. Proc Natl Acad Sci USA 1994, 91: 858–862.
19. Khan I U, Wallin R, Gupta R S and Kammer G M. Protein kinase A-catalyzed phosphorylation of heat shock protein 60 chaperone regulates its attachment to histone 2B in the T lymphocyte plasma membrane. Proc Natl Acad Sci USA 1998, 95: 10425–10430.
20. Gupta R S, Ho T K, Moffat M R and Gupta R. Podophyllotoxin-resistant mutants of Chinese hamster ovary cells. Alteration in a microtubule-associated protein. J Biol Chem 1982, 257: 1071–1078.
21. Jindal S, Dudani A K, Singh B, Harley C B and Gupta R S. Primary structure of a human mitochondrial protein homologous to the bacterial and plant chaperonins and to the 65-kilodalton mycobacterial antigen. Mol Cell Biol 1989, 9: 2279–2283.
22. Picketts D J, Mayanil C S and Gupta R S. Molecular cloning of a Chinese hamster mitochondrial protein related to the 'chaperonin' family of bacterial and plant proteins. J Biol Chem 1989, 264: 12001–12008.
23. Gupta R S and Austin R C. Mitochondrial matrix localization of a protein altered in mutants resistant to the microtubule inhibitor podophyllotoxin. Eur J Cell Biol 1987, 45: 170–176.
24. Gupta R S and Dudani A K. Mitochondrial binding of a protein affected in mutants resistant to the microtubule inhibitor podophyllotoxin. Eur J Cell Biol 1987, 44: 278–285.
25. Gupta R S. Mitochondria, molecular chaperone proteins and the *in vivo* assembly of microtubules. Trends Biochem Sci 1990, 15: 415–418.
26. Fisch P, Malkovsky M, Kovats S, Sturm E, Braakman E, Klein B S, Voss S D, Morrissey L W, DeMars R, Welch W J, Bolhuis R L H and Sondel P M. Recognition by human Vγ9/Vδ2 T cells of a GroEL homolog on Daudi Burkitt's lymphoma cells. Science 1990, 250: 1269–1273.
27. Koga T, Wand-Wurttenberger A, DeBruyn J, Munk M E, Schoel B and Kaufmann S H E. T cells against a bacterial heat shock protein recognize stressed macrophages. Science 1989, 245: 1112–1115.
28. Kaur I, Voss S D, Gupta R S, Schell K, Fisch P and Sondel P M. Human peripheral γδ T cells recognize hsp60 molecules on Daudi Burkitt's lymphoma cells. J Immunol 1993, 150: 2046–2055.
29. Soltys B J and Gupta R S. Cell surface localization of the 60 kDa heat shock chaperonin protein (hsp60) in mammalian cells. Cell Biol Int 1997, 21: 315–320.

30. Bocharov A V, Vishnyakova T G, Baranova I N, Remaley A T, Patterson A P and Eggerman T L. Heat shock protein 60 is a high-affinity high-density lipoprotein binding protein. Biochem Biophys Res Commun 2000, 277: 228–235.
31. Lukacs K V, Lowrie D B, Stokes R W and Colston M J. Tumor cells transfected with a bacterial heat-shock gene lose tumorigenicity and induce protection against tumors. J Exp Med 1993, 178: 343–348.
32. Wells A D and Malkovsky M. Heat shock proteins, tumor immunogenicity and antigen presentation: an integrated view. Immunol Today 2000, 21: 129–132.
33. Gupta R S and Gupta R. Mutants of chinese hamster ovary cells affected in two different microtubule-associated proteins. Genetic and biochemical studies. J Biol Chem 1984, 259: 1882–1890.
34. Soltys B J and Gupta R S. Mitochondrial molecular chaperones Hsp60 and mHsp70: Are their roles restricted to mitochondria? In Abe, H. and Latchman, D. S. (Eds.) Handbook of Experimental Pharmacology: Heat Shock Proteins. Springer-Verlag New York, Inc., New York: 1998, pp 69–100.
35. Xu Q, Schett G, Seitz C S, Hu Y, Gupta R S and Wick G. Surface staining and cytotoxic activity of heat-shock protein 60 in stressed aortic endothelial cells. Circ Res 1994, 75: 1078–1085.
36. Schett G, Metzler B, Mayr M, Amberger A, Niederwieser D, Gupta R S, Mizzen L, Xu Q and Wick G. Macrophage-lysis mediated by autoantibodies to heat shock protein 65/60. Atherosclerosis 1997, 128: 27–38.
37. Poccia F, Piselli P, Vendetti S, Bach S, Amendola A, Placido R and Colizzi V. Heat-shock protein expression on the membrane of T cells undergoing apoptosis. Immunology 1996, 88: 6–12.
38. Gupta S and Knowlton A A. Cytosolic heat shock protein 60, hypoxia, and apoptosis. Circulation 2002, 106: 2727–2733.
39. Soltys B J and Gupta R S. Immunoelectron microscopic localization of the 60-kDa heat shock chaperonin protein (Hsp60) in mammalian cells. Exp Cell Res 1996, 222: 16–27.
40. Brudzynski K, Martinez V and Gupta R S. Immunocytochemical localization of heat-shock protein 60-related protein in beta-cell secretory granules and its altered distribution in non-obese diabetic mice. Diabetologia 1992, 35: 316–324.
41. Cechetto J D, Soltys B J and Gupta R S. Localization of mitochondrial 60-kD heat shock chaperonin protein (Hsp60) in pituitary growth hormone secretory granules and pancreatic zymogen granules. J Histochem Cytochem 2000, 48: 45–56.
42. Velez-Granell C S, Arias A E, Torres-Ruiz J A and Bendayan M. Molecular chaperones in pancreatic tissue: the presence of cpn10, cpn60 and hsp70 in distinct compartments along the secretory pathway of the acinar cells. J Cell Sci 1994, 107: 539–549.
43. Velez-Granell C S, Arias A E, Torres-Ruiz J A and Bendayan M. Presence of Chromatium vinosum chaperonins 10 and 60 in mitochondria and peroxisomes of rat hepatocytes. Biol Cell 1995, 85: 67–75.
44. Rothman S S. Protein transport by the pancreas. Science 1975, 190: 747–753.
45. Sadacharan S K, Cavanagh A C and Gupta R S. Immunoelectron microscopy provides evidence for the presence of mitochondrial heat shock 10-kDa protein (chaperonin 10) in red blood cells and a variety of secretory granules. Histochem Cell Biol 2001, 116: 507–517.

46. Jamieson J D and Palade G E. Intracellular transport of secretory proteins in the pancreatic exocrine cell. II. Transport to condensing vacuoles and zymogen granules. J Cell Biol 1967, 34: 597–615.
47. Bassan M, Zamostiano R, Giladi E, Davidson A, Wollman Y, Pitman J, Hauser J, Brenneman D E and Gozes I. The identification of secreted heat shock 60-like protein from rat glial cells and a human neuroblastoma cell line. Neurosci Lett 1998, 250: 37–40.
48. Orci L, Vassalli J-D and Perrelet A. The insulin factory. Scientific American 1988, 259: 85–94.
49. Hendrick J P and Hartl F U. The role of molecular chaperones in protein folding. FASEB J 1995, 9: 1559–1569.
50. Jarvis J A, Ryan M T, Hoogenraad N J, Craik D J and Hoj P B. Solution structure of the acetylated and noncleavable mitochondrial targeting signal of rat chaperonin 10. J Biol Chem 1995, 270: 1323–1331.
51. Cavanagh A C and Morton H. The purification of early-pregnancy factor to homogeneity from human platelets and identification as chaperonin 10. Eur J Biochem 1994, 222: 551–560.
52. Quinn K A, Cavanagh A C, Hillyard N C, McKay D A and Morton H. Early pregnancy factor in liver regeneration after partial hepatectomy in rats: relationship with chaperonin 10. Hepatology 1994, 20: 1294–1302.
53. Cavanagh A C. Identification of early pregnancy factor as chaperonin 10: implications for understanding its role. Rev Reprod 1996, 1: 28–32.
54. Fletcher B H, Cassady A I, Summers K M and Cavanagh A C. The murine chaperonin 10 gene family contains an intronless, putative gene for early pregnancy factor, Cpn10-rs1. Mamm Genome 2001, 12: 133–140.
55. Alberts B, Johnson A, Lewis J, Raff M, Roberts K and Walter P. *Molecular Biology of the Cell*. Garland Publishing, Inc., New York: 2002.
56. Weiss L. *Cell and Tissue Biology: A Textbook of Histology*. Urban and Schwarzenberg, Baltimore: 1988.
57. Domanico S Z, DeNagel D C, Dahlseid J N, Green J M and Pierce S K. Cloning of the gene encoding peptide-binding protein 74 shows that it is a new member of the heat shock protein 70 family. Mol Cell Biol 1993, 13: 3598–3610.
58. Mizukoshi E, Suzuki M, Loupatov A, Uruno T, Hayashi H, Misono T, Kaul S C, Wadhwa R and Imamura T. Fibroblast growth factor-1 interacts with the glucose-regulated protein GRP75/mortalin. Biochem J 1999, 343: 461–466.
59. Bhattacharyya T, Karnezis A N, Murphy S P, Hoang T, Freeman B C, Phillips B and Morimoto R I. Cloning and subcellular localization of human mitochondrial hsp70. J Biol Chem 1995, 270: 1705–1710.
60. Singh B, Soltys B J, Wu Z C, Patel H V, Freeman K B and Gupta R S. Cloning and some novel characteristics of mitochondrial Hsp70 from Chinese hamster cells. Exp Cell Res 1997, 234: 205–216.
61. Ungermann C, Neupert W and Cyr D M. The role of Hsp70 in conferring unidirectionality on protein translocation into mitochondria. Science 1994, 266: 1250–1253.
62. Caplan A J, Cyr D M and Douglas M G. Eukaryotic homologues of *Escherichia coli* DnaJ: a diverse protein family that functions with hsp70 stress proteins. Mol Biol Cell 1993, 4: 555–563.

63. VanBuskirk A M, DeNagel D C, Guagliardi L E, Brodsky F M and Pierce S K. Cellular and subcellular distribution of PBP72/74, a peptide-binding protein that plays a role in antigen processing. J Immunol 1991, 146: 500–506.
64. Sacht G, Brigelius-Flohe R, Kiess M, Sztajer H and Flohe L. ATP-sensitive association of mortalin with the IL-1 receptor type I. Biofactors 1999, 9: 49–60.
65. Kaul S C, Taira K, Pereira-Smith O M and Wadhwa R. Mortalin: present and prospective. Exp Gerontol 2002, 37: 1157–1164.
66. Wadhwa R, Kaul S C, Ikawa Y and Sugimoto Y. Identification of a novel member of mouse hsp70 family. Its association with cellular mortal phenotype. J Biol Chem 1993, 268: 6615–6621.
67. Choglay A A, Chapple J P, Blatch G L and Cheetham M E. Identification and characterization of a human mitochondrial homologue of the bacterial co-chaperone GrpE. Gene 2001, 267: 125–134.
68. Syken J, Macian F, Agarwal S, Rao A and Münger K. TID1, a mammalian homologue of the drosophila tumor suppressor lethal(2) tumorous imaginal discs, regulates activation-induced cell death in Th2 cells. Oncogene 2003, 22: 4636–4641.
69. Cechetto J D and Gupta R S. Immunoelectron microscopy provides evidence that tumor necrosis factor receptor-associated protein 1 (TRAP-1) is a mitochondrial protein which also localizes at specific extramitochondrial sites. Exp Cell Res 2000, 260: 30–39.
70. Gupta R S. Phylogenetic analysis of the 90 kD heat shock family of protein sequences and an examination of the relationship among animals, plants, and fungi species. Mol Biol Evol 1995, 12: 1063–1073.
71. Chen C F, Chen Y, Dai K, Chen P L, Riley D J and Lee W H. A new member of the hsp90 family of molecular chaperones interacts with the retinoblastoma protein during mitosis and after heat shock. Mol Cell Biol 1996, 16: 4691–4699.
72. Simmons A D, Musy M M, Lopes C S, Hwang L Y, Yang Y P and Lovett M. A direct interaction between EXT proteins and glycosyltransferases is defective in hereditary multiple exostoses. Hum Mol Genet 1999, 8: 2155–2164.
73. Herwig S and Strauss M. The retinoblastoma protein: a master regulator of cell cycle, differentiation and apoptosis. Eur J Biochem 1997, 246: 581–601.
74. Morita T, Amagai A and Maeda Y. Unique behavior of a dictyostelium homologue of TRAP-1, coupling with differentiation of D. discoideum cells. Exp Cell Res 2002, 280: 45–54.
75. Ledgerwood E C, Prins J B, Bright N A, Johnson D R, Wolfreys K, Pober J S, O'Rahilly S and Bradley J R. Tumor necrosis factor is delivered to mitochondria where a tumor necrosis factor-binding protein is localized. Lab Invest 1998, 78: 1583–1589.
76. Glick B and Schatz G. Import of proteins into mitochondria. Ann Rev Genet 1991, 25: 21–44.
77. Bradbury M W and Berk P D. Mitochondrial aspartate aminotransferase: direction of a single protein with two distinct functions to two subcellular sites does not require alternative splicing of the mRNA. Biochem J 2000, 345: 423–427.
78. Cechetto J D, Sadacharan S K, Berk P D and Gupta R S. Immunogold localization of mitochondrial aspartate aminotransferase in mitochondria and on the cell surface in normal rat tissues. Histol Histopathol 2002, 17: 353–364.
79. Ghebrehiwet B and Peerschke E I. Structure and function of gC1q-R: a multiligand binding cellular protein. Immunobiology 1998, 199: 225–238.

80. Soltys B J, Kang D and Gupta R S. Localization of P32 protein (gC1q-R) in mito-chondria and at specific extramitochondrial locations in normal tissues. Histochem Cell Biol 2000, 114: 245–255.

81. Green D R and Reed J C. Mitochondria and apoptosis. Science 1998, 281: 1309–1312.

82. Soltys B J, Andrews D A, Jemmerson R and Gupta R S. Cytochrome c localizes in secretory granules in pancreas and anterior pituitary. Cell Biol Int 2001, 25: 331–338.

83. Poyton R O, Duhl D M J and Clarkson G H D. Protein export from the mitochondrial matrix. Trends Cell Biol 1992, 2: 369–375.

84. Smalheiser N R. Proteins in unexpected locations. Mol Biol Cell 1996, 7: 1003–1014.

Changing Paradigms of Protein Trafficking and Protein Function

3

Novel Pathways of Protein Secretion

Giovanna Chimini and Anna Rubartelli

3.1. Introduction

Intercellular communications are fundamental for many of the biological processes that are involved in the survival of living organisms, and secretory proteins are among the most important messengers in this network of information. Proteins destined for this function are endowed with a hydrophobic signal peptide which targets them to the endoplasmic reticulum (ER) and are released in the extracellular environment by a 'classical' pathway of constitutive or regulated secretion. However, in the early 1990s it became evident that non-classical mechanisms must exist for the secretion of some proteins which, despite their extracellular localisation and function, lack a signal peptide. Indeed, the family of these leaderless secretory proteins continues to grow and comprises proteins that, although apparently unrelated, share both structural and functional features. This chapter will review current hypotheses on the mechanisms underlying non-classical secretion and discuss their implications in the regulation of the inflammatory and immune response. The relevance of non-classical secretion pathways to molecular chaperone biology is also discussed in Chapters 2 and 12.

3.2. Leaderless secretory proteins

Secretory mechanisms that are discrete to the classical pathways appear early in evolution. Gram negative bacteria are endowed with many (up to six) types of secretion mechanisms that are, at least in part, independent of the general secretory pathway, the prototype being the haemolysin secretion system [1]. In addition, two pathways of secretion that avoid the ER exist in yeast. Whereas one of these seems to be activated only as a detoxifying tool [2], the other is essential because it allows the release of the a-factor, a key mating factor [3]. In higher eukaryotes, leaderless secretory proteins display some common structural features, such as a relatively low molecular mass (12–45 kDa with few exceptions), the absence of N-linked glycosylation, even if potential sites are present, and the presence of free cysteines that are not engaged in disulphide bridges [4]. These characteristics suggest that leaderless secretory proteins avoid the ER, in which post-translational modifications such as N-linked glycosylation and formation of disulphide bridges take place. Moreover, the pharmacological evidence that brefeldin A, a drug that blocks secretion of classical secretory proteins, does not affect the release of leaderless proteins [5, 6] further supports the hypothesis that their secretion must follow non-classical export routes.

Due to the relatively large cytosolic accumulation of these proteins, it was originally proposed that their presence outside the cell could be the consequence of a passive release of cytoplasmic content, as might occur in the case of cell death. This possibility was ruled out by observations that the presence of leaderless proteins in the extracellular environment is selective and does not correlate with the presence of other cytoplasmic proteins such as lactate dehydrogenase, a marker for cell lysis [4]. Leaderless secretion has also been shown to be energy- or temperature-dependent and can be blocked by a number of treatments or drugs. On the basis of this evidence, it is now currently accepted that the secretion of leaderless proteins is controlled by an active mechanism that excludes the ER–Golgi apparatus.

Table 3.1 lists the most studied leaderless secretory proteins. Some of these belong to the interleukin family (IL-1, IL-16, IL-18), others are cytosolic enzymes or derive from proteins that have a well-defined intracellular function (thioredoxin, thioredoxin reductase, phosphoglucose isomerase/AMF, EMAP II, caspase I, transglutaminase), and others are nuclear proteins which may be readdressed to the extracellular compartment (HMGB-1, engrailed-2). An unexpected nuclear localisation has also been reported. Although in some cases this localisation is associated with a function (galectin-3 is involved in pre-mRNA splicing [33] and macrophage migration inhibitory factor (MIF) might work as a transcription factor modulator [34]), in the case of fibroblast growth factor (FGF)-1 and -2 and IL-1α a nuclear function is still debated.

The list of leaderless secretory proteins continues to grow. However, some proteins that have previously been considered to belong to this class have turned out to function mostly, if not exclusively, intracellularly. For instance, pro- and para-thymosine alpha, previously regarded as thymic hormones, now appear to be essential nuclear factors [35] and their extracellular role is debated [36]; platelet-derived endothelial cell growth factor turned out to be identical to the well-known enzyme thymidine phosphorylase and to play an angiogenic role as a result of its intracellular enzymatic activity [37].

Table 3.1. Leaderless secretory proteins

	Extracellular function	Intracellular function	Key reference
IL-1α	Pro-inflammatory immune mediator	Activator of transcription	[7, 8]
IL-1β	Pro-inflammatory immune mediator	–	[7]
IL-18	Pro-inflammatory immune mediator	–	[9]
caspase I (ICE)	?	IL-1/IL-18 converting enzyme	[10, 11]
IL-16	Pro-inflammatory immune mediator, T-lymphocyte chemoattractant	–	[12]
FGF-1	Growth, angiogenic and motility factor	Growth regulator (nuclear)?	[13]
FGF-2	Growth, angiogenic and motility factor	Growth regulator (nuclear)?	[14]
FGF-9	Neurotrophin, Growth survival factor	–	[15]
CNTF	Neurotrophin	–	[16]
MIF	Pro-inflammatory mediator, shock factor	Transcription factor modulator	[17]
EMAP II	Pro-inflammatory immune mediator	Apoptosis mediator	[18, 19]
Annexin I	Anti-inflammatory immune mediator	Vesicular traffic	[20–22]
AMF	Growth motility factor	Glycolytic enzyme	[23]
TRX/ADF	Immune mediator, chemotactic factor, involved in implantation and establishment of pregnancy	Major cellular disulfide reductase	[24]
TRX reductase	Enzyme reducing oxidised TRX?	Enzyme reducing oxidised TRX	[24]
Galectines	Pro/anti-inflammatory factors	Phagocytosis	[25–27]
Transglutaminase	Protein cross-linking	Apoptosis (intracellular protein cross-linking)	[28]
Coagulation factor XIII	Coagulation	–	[29]
HMGB1	Pro-inflammatory, shock factor	Chromatin component	[30]
Engrailed-2	Transcription factor by intercellular transfer?	Homeodomain transcription factor	[31, 32]

Finally, it is worth stressing that, from a functional point of view, most of these proteins display functions related to the regulation of inflammatory processes [37]. The finding that molecular chaperones also act to control inflammation is, therefore, suggestive.

3.3. Mechanisms of leaderless secretion

3.3.1. Targeting motifs

Although a considerable amount of effort has been dedicated to the understanding of leaderless secretion, the molecular mechanism is still only partially defined. It is likely that this mechanism involves a sequential series of events. First of all, recognition is required and a given leaderless secretory protein must be selected from amongst a myriad of cytosolic macromolecules. A common sorting motif has not yet been identified, although in a number of instances a requirement for a specific primary structure has been determined. In two cases, a single cysteine is crucial for secretion and the mutation of Cys 30 in FGF-1 [38, 39] and Cys 277 in tissue transglutaminase [40] is sufficient to impair their externalisation. Short sequences are required for secretion of chick ciliary neurotrophic factor (CNTF) and galectin-3, and in both proteins these lie in the N-terminal domain. Whereas for CNTF this sequence (AA 46–53) contains six hydrophobic aminoacids [41], in the case of galectin-3 two proline residues (Pro 90 and Pro 93) in the context of AA 89–96 seem crucial [42, 43].

Post-translational modifications play an important role in the secretion of two nuclear factors: acetylation for the chromatin component high mobility group box 1 protein (HMGB-1) [44] and phosphorylation for the homeoprotein engrailed-2 [31]. In both cases, the protein shuttles continuously from the nucleus to the cytoplasm due to nuclear export signals, but the equilibrium is almost completely shifted towards nuclear accumulation. Hyperacetylation of HMGB-1 and de-phosphorylation of engrailed-2 relocate the proteins to the cytoplasm. In turn, cytoplasmic availability allows secretion. Phosphorylation also appears to be required for the secretion of the glycolytic enzyme phosphohexose isomerase/autocrine motility factor (AMF). In this case, however, the modification induces secretion, whereas the non-phosphorylated enzyme remains within the cell [45]. Phosphohexose isomerase/AMF is the archetypal moonlighting protein and is discussed again in Chapters 4 and 5.

Other post-translational modifications such as mirystoylation or farnesylation might positively modulate the secretion of certain leaderless secretory proteins such as IL-1α [46] and galectin-3 [42] by enhancing their recruitment to cell membranes. Nevertheless, these modifications do not represent a bona fide secretory signal, because other cytosolic proteins undergo lipid modification but are not secreted.

3.3.2. Reaching the extracellular space: vesiculation or translocation?

Once recognised, the proteins must be externalised. This can be accomplished by two different mechanisms, namely vesiculation or translocation.

3.3.2.1. Vesiculation

Vesicles may result from either outward or inward membrane bending. In the first case, evagination of the membrane leads to bleb formation followed by the release of extracellular vesicles. Their membrane instability allows the rapid solubilisation of the protein. This mechanism, which has been first documented for the lectin L-14/galectin-1 [47], implies a prior concentration of the protein in patches beneath the plasma membrane. Interestingly, bleb formation can be restricted to the apical membrane in epithelial cells, resulting in polarised secretion [48]. This is also the case of the apocrine secretion described in glandular cells of the male reproductive system which leads to the apical release of a number of leaderless proteins [49] including tissue transglutaminase [50] and MIF [51]. Moreover, microvesicle shedding has been proposed to mediate IL-1β release from a monocytic cell line [52].

In the second case, repeated invaginations of the plasma membrane generate multivesicular bodies. These are multilamellar complexes containing vesicles called exosomes that are released upon the fusion of the multivesicular bodies with the plasma membrane [53]. Exosomes have been observed primarily in haemopoietic cells in which they can serve a variety of functions, including shedding of transferrin receptor during reticulocyte maturation [54] or antigen presentation [55]. In addition, exosomes may mediate externalisation of leaderless proteins. Indeed, galectin-3 and several annexins have been detected by proteomic analysis in exosomes from dendritic cells (DCs) [56], suggesting that they are selectively recruited in this compartment. The potential role of exosomes in the release of molecular chaperones is discussed in Chapter 12.

3.3.2.2. Translocation

Leaderless proteins that are not secreted by vesiculation must cross a cell membrane to reach the extracellular space. Membrane translocation has to meet a number of requirements. Firstly, as a general rule, proteins must unfold in order to acquire a 'translocation competent conformation.' Unfolding is usually assisted by cytoplasmic chaperones [57]. This has been clearly demonstrated for secretion of leaderless proteins in prokaryotes [58], whereas it has not been directly addressed in mammalian leaderless secretion. Interestingly, in the case of FGF-1, unfolding does not seem crucial for translocation. Indeed, Prudovsky et al. [59]

suggest that FGF-1 is secreted as a dimer assembled in a multi-molecular complex with other cytosolic proteins. However, this apparent inconsistency is not the only exception to the rule because peroxisomal proteins may be imported into peroxisomes as heterotrimers [60]. Secondly, protein translocation implies the presence of dedicated membrane transporters. Several specific transporters have been identified as mediators of membrane crossing in most intracellular organelles, from the ER to mitochondria and peroxisomes. Also, the lysosomal membrane is equipped with a transporter (Lamp-2A) which is able to import up to 30% of cytosolic proteins under stress conditions [61].

In yeast and bacteria, the secretion of leaderless proteins [1, 3] is, in most cases, dependent on membrane proteins belonging to the family of ATP-binding cassette (ABC) transporters. This family of transporters is conserved in mammals, in which 48 transporters organised into 7 structural classes (A to G) have been identified [62]. They include, amongst others, the multi-drug resistance protein, responsible for tumour resistance to chemotherapeutic drugs; the cystic fibrosis gene product; the ER proteins Tap-1 and Tap-2, involved in translocation of antigenic peptides; and ABCA1, a crucial regulator of clearance of apoptotic cells also implicated in lipid homeostasis.

By analogy with yeast and bacteria, the implication of ABC transporters in leaderless secretion has also been investigated in mammals. Although a direct demonstration has not been provided, pharmacological evidence strongly suggests that member(s) of the ABCA class mediate translocation of a number of leaderless proteins. Specifically, glybenclamide, a known inhibitor of the activity of ABCA1, has been reported to impair the secretion of IL-1β [63], Annexin I [64], MIF [65] and HMGB-1 (Rubartelli, unpublished observation).

In principle, translocation of a leaderless protein may occur at the plasma membrane or at the membrane of any intracellular organelle able to undergo exocytosis. Despite extensive study, direct evidence for protein translocation at the plasma membrane was lacking until recently, when *in vitro* translocation of FGF-2 and galectin-1 across inside-out membrane vesicles was demonstrated [66]. In contrast, a large body of evidence suggests that leaderless proteins may be imported into cytoplasmic organelles associated with the lysosomal compartment. Immunohistochemical studies have shown the presence of Annexin I in eosinophil granules [67], of chicken CNTF in an endosomal compartment of transfected cells [41], of FGF-2 in mast cell secretory granules [68], and of MIF within secretory granules of pituitary cells [69]. Morphological and biochemical evidence has reported the localisation of engrailed-2 [70] and HMGB-1 [71] in intracellular vesicles belonging to the early endocytic pathway and of IL-1β [72], IL-18 [73], caspase-I and Annexin I (Rubartelli, unpublished results) in the endolysosomes of activated monocytes.

Taken together, these results suggest that the lysosomal compartment is involved in leaderless secretion. Because secretory lysosomes are particularly abundant in haemopoietic cells, in which they are implicated in immune inflammatory processes [74], lysosome-mediated secretion of leaderless proteins is consistent with the role played by these proteins in the modulation of inflammation. Interestingly, the involvement of acidic vesicles in the export of leaderless proteins is evolutionarily conserved; as in *Dictyostelium discoideum*, translocation into exocytic contractile vacuoles of DdCAD-1, a leaderless adhesion protein, is necessary for its externalisation [75].

As indicated earlier, the ABCA inhibitor glybenclamide blocks secretion of several leaderless proteins: more specifically, glybenclamide prevents the appearance in secretory lysosomes of IL-1β [72] and other leaderless proteins (Rubartelli, unpublished results). This implies a lysosomal localisation of the putative ABCA protein responsible for translocation, in line with the presumption that all intracellular membranes may be endowed with transporters of the ABC family [76].

It is to be noted that, depending on the cell system, the same leaderless protein seems to use different pathways of secretion. For example, IL-1β has been found to be released by a myelomonocytic cell line via vesiculation [52], whereas in primary monocytes it undergoes lysosome-mediated exocytosis [72]. Galectin-1, one of the first examples of secretion by vesicle shedding [47], has been found to be capable of translocating directly at the plasma membrane [66]. Similarly, FGF-2 is able to cross plasma membrane vesicles [66], despite the fact that it was previously reported to accumulate in mast cell granules and be secreted after degranulation [68]. Whether these discrepancies are due to the experimental model or reflect physiological modulation remains to be elucidated.

3.4. Lysosome-mediated polarised secretion

In general, lysosome-mediated secretion is a regulated process in that a triggering signal is required to induce exocytosis [74]. In the case of IL-1β, two steps are needed for secretion by monocytes. Firstly, an inflammatory stimulus such as lipopolysaccharide (LPS) induces synthesis which results in cytosolic accumulation and lysosomal translocation; then a second extracellular signal triggers exocytosis resulting in IL-1β release [72, 77]. A similar two-step mechanism seems to account for the regulated secretion of other pro-inflammatory leaderless cytokines such as IL-18 [73] and HMGB-1 [71]. In all these cases, the signal triggering secretion is generated during the process of inflammation. ATP, promoting IL-1β and IL-18 secretion [78], is released by monocytes themselves and by other cells involved in inflammation (i.e., platelets) soon after

LPS stimulation [79]. In contrast, active phospholipids such as phosphatidyl-choline, which are responsible for secretion of HMGB-1, appear later in the inflammatory microenvironment [71].

Interestingly, in addition to inflammatory cells such as monocytes, DCs, the professional antigen presenting cells, also express inflammatory leaderless cytokines upon activation by maturational stimuli such as LPS or the engagement of CD40. However, in these cells, soluble signals seem unable to drive secretion; rather secretion occurs following interaction of DCs with antigen-specific T cells [73, 80]. Morphological approaches have demonstrated that the interaction between DCs and CD8$^+$ T cells is associated with the recruitment of IL-1β– or IL-18–containing secretory lysosomes in the areas of contact among the cells. This results in a polarisation of these organelles and evidence of lysosome exocytosis at the intercellular space – the so-called 'immunological synapse' [81]. These findings warrant two considerations.

On the one hand, they underlie the existence of a bi-directional cross-talk between T cells and DCs, in which a T cell induces the functional polarisation of a DC and the DC responds by a degranulation that is orientated towards the same T cell, with obvious relevance for the control of the immune response. On the other hand, the different ways by which monocytes and DCs regulate secretion may account for the different function of IL-1β and IL-18 in inflammation and the immune response (Fig. 3.1). Monocytes respond to soluble signals with generalised exocytosis, thus allowing the spreading of inflammatory cytokines in the microenvironment, whereas DCs respond to the localised signal provided by the interacting T cell. This restricts the area of release to the immunological synapse and allows the activation of target cells without a wider distribution of the cytokine, thus controlling inflammation. Thus, lysosome-mediated secretion of inflammatory leaderless proteins allows polarised secretion in non-polarised cells.

3.5. Advantages of a leaderless secretory pathway

Leaderless secretion is generally inefficient and of a relatively 'short range' compared to the classical pathway, and this questions the rationale of its conservation. We can envisage several explanations for this. First of all, the vast majority of leaderless secretory proteins are cytokines, as characterised by a high biological activity and their involvement in the regulation of inflammatory processes, both as inducers and as silencers. The cytokine network needs to be perfectly balanced in order to control the onset, progression and resolution of inflammation; indeed, its dysregulation might lead to pathological conditions such as chronic inflammation and autoimmune diseases [7]. In this respect, slow and inefficient

Non-Polarized   Polarized

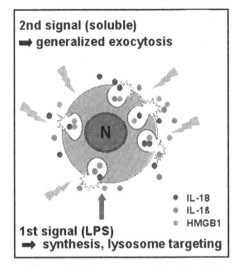
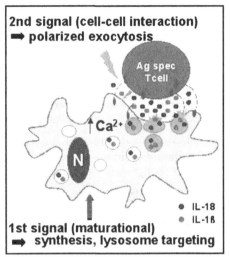

Figure 3.1. Lysosome-mediated leaderless secretion in inflammation and immune response. *Non-polarised:* A first inflammatory stimulus (e.g., LPS) induces monocytes to actively synthesise IL-1ß [7] and to hyperacetylate HMGB-1 that is readdressed from the nucleus to the cytoplasm [44]. Both proteins accumulate into the cytosol and in part into secretory lysosomes [71, 72]. A second extracellular soluble signal (e.g., ATP for IL-1ß [79], lysophosphatydylcholine for HMGB-1 [71]) triggers generalised lysosome exocytosis. Although IL-18 is constitutively expressed by monocytes, both signals (LPS and ATP) are needed for secretion to occur [78]. *Polarised:* In DCs, a first maturational stimulus, soluble (LPS, TNF-α) or cell-mediated (CD40 triggering by CD40 ligand expressing activated CD4[+] T cells), induces IL-1β synthesis. The second signal is provided by antigen-specific T cells, which induces a $[Ca^{2+}]_i$ rise, followed by the recruitment of IL-1β–containing secretory lysosomes towards the interacting T cell, and exocytosis which is restricted to the intercellular space (immunological synapse) [81]. A similar mechanism undergoes IL-18 secretion [73, 80].

secretion may become an advantage in that it self-limits the protein activity in a restricted environment and allows the control of potentially toxic effects. In addition, many leaderless proteins require reducing conditions to maintain their function.

Mutation assays have demonstrated the requirement of one or more free cysteines for IL-1β [82], IL-18 [83] and FGF-1 [84]. Galectins are sensitive to oxidation if not bound to their appropriate glycoconjugates [85], and at least some of the cytokine functions of thioredoxin (TRX) [24] and MIF [86] require the active enzymatic redox site CXXC. These observations suggest that in principle these proteins should rapidly lose their activity in the oxidising extracellular milieu. On the one hand, this can be a further way of regulating the potentially dangerous hyperactivity of these proteins; however, on the other hand, it is tempting to speculate that the parallel secretion of oxidation-sensitive cytokines and of TRX and TRX reductase in the local microenvironment of the immune response serves to lengthen their survival span and, for instance, to prolong immunostimulation [87].

Another advantage may be the prevention of intracellular autocriny via the compartmentalisation of receptor and ligand. Indeed, in physiological situations, co-expression of leaderless cytokines and their specific receptors is a common event. Thus, during their transit to the extracellular milieu, leaderless proteins should avoid the interaction with their own receptors, because this might lead to undue early activation with unwanted consequences, such as uncontrolled proliferation or even cell transformation [4]. In the case of the leaderless protein FGF-2, the insertion of a secretory leader sequence at the N terminus has been shown to induce cell transformation in the cell line expressing the FGF-2 receptor [88]. Exclusion from the classical secretory pathway may also prevent misfolding of some leaderless proteins.

As mentioned before, many of these proteins bear free sulphhydryl groups which must be maintained in the reduced state in order to guarantee folding and bioactivity. The ER lumen is highly oxidising and favours the formations of disulphide bridges. For proteins that present free thiols, the transit through the ER might thus result in either retention or secretion in a non-functional folding. Similarly, avoiding the Golgi compartment can be an advantage for galectins, because sugars, which are highly abundant in the Golgi lumen, may trap these proteins and thereby impair their transport.

For some proteins, leaderless secretion occurs only under non-physiological conditions. For instance, in yeast, over-expression of some endogenous proteins, the intracellular accumulation of which may be toxic, induces their secretion through a non-classical export system [2]. A similar mechanism also seems to exist in mammals, an example being the mitochondrial sulphotransferase rhodanese, which in physiological conditions accumulates in mitochondria, but when over-expressed is also efficiently secreted [89]. Similarly, heterologous expression of green fluorescent protein (GFP) leads to its cytosolic accumulation but also activates the secretion of improperly folded molecules [90]. Non-classical secretion may thus act as a safety valve, maintaining cellular homeostasis when the cytoplasmic degradative pathways are overloaded. In this context, it is worth stressing that results obtained from studies in which chimaeric proteins, bearing putative sequences for leaderless secretion, are over-expressed must be interpreted with caution, because secretion might result from misfolding rather than from true recognition.

Finally, it must be remembered that many leaderless secretory proteins also have an intracellular function. Some of these are cytosolic enzymes: TRX and MIF are oxide reductases [24, 91]; AMF is the ubiquitous glycolytic enzyme phosphohexose isomerase [23] and tissue transglutaminase catalises the cross-linking of intracellular proteins [28]. Others such as HMGB-1 and engrailed-2 are nuclear factors [30, 32], and Annexin I regulates vesicular traffic [20]. It is possible that, on the basis of its physiological and developmental state, a cell addresses a cytosolic protein towards an additional extracellular function – the non-classical secretory route would guarantee the likelihood of this double function.

REFERENCES

1. de Lima Pimenta A, Blight M A, Chervaux C and Holland I B. Protein secretion in Gram negative bacteria. In Kuchler, K., Rubartelli, A. and Holland, B. (Eds.) *Unusual Secretory Pathways: From Bacteria to Man*. Chapman & Hall Landes Bioscience, New York/Austin, TX 1997, pp 1–48.
2. Cleves A E, Cooper D N, Barondes S H and Kelly R B. A new pathway for protein export in *Saccharomyces cerevisiae*. J Cell Biol 1996, 133: 1017–1026.
3. Kuchler K and Egner R. Unusual protein secretion and translocation pathways in yeast: implication of ABC transporters. In Kuchler, K., Rubartelli, A. and Holland, B. (Eds.) *Unusual Secretory Pathways: From Bacteria to Man*. Chapman & Hall Landes Bioscience, New York/Austin, TX 1997, pp 49–86.
4. Rubartelli A and Sitia R. Secretion of mammalian proteins that lack a signal sequence. In Kuchler, K., Rubartelli, A. and Holland, B. (Eds.) *Unusual Secretory Pathways: From Bacteria to Man*. Chapman & Hall Landes Bioscience, New York/Austin, TX 1997, pp 87–115.
5. Rubartelli A, Cozzolino F, Talio M and Sitia R. A novel secretory pathway for interleukin-1 beta, a protein lacking a signal sequence. EMBO J 1990, 9: 1503–1510.
6. Rubartelli A, Bajetto A, Allavena G, Wollman E and Sitia R. Secretion of thioredoxin by normal and neoplastic cells through a leaderless secretory pathway. J Biol Chem 1992, 267: 24161–24164.
7. Dinarello C A. Proinflammatory cytokines. Chest 2000, 118: 503–508.
8. Werman A, Werman-Venkent R, White R, Lee J K, Werman B, Krelin Y, Voronov E, Dinarello C A, and Apte R N. The precursor form of IL1alpha is an intracrine proinflammatory activator of transcription. Proc Natl Acad Sci U S A, 2004, 101: 2434–2439.
9. Dinarello C A and Fantuzzi G. Interleukin-18 and host defense against infection. J Infect Dis 2003, 187: S370–S384.
10. Fantuzzi G and Dinarello C A. Interleukin-18 and interleukin-1β: two cytokine substrates for ICE (caspase-1). J Clin Immunol 1999, 19: 1–11.
11. Laliberte R E, Eggler J and Gabel C A. ATP treatment of human monocytes promotes caspase-1 maturation and externalization. J Biol Chem 1999, 274: 36944–36951.
12. Cruikshank W W, Kornfeld H and Center D M. Interleukin-16. J Leuk Biol 2000, 67: 757–766.

13. Christofori G. The role of fibroblast growth factors in tumour progression and angiogenesis. In Bicknell, R., Lewis, C. E. and Ferrara, N. (Eds.) *Tumour Angiogenesis*. Oxford University Press, Oxford, UK 1997, pp 201–238.

14. Bikfalvi A, Savona C, Perollet C and Javerzat S. New insights in the biology of fibroblast growth factor-2. Angiogenesis 1998, 1: 155–173.

15. Tsai S J, Wu M H, Chen H M, Chuang P C and Wing L Y. Fibroblast growth factor-9 is an endometrial stromal growth factor. Endocrinology 2002, 143: 2715–2721.

16. Sleeman M W, Anderson K D, Lambert P D, Yancopoulos G D and Wiegand S J. The ciliary neurotrophic factor and its receptor, CNTFR alpha. Pharm Acta Helv 2000, 74: 265–272.

17. Calandra T. Macrophage migration inhibitory factor and host innate immune responses to microbes. Scand J Infect Dis 2003, 35: 573–576.

18. Ko Y G, Park H, Kim T, Lee J W, Park S G, Seol W, Kim J E, Lee W H, Kim S H, Park J E and Kim S. A cofactor of tRNA synthetase, p43, is secreted to up-regulate proinflammatory genes. J Biol Chem 2001, 276: 23028–23033.

19. Berger A C, Alexander H R, Tang G, Wu P S, Hewitt S M, Turner E, Kruger E, Figg W D, Grove A, Kohn E, Stern D and Libutti S K. Endothelial monocyte activating polypeptide II induces endothelial cell apoptosis and may inhibit tumor angiogenesis. Microvasc Res 2000, 60: 70–80.

20. Rescher U, Zobiack N and Gerke V. Intact Ca^{2+}-binding sites are required for targeting of annexin 1 to endosomal membranes in living HeLa cells. J Cell Sci 2000, 113: 3931–3938.

21. Solito E, Nuti S and Parente L. Dexamethasone-induced translocation of lipocortin (annexin) 1 to the cell membrane of U-937 cells. Br J Pharmacol 1994, 112: 347–348.

22. Christmas P, Callaway J, Fallon J, Jones J and Haigler H T. Selective secretion of annexin 1, a protein without a signal sequence, by the human prostate gland. J Biol Chem 1991, 266: 2499–2507.

23. Tsutsumi S, Yanagawa T, Shimura T, Fukumori T, Hogan V, Kuwano H and Raz A. Regulation of cell proliferation by autocrine motility factor/phosphoglucose isomerase signaling. J Biol Chem 2003, 278: 32165–32172.

24. Arner E S and Holmgren A. Physiological functions of thioredoxin and thioredoxin reductase. Eur J Biochem 2000, 267: 6102–6109.

25. Rabinovich G A, Baum L G, Tinari N, Paganelli R, Natoli C, Liu F T and Iacobelli S. Galectins and their ligands: amplifiers, silencers or tuners of the inflammatory response? Trends Immunol 2002, 23: 313–320.

26. Rabinovich G A, Rubinstein N and Toscano M A. Role of galectins in inflammatory and immunomodulatory processes. Biochim Biophys Acta 2002, 1572: 274–284.

27. Sano H, Hsu D K, Apgar J R, Yu L, Sharma B B, Kuwabara I, Izui S and Liu F T. Critical role of galectin-3 in phagocytosis by macrophages. J Clin Invest 2003, 112: 389–397.

28. Griffin M, Casadio R and Bergamini C M. Transglutaminases: Nature's biological glues. Biochem J 2002, 368: 377–396.

29. Grundmann U, Amann E, Zettlmeissl G and Kupper H A. Characterization of cDNA coding for human factor XIIIa. Proc Natl Acad Sci USA 1986, 83: 8024–8028.

30. Wang H, Bloom O, Zhang M, Vishnubhakat J M, Ombrellino M, Che J, Frazier A, Yang H, Ivanova S, Borovikova L, Manogue K R, Faist E, Abraham E, Andersson J,

Andersson U, Molina P E, Abumrad N N, Sama A and Tracey K J. HMG-1 as a late mediator of endotoxin lethality in mice. Science 1999, 285: 248–251.

31. Maizel A, Tassetto M, Filhol O, Cochet C, Prochiantz A and Joliot A. Engrailed homeoprotein secretion is a regulated process. Development 2002, 129: 3545–3553.

32. Maizel A, Bensaude O, Prochiantz A and Joliot A. A short region of its homeodomain is necessary for engrailed nuclear export and secretion. Development 1999, 126: 3183–3190.

33. Dagher S F, Wang J L and Patterson R J. Identification of galectin-3 as a factor in pre-mRNA splicing. Proc Natl Acad Sci USA 1995, 92: 1213–1217.

34. Kleemann R, Hausser A, Geiger G, Mischke R, Burger-Kentischer A, Flieger O, Johannes F J, Roger T, Calandra T, Kapurniotu A, Grell M, Finkelmeier D, Brunner H and Bernhagen J. Intracellular action of the cytokine MIF to modulate AP-1 activity and the cell cycle through Jab1. Nature 2000, 408: 211–216.

35. Vareli K, Frangou-Lazaridis M, van der Kraan I, Tsolas O and van Driel R. Nuclear distribution of prothymosin alpha and parathymosin: evidence that prothymosin alpha is associated with RNA synthesis processing and parathymosin with early DNA replication. Exp Cell Res 2000, 257: 152–161.

36. Hannappel E and Huff T. The thymosins. Prothymosin alpha, parathymosin, and beta-thymosins: structure and function. Vit Horm 2003, 66: 257–296.

37. Focher F and Spadari S. Thymidine phosphorylase: a two-face Janus in anticancer chemotherapy. Cur Cancer Drug Targets 2001, 1: 141–153.

38. Jackson A, Tarantini F, Gamble S, Friedman S and Maciag T. The release of fibroblast growth factor-1 from NIH 3T3 cells in response to temperature involves the function of cysteine residues. J Biol Chem 1995, 270: 33–36.

39. Tarantini F, Gamble S, Jackson A and Maciag T. The cysteine residue responsible for the release of fibroblast growth factor-1 residues in a domain independent of the domain for phosphatidylserine binding. J Biol Chem 1995, 270: 29039–29042.

40. Balklava Z, Verderio E, Collighan R, Gross S, Adams J and Griffin M. Analysis of tissue transglutaminase function in the migration of Swiss 3T3 fibroblasts: the active-state conformation of the enzyme does not affect cell motility but is important for its secretion. J Biol Chem 2002, 277: 16567–16575.

41. Reiness C G, Seppa M J, Dion D M, Sweeney S, Foster D N and Nishi R. Chick ciliary neurotrophic factor is secreted via a nonclassical pathway. Mol Cell Neurosci 2001, 17: 931–944.

42. Menon R P and Hughes R C. Determinants in the N-terminal domains of galectin-3 for secretion by a novel pathway circumventing the endoplasmic reticulum-Golgi complex. Eur J Biochem 1999, 264: 569–576.

43. Gong H C, Honjo Y, Nangia-Makker P, Hogan V, Mazurak N, Bresalier R S and Raz A. The NH2 terminus of galectin-3 governs cellular compartmentalization and functions in cancer cells. Cancer Res 1999, 59: 6239–6245.

44. Bonaldi T, Talamo F, Scaffidi P, Ferrera D, Porto A, Bachi A, Rubartelli A, Agresti A and Bianchi M E. Monocytic cells hyperacetylate chromatin protein HMGB1 to redirect it towards secretion. EMBO J 2003, 22: 5551–5560.

45. Haga A, Niinaka Y and Raz A. Phosphohexose isomerase/autocrine motility factor/neuroleukin/maturation factor is a multifunctional phosphoprotein. Biochim Biophys Acta 2000, 1480: 235–244.

46. Stevenson F T, Bursten S L, Fanton C, Locksley R M and Lovett D H. The 31-kDa precursor of interleukin 1 alpha is myristoylated on specific lysines within the 16-kDa N-terminal propiece. Proc Natl Acad Sci USA 1993, 90: 7245–7249.

47. Cooper D N and Barondes S H. Evidence for export of a muscle lectin from cytosol to extracellular matrix and for a novel secretory mechanism. J Cell Biol 1990, 110: 1681–1691.

48. Lindstedt R, Apodaca G, Barondes S H, Mostov K E and Leffler H. Apical secretion of a cytosolic protein by Madin-Darby canine kidney cells. Evidence for polarized release of an endogenous lectin by a nonclassical secretory pathway. J Biol Chem 1993, 268: 11750–11757.

49. Hermo L and Jacks D. Nature's ingenuity: bypassing the classical secretory route via apocrine secretion. Mol Reprod Dev 2002, 63: 394–410.

50. Steinhoff M, Eicheler W, Holterhus P M, Rausch U, Seitz J and Aumuller G. Hormonally induced changes in apocrine secretion of transglutaminase in the rat dorsal prostate and coagulating gland. Eur J Cell Biol 1994, 65: 49–59.

51. Eickhoff R, Wilhelm B, Renneberg H, Wennemuth G, Bacher M, Linder D, Bucala R, Seitz J and Meinhardt A. Purification and characterization of macrophage migration inhibitory factor as a secretory protein from rat epididymis: evidences for alternative release and transfer to spermatozoa. Mol Med 2001, 7: 27–35.

52. MacKenzie A, Wilson H L, Kiss-Toth E, Dower S K, North R A and Surprenant A. Rapid secretion of interleukin-1β by microvesicle shedding. Immunity 2001, 15: 825–835.

53. Murk J L, Stoorvogel W, Kleijmeer M J and Geuze H J. The plasticity of multivesicular bodies and the regulation of antigen presentation. Sem Cell Dev Biol 2002, 13: 303–311.

54. Johnstone R M, Adam M, Hammond J R, Orr L and Turbide C. Vesicle formation during reticulocyte maturation. Association of plasma membrane activities with released vesicles (exosomes). J Biol Chem 1987, 262: 9412–9420.

55. Théry C, Zitvogel L and Amigorena S. Exosomes: composition, biogenesis and function. Nat Rev Immunol 2002, 2: 569–579.

56. Théry C, Boussac M, Véron P, Ricciardi-Castagnoli P, Raposo G, Garin G and Amigorena S. Proteomic analysis of dendritic cell-derived exosomes: A secreted subcellular compartment distinct from apoptotic vesicles. J Immunol 2001, 166: 7309–7318.

57. Fink A L. Chaperone-mediated protein folding. Physiol Rev 1999, 79: 425–449.

58. Holland I B, Benhabdelhak H, Young J, de Lima Pimenta A, Schmitt L and Blight M A. Bacterial ABC transporters involved in protein translocation. In Holland, I. B. (Ed.) ABC Proteins: from Bacteria to Man. Academic Press, London/San Diego 2003, pp 209–242.

59. Prudovsky I, Bagala C, Tarantini F, Mandinova A, Soldi R, Bellum S and Maciag T. The intracellular translocation of the components of the fibroblast growth factor 1 release complex precedes their assembly prior to export. J Cell Biol 2002, 158: 201–208.

60. McNew J A and Goodman J M. An oligomeric protein is imported into peroxisomes in vivo. J Cell Biol 1994, 127: 1245–1257.

61. Cuervo A M, Mann L, Bonten E J, d'Azzo A and Dice J F. Cathepsin A regulates chaperone-mediated autophagy through cleavage of the lysosomal receptor. EMBO J 2003, 22: 47–59.
62. Dean M, Hamon Y and Chimini G. The human ATP-binding cassette (ABC) transporter superfamily. J Lipid Res 2001, 42: 1007–1017.
63. Hamon Y, Luciani M F, Becq F, Verrier B, Rubartelli A and Chimini G. Interleukin-1β secretion is impaired by inhibitors of the ATP binding cassette transporter, ABC1. Blood 1997, 90: 2911–2915.
64. Chapman L P, Epton M J, Buckingham J C, Morris J F and Christian H C. Evidence for a role of the adenosine 5'-triphosphate-binding cassette transporter A1 in the externalization of annexin I from pituitary folliculo-stellate cells. Endocrinology 2003, 144: 1062–1073.
65. Flieger O, Engling A, Bucala R, Lue H, Nickel W and Bernhagen J. Regulated secretion of macrophage migration inhibitory factor is mediated by a non-classical pathway involving an ABC transporter. FEBS Lett 2003, 551: 78–86.
66. Schafer T, Zentgraf H, Zehe C, Brugger B, Bernhagen J and Nickel W. Unconventional protein secretion: direct translocation of fibroblast growth factor 2 across the plasma membrane of mammalian cells. J Biol Chem 2003, 279: 6244–6251.
67. Oliani S M, Damazo A S and Perretti M. Annexin 1 localisation in tissue eosinophils as detected by electron microscopy. Med Inflamm 2002, 11: 287–292.
68. Qu Z, Kayton R J, Ahmadi P, Liebler J M, Powers M R, Planck S R and Rosenbaum J T. Ultrastructural immunolocalization of basic fibroblast growth factor in mast cell secretory granules. Morphological evidence for bfgf release through degranulation. J Histochem Cytochem 1998, 46: 1119–1128.
69. Nishino T, Bernhagen J, Shiiki H, Calandra T, Dohi K and Bucala R. Localization of macrophage migration inhibitory factor (MIF) to secretory granules within the corticotrophic and thyrotrophic cells of the pituitary gland. Mol Med 1995, 1: 781–788.
70. Joliot A, Maizel A, Rosenberg D, Trembleau A, Dupas S, Volovitch M and Prochiantz A. Identification of a signal sequence necessary for the unconventional secretion of Engrailed homeoprotein. Cur Biol 1998, 8: 856–863.
71. Gardella S, Andrei C, Ferrera D, Lotti L V, Torrisi M R, Bianchi M E and Rubartelli A. The nuclear protein HMGB1 is secreted by monocytes via a non-classical, vesicle-mediated secretory pathway. EMBO Rep 2002, 3: 995–1001.
72. Andrei C, Dazzi C, Lotti L, Torrisi M R, Chimini G and Rubartelli A. The secretory route of the leaderless protein interleukin 1beta involves exocytosis of endolysosome-related vesicles. Mol Biol Cell 1999, 10: 1463–1475.
73. Gardella S, Andrei C, Poggi A, Zocchi M R and Rubartelli A. Control of interleukin-18 secretion by dendritic cells: role of calcium influxes. FEBS Lett 2000, 481: 245–248.
74. Blott E J and Griffiths G M. Secretory lysosomes. Nat Rev Mol Cell Biol 2002, 3: 122–131.
75. Sesaki H, Wong E F and Siu C H. The cell adhesion molecule DdCAD-1 in Dictyostelium is targeted to the cell surface by a nonclassical transport pathway involving contractile vacuoles. J Cell Biol 1997, 138: 939–951.
76. Holland I B. ABC Proteins: from Bacteria to Man. Academic Press, London/San Diego: 2003.

77. Andrei C, Margiocco P, Poggi A, Lotti L V, Torrisi M R, Rubartelli A. Phospholipases C and A_2 control lysosome – mediated IL-1beta secretion: Implications for inflammatory processes. Proc Natl Acad Sci USA 2004, 101: 9745–9750.

78. Perregaux D G, McNiff P, Laliberte R, Conklyn M and Gabel C A. ATP acts as an agonist to promote stimulus-induced secretion of IL-1β and IL-18 in human blood. J Immunol 2000, 165: 4615–4623.

79. Di Virgilio F, Chiozzi P, Ferrari D, Falzoni S, Sanz J M, Morelli A, Torboli M, Bolognesi G and Baricordi O R. Nucleotide receptors: an emerging family of regulatory molecules in blood cells. Blood 2001, 97: 587–600.

80. Gardella S, Andrei C, Costigliolo S, Poggi A, Zocchi M R and Rubartelli A. Interleukin-18 synthesis and secretion by dendritic cells are modulated by interaction with antigen-specific T cells. J Leuk Biol 1999, 66: 237–241.

81. Gardella S, Andrei C, Lotti L V, Poggi A, Torrisi M R, Zocchi M R and Rubartelli A. CD8+ T lymphocytes induce polarized exocytosis of secretory lysosomes by dendritic cells with release of interleukin-1β and cathepsin D. Blood 2001, 98: 2152–2159.

82. Kamogashira T, Masui Y, Ohmoto Y, Hirato T, Nagamura K, Mizuno K, Hong Y M, Kikumoto Y, Nakai S and Hirai Y. Site-specific mutagenesis of the human interleukin-1β gene: structure-function analysis of the cysteine residues. Biochem Biophys Res Comm 1988, 150: 1106–1114.

83. Pei D S, Fu Y, Sun Y F and Zhao H R. Site-directed mutagenesis of the cysteines of human IL-18 and its effect on IL-18 activity. Sheng Wu Hua Xue Yu Sheng Wu Wu Li Xue Bao (Shanghai) 2002, 34: 57–61.

84. Ortega S, Schaeffer M T, Soderman D, DiSalvo J, Linemeyer D L, Gimenez-Gallego G and Thomas K A. Conversion of cysteine to serine residues alters the activity, stability, and heparin dependence of acidic fibroblast growth factor. J Biol Chem 1991, 266: 5842–5846.

85. Cho M and Cummings R D. Galectin-1, a β-galactoside-binding lectin in Chinese hamster ovary cells. I. Physical and chemical characterization. J Biol Chem 1995, 270: 5198–5206.

86. Kleemann R, Kapurniotu A, Mischke R, Held J and Bernhagen J. Characterization of catalytic centre mutants of macrophage migration inhibitory factor (MIF) and comparison to Cys81Ser MIF. Eur J Biochem 1999, 261: 753–766.

87. Angelini G, Gardella S, Ardy M, Ciriolo M R, Filomeni G, Di Trapani G, Clarke F, Sitia R and Rubartelli A. Antigen-presenting dendritic cells provide the reducing extracellular microenvironment required for T lymphocyte activation. Proc Nat Acad Sci USA 2002, 99: 1491–1496.

88. Rogelj S, Weinberg R A, Fanning P and Klagsbrun M. Basic fibroblast growth factor fused to a signal peptide transforms cells. Nature 1988, 331: 173–175.

89. Sloan I S, Horowitz P M and Chirgwin J M. Rapid secretion by a nonclassical pathway of overexpressed mammalian mitochondrial rhodanese. J Biol Chem 1994, 269: 27625–27630.

90. Tanudji M, Hevi S and Chuck S L. Improperly folded green fluorescent protein is secreted via a non-classical pathway. J Cell Sci 2002, 115: 3849–3857.

91. Lue H, Kleemann R, Calandra T, Roger T and Bernhagen J. Macrophage migration inhibitory factor (MIF): mechanisms of action and role in disease. Microbes Infect 2002, 4: 449–460.

4

Moonlighting Proteins: Proteins with Multiple Functions

Constance J. Jeffery

4.1. Introduction

Moonlighting proteins, also referred to as 'gene sharing', refer to a subset of multifunctional proteins in which two or more different functions are performed by one polypeptide chain, and the multiple functions are not a result of splice variants, gene fusions, or multiple isoforms [1]. In addition, they do not include proteins with the same function in multiple locations or protein families in which different members have different functions, if each individual member has only one function. A single protein with multiple functions may seem surprising, but there are actually many cases of proteins that 'moonlight'.

4.2. Examples and mechanisms of combining two functions in one protein

The current examples of moonlighting proteins include enzymes, DNA binding proteins, receptors, transmembrane channels, chaperones and ribosomal proteins (Table 4.1). In general, there are several different methods by which a moonlighting protein can combine two functions within one polypeptide chain. A single protein can have a second function when it moves to a different cellular location; when it is expressed in a different cell type; when it binds a substrate, product, or cofactor; when it interacts with another protein to form a multimer, or when it interacts with a large multiprotein complex. In addition, a few enzymes have two active sites for different substrates (Figure 4.1). The methods are not mutually exclusive and sometimes a combination of methods is employed.

Cellular location: Several cytosolic or nuclear enzymes have a second function outside of the cell. Phosphoglycerate kinase and phosphoglucose isomerase

Table 4.1. Moonlighting proteins

One function	Another function	Reference
Plasmin reductase	Phosphoglycerate kinase	[2]
Phosphoglucose isomerase	Neuroleukin, autocrine motility factor, differentiation and maturation mediator	[3–6]
Thymidine phosphorylase	Platelet-derived endothelial cell growth factor	[7]
Thymosin β4 (sequester actin)	Secreted chemotaxis ligand	[8]
SMC3 (sister chromatin cohesion)	Basement membrane bamacam	[9, 10]
Histone H1	Thyroglobulin receptor	[11]
Neuropilin (VEGF receptor)	Receptor for semaphorin III (nerve axons)	[12]
Thymidylate synthase	Translation inhibitor	[13]
birA biotin sythetase	bio operon repressor	[14]
PutA proline dehydrogenase	Transcriptional repressor	[15]
Aconitase	Iron responsive element binding protein (IRE-BP)	[16]
Paramyxovirus hemaglutinin	Neuraminidase	[17]
4a-Carbinolamine dehydratase	Dimerization cofactor (DCoH)	[18]
δ-Aminolevulinic acid dehydratase	Proteasome inhibitory subunit CF-2	[19]
Ribosomal proteins	DNA repair, translational regulators, etc.	[20]
Clf1p pre-mRNA splicing factor	Initiation of DNA replication	[21–24]
Proteasome base complex	RNA polIII transcription	[25]
Cyclooxygenase-1	Heme-dependent peroxidase	[26]
Lysyl hydroxylase isoform 3	Collagen glucosyltransferase	[27]
CFTR chloride channel	Regulator of other epithelial anion channels	[28]
Mitochondrial Lon protease	Chaperone	[29]
Bacterial FtsH chaperone	Metalloprotease	[29]
Lens crystallins	Heat shock proteins, lactate dehydrogenase, argininosuccinate lyase, retinaldehyde dehydrogenase, enolase, quinone oxidoreductase, glyceraldehyde-3-phosphate dehydrogenase	[30]
PHGPx (glutathione peroxidase)	Sperm structural protein	[31]
E. coli thioredoxin	Subunit of T7 DNA polymerase	[38]
PMS2 mismatch repair enzyme	Hypermutation of antibody variable chains	[39]
Leukotriene A4 hydrolase	Aminopeptidase	[40]
1-cys peroxiredoxin (peroxidase)	Phospholipase aiPLA2	[41]
Tetrahymena citrate synthase	14-nm cytoskeletal protein	[42]
Transferrin receptor	Glyceraldehyde-3-phosphate dehydrogenase	[43]
Lactose synthetase	Galactosyltransferase	[44]
Homing endonuclease	Intron splicing factor	[45]
N. crassa tyrosyl tRNA synthetase	Promotes folding of group I introns	[46]
Cytochrome c (electron transport)	Apoptosis	[47]

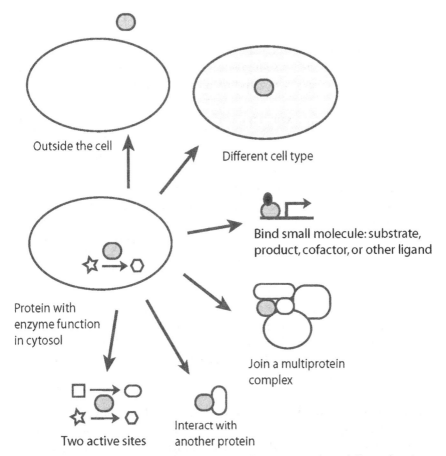

Figure 4.1. Methods of switching functions. A single protein can have different functions in different cellular locations, when expressed in different cell types; when it binds a ligand, substrate, product, or cofactor; when it interacts with another protein to form a multimer; when it interacts with a large multiprotein complex; or by having two binding sites for different substrates or ligands.

catalyse the seventh and second steps, respectively, in glycolysis in the cytosol of most cells. Both are also found to have a second function outside of the cell. Phosphoglycerate kinase is a disulphide reductase that reduces plasmin [2]. The reduced plasmin undergoes proteolysis to produce angiostatin, an angiogenesis inhibitor. Phosphoglucose isomerase binds to cell surface receptors on target cells and causes a variety of effects, including differentiation of pre-B cells to antibody secreting cells, an increase in motility of some tumour cells, and differentiation of HL-60 leukaemia cells to monocytes [3–6].

Thymidine phosphorylase, which is also called platelet-derived endothelial cell growth factor, removes the phosphoryl group from thymidine and deoxyuridine in the cytoplasm and stimulates chemotaxis of endothelial cells outside of the cell [7]. Thymosin beta 4 sulphoxide is an inhibitor of actin polymerisation in the cytosol and serves as a negative modulator of the inflammatory response outside the cell [8].

Whereas the preceding examples include cytosolic proteins that serve as soluble growth factors, enzymes, or cytokines outside of the cell, other cytoplasmic or nuclear proteins have an extracellular second function in which they are not soluble. The mouse SMC3 protein (structural maintenance of chromosome 3), also known as bamacam, functions in sister chromatid cohesion in the nucleus and is also a component of the basement membrane [9, 10]. Histone H1 is another nuclear protein with a function outside the cell, but it remains attached to the extracellular surface of the cell membrane and serves as a receptor for thyroglobulin [11].

Different cell types: Expression by multiple cell types can also result in a protein having multiple functions. For example, neuropilin is a cell surface receptor in neurons and endothelial cells [12]. When expressed in neurons, neuropilin binds semaphorin III and plays a role in axonal guidance. When expressed in endothelial cells, it binds vascular endothelial growth factor (VEGF) and helps signal the need for new blood cells.

Binding substrate, product, cofactor, or other ligand: Binding to a substrate, product, cofactor, or other ligand can cause a change in the function of a protein. The enzymes thymidylate synthase [13], biotin synthetase (*Escherichia coli birA*) [14], PutA proline dehydrogenase [15], and aconitase [16] (also called iron responsive binding protein, IRE-BP) are three cytosolic or membrane-bound enzymes that detect changes in the cellular concentration of a ligand and then bind to DNA or RNA and regulate transcription or translation.

Paramyxovirus hemagglutinin-neuraminidase responds to changes in the pH of its environment by changing conformation of several amino acid side chains and a loop in the active site. These movements may enable a switch between the sialic acid binding and hydrolysis functions of the protein [17].

Forming a complex with other proteins: Entering into multiprotein complexes is another method by which a protein can exhibit a moonlighting function. In some cases, the moonlighting protein interacts with only one or a few other proteins. 4α-carbinolamine dehydratase (also called DCoH) is an enzyme in liver cells. It also binds to the transcription factor HNF1 δ (hepatic nuclear factor

1δ). By influencing the dimerisation of HNF1 δ, DCoH regulates the binding of the transcription factor to DNA [18].

In other cases a protein becomes part of a large multiprotein complex, such as the proteasome or the ribosome, which is composed of many different polypeptide chains. Delta-aminolevulinic acid dehydratase, and enzyme in heme biosynthesis, is the same protein as the 240-kDA inhibitory component of the proteasome [19]. Several other cytoplasmic or nuclear enzymes have been found to be identical to proteins in the ribosome (reviewed in [20]). In addition, there are a few examples of moonlighting proteins that participate in multiple multiprotein complexes, changing roles with the different polypeptide partners. *Saccharomyces cerevisiae* Clf1p apparently performs different functions by interacting with different proteins in two multiprotein complexes for pre-mRNA splicing and the initiation of DNA replication. It interacts with the U5 and U6 small nuclear ribonucleoprotein particles, pre-mRNA, and other components of pre-mRNA splicing reactions. It also interacts with the DNA replication initiation protein Orc2p in the origin of replication complex [21–24].

Sug1/Rpt6 and Sug2/Rpt4 are AAA proteins (an ATP-dependent protein superfamily including molecular chaperones involved in protein assembly/disassembly) that form part of the base complex of the proteasome. The base complex and lid complex, which make up the 19S particle, join with the 20S proteolytic complex to catalyse proteolysis. However, the base complex also plays a role in RNA polIII transcription, without the lid complex or the 20S particle. In response to galactose induction, the base complex moves to the GAL1-10 promoter, and the Sug1/Rpt6 and Sug2/Rpt4 proteins interact directly with the Gal4 transactivator to alter transcription levels from the GAL1-10 promoter [25].

Multiple binding sites for different substrates or ligands: Other proteins do not necessarily have a switch mechanism to change functions; they simply have multiple binding sites or active site pockets for different ligands or substrates. The enzyme prostaglandin H2 synthase-1 has two active sites: a heme-dependent peroxidase active site and a cyclooxygenase active site. The two active site pockets are found near each other in the enzyme structure [26], and both catalyse reactions in the synthesis of prostaglandin H2. Similarly, the enzyme lysyl hydroxylase 3 catalyses two steps in collagen biosynthesis [27].

Overall, there are a number of ways in which a protein can combine two functions within one polypeptide chain. The methods are not mutually exclusive, and, in some cases, a combination of factors contribute to switching between functions.

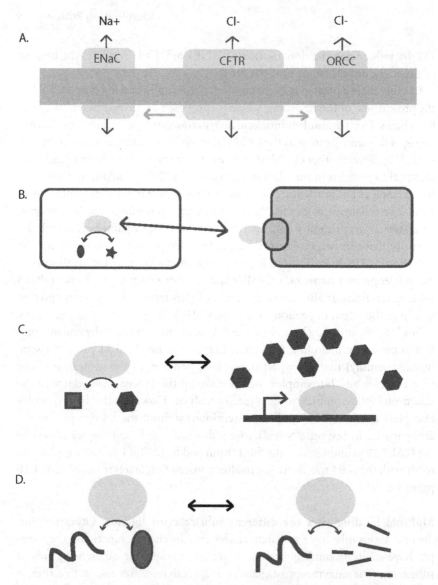

Figure 4.2. Examples of benefits provided by proteins that moonlight: (A) Coordination of functions within a cell. The CFTR is a chloride channel and it regulates (horizontal grey arrows) the activity of the EnaC sodium channel and the ORCC chloride channel. (B) Coordination between different cell types. Some proteins with catalytic activity in the cytoplasm of one cell type (black oval to star) also have an extracellular function in which they bind to a receptor on other cells types. (C) Feedback. Some enzymes that catalyse a chemical reaction (square to hexagon) can sense when the accumulation of product (hexagon) is high and bind to DNA to inhibit the synthesis of more copies of the enzyme. (D) Switch between pathways. Some chaperones that help a protein fold (curved line to black oval) also have a protease activity that can degrade a protein (curved line to short line fragments), depending on cellular conditions.

4.3. Why have moonlighting proteins?

The wide variety of moonlighting proteins and combinations of functions suggests that moonlighting evolved independently many times. This suggests that there are benefits to having moonlighting proteins, or that it is relatively easy for a second function to evolve. Analysis of the examples in Table 4.1 suggests that both are true. In fact, moonlighting proteins appear to provide several kinds of benefits to the organism. Also, there are two proposed mechanisms for moonlighting proteins to have evolved that make use of general physical properties of many protein structures.

4.3.1. Benefits to the organism

Having moonlighting proteins can provide a means to coordinate different biochemical pathways, a means to respond to stress or changes in the environment, and as a feedback mechanism (Figure 4.2).

Coordination: As the complex modern cell evolved, a need arose for methods to coordinate the many intracellular biochemical pathways for signalling, transport, biosynthesis and other functions, and moonlighting proteins provide one such mechanism. For example, combining the two enzymatic functions of lysyl hydroxylase 3, described earlier, within one protein might help coordinate two steps involved in collagen maturation. The cystic fibrosis transmembrane conductance regulator (CFTR) is a chloride channel and also regulates the activity of the outwardly rectifying chloride channel (ORCC) and a sodium channel (ENaC) (Figure 4.2) [28]. The ability of one transmembrane channel to coordinate the activity of several kinds of channels helps to maintain ion homeostasis within epithelial cells. In addition, as multicellular organisms developed, the need for coordination of activities between different cells, cell types, and organs arose, which might be one reason there are multiple examples of intracellular enzymes with a second function as a cytokine or growth factor (Figure 4.2).

Switch between pathways: The combination of two alternative functions within one protein might also provide an efficient method to switch between two pathways in response to changing conditions in the environment, such as changes in food supply or the introduction of a stress (Figure 4.2). Changes in cellular iron concentrations cause a decrease in the catalytic activity of aconitase, a cytosolic enzyme. Aconitase, also known as the iron-responsive element binding protein, then binds to DNA to cause changes in transcription of proteins involved in iron accumulation [16]. The two functions of delta-aminolevulinic acid dehydratase, as an enzyme in the heme biosynthesis pathway and as the proteasome inhibitor CF-2, might provide a method to switch between protein degradation and heme biosynthesis [19]. Mitochondrial Lon protease, which is both a protease and a chaperone (reviewed in [29]), provides a more general switch between protein degradation and protein biosynthesis.

Feedback: Some enzymes that catalyse one step of a biochemical pathway moonlight as a sensor for the overall level of activity of the pathway by measuring the concentration of substrates or products and then binding to DNA or RNA in order to regulate transcription or translation of enzymes within the pathway (Figure 4.2). This combination of functions provides a feedback mechanism to regulate the activity of the pathway. Thymidylate synthase 3, biotin synthetase (birA), and PutA proline dehydrogenase are three examples of enzymes that also bind DNA or RNA to regulate transcription or translation in response to changing levels of substrate, product, or cofactor [13–15].

No clear benefit: Although in many cases it appears there is indeed a benefit to having two functions within one polypeptide chain, it is not always clear if there is a connection between the two functions in some moonlighting proteins. As described earlier, phosphoglucose isomerase is both an enzyme in glycolysis in the cell cytosol and an extracellular cytokine [3–6]. It is not clear why an organism would make use of a glycolytic enzyme in this way. One possibility is that after the second function evolved both functions benefit the organism independently and there was no selective pressure to remove it.

4.4. Models for the evolution of moonlighting proteins

But how can a protein develop a second function within the same polypeptide chain? From consideration of the examples listed in Table 4.1, it appears that there are two general mechanisms for a protein to evolve a second function (Figure 4.3).

Recruitment of a protein without significant change in protein structure: The first method involves recruitment of the protein for a new function, perhaps as a new organ or cell type evolves, without major changes in protein structure (Figure 4.3). The crystallins, members of the small heat shock protein family, are a classic example of this method. Several crystallins are ubiquitous, soluble, cytosolic enzymes that were recruited for a second function in the lens when

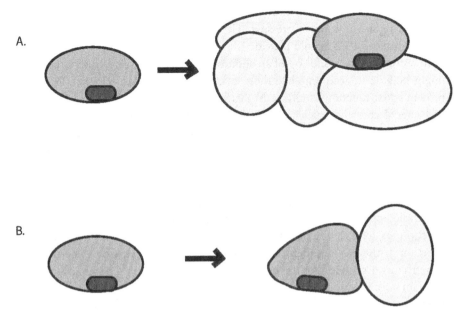

Figure 4.3. Methods of evolution. A protein can evolve a second function (A) by recruitment without a significant change in protein structure or (B) by modification of 'unused' solvent exposed surface area. In both cases, the original function, as represented by the black oval 'active site' remains.

the eye evolved [30]. Two other proteins that appear to have evolved a second function by recruitment into a multiprotein complex are PHGPx (glutathione peroxidase), a soluble enzyme that is also a sperm structural protein [31], and *E. coli* thioredoxin, which was adopted by the T7 phage as a subunit of its DNA polymerase [32].

Modification of solvent-exposed surface area: The other general method for a protein to develop a moonlighting function is based on the observation that several ubiquitous enzymes appear to have evolved a protein-binding site on the protein surface in addition to their catalytic sites (Figure 4.3). One example is phosphoglucose isomerase (PGI). PGI is found in almost all species and so must have evolved over three billion years ago. Throughout all that time, the enzyme active site has been conserved, but the protein surface has undergone many changes. As is seen in many enzymes, the active site pocket of PGI is actually rather small compared to the total surface area of the protein. In this relatively large protein – a dimer with 557 amino acids in each subunit – there is a large amount of apparently unused solvent exposed surface area that might not be under tight evolutionary pressure. The random accumulation of mutations on the surface of PGI could provide the material and opportunity that might have resulted in the evolution of an additional binding site. The protein surface might be able to accommodate these changes without adversely affecting the protein's first function: catalysis. In fact, a comparison between the structures of bacterial and mammalian PGI indicates that several alpha helices, surface loops, and surface pockets have undergone considerable change through evolution, whereas the active site pockets of the two enzymes have remained almost identical [33].

Both the recruitment and modification of surface area methods of evolution of a moonlighting function make use of general features of protein structure and could apply to many proteins. Whichever way that a protein evolves a new function, as long as the new function does not adversely affect the original function of the protein, then the second function might provide some of the benefits just described. It is therefore possible that there would be an advantage to retain both functions during further evolution, or at least no selective pressure to eliminate either.

4.5. How many proteins moonlight?

The two proposed methods of evolving moonlighting functions could have happened to many proteins. In addition, the examples of moonlighting proteins identified to date include many types of proteins, including enzymes, transcription factors, channels, and receptors, and proteins from many diverse organisms and cell types. This wide variety of proteins and combinations of functions suggest that many types of proteins can moonlight. While we cannot put a distinct value on the number of proteins that moonlight, it is possible that moonlighting proteins might be common, and we could find that many other proteins also have additional functions that have not yet been found.

4.6. Identifying more moonlighting proteins

Although it is not yet clear how many proteins might moonlight, moonlighting proteins that have different functions in different locations or within different multiprotein complexes might be especially common because they could provide the key benefits listed earlier: coordinating cellular activities, response to changes in the environment, and feedback mechanisms. The current interest in large-scale proteomics studies that apply biochemical or genetic methods to characterise the locations, protein–protein interactions, or expression levels of thousands of proteins is likely to lead to the identification of more examples of moonlighting proteins.

4.6.1. Methods employing protein locations

Mass spectrometry and two-dimensional gel electrophoresis can be used to identify proteins in complex mixtures such as whole cells, organelles, or multiprotein complexes. Similarly, analysis of RNA levels using micro-array methods can suggest which proteins are being expressed in a chosen cell type or tissue. The resulting protein expression profiles can be repeated for different cell types, at different times in development, before and after application of a signal, or in diseased and healthy tissues. The differences in protein expression patterns are then used to help deduce the function of that protein. For example, in general, groups of proteins that function together in a biochemical pathway, multiprotein complex or signalling pathway are often expressed in the same cell types and under the same growth conditions, whereas in other cell types none of the proteins in those complexes or pathways are expressed – in other words, an 'all or none' pattern.

However, a moonlighting protein might have an unusual pattern of expression which is inconsistent with its function in a single biochemical pathway or protein complex. For example, it might be expressed along with the other enzymes in a multiprotein complex or multienzyme pathway in some cell types or under some conditions, but it might also be expressed in other cell types without the other proteins, in which it performs its second function. When protein expression profiles identify a protein in an unexpected cell type, organelle, or multiprotein complex, it is possible that the protein might have a moonlighting function.

4.6.2. Interactions with different binding partners

Identifying binding partners, whether other proteins or small molecules, might also help to identify moonlighting proteins, because the development of a new binding site is one way by which moonlighting proteins are proposed to evolve. Micro-array technology can be used to measure the interaction of a protein with other proteins, small molecules, antibodies, or peptides *in vitro*. The yeast two-hybrid method is used to determine *in vivo* physical interactions between proteins; for example, in protein complexes or in signalling networks. In fact, yeast two-hybrid screens are notorious for identifying many 'false positive', protein–protein interactions that do not seem to be involved in the protein complex or signalling network being studied. Moonlighting proteins can provide a possible explanation for some of the apparent 'false positives' that are observed in yeast two-hybrid experiments.

4.6.3. Unexpected protein expression levels

An unusually high expression level of a particular protein might also suggest the protein has a second function. In general, many proteins in biochemical and signalling pathways are not needed at high concentrations in the cell because they can be used repeatedly, function in a highly specialised pathway, or are involved in a cascade that amplifies a signal. However, a second function, perhaps as part of a protein complex, might require much higher levels of protein expression. Delta-aminolevulinic acid dehydratase is expressed at levels far more than is needed to catalyse a step in heme biosynthesis (up to 1% of total soluble protein). However, the surprisingly high level of protein expression makes more sense when we consider that the protein is also a proteasome inhibitory subunit.

4.7. Moonlighting proteins in disease and rational drug design

Whereas the preceding discussion included many examples of proteins in basic physiological pathways, moonlighting proteins can also be important in studies of disease. Already, moonlighting proteins have been found to be involved in tumour cell motility, angiogenesis, DNA synthesis or repair, chromatin and cytoskeleton structure, and cystic fibrosis (reviewed in [34]). The ability of a protein to moonlight can complicate the elucidation of molecular mechanisms of disease, the identification of biomarkers of disease progression, and the development of novel therapeutics.

The presence of moonlighting proteins can complicate understanding of the molecular mechanisms of disease development. Even in the case where the key proteins involved have been identified, the observed role of a particular protein in disease progression might be difficult to explain if only one function of a moonlighting protein is known. In fact, even if both functions are known, the identification of a molecular mechanism by which the proteins cause the observed symptoms can be complicated, and it might not be clear which function (or both) is responsible for the observed symptoms.

Even in the case of a genetic disease caused by altered levels or activity of a single protein, the effects of mutations on one function of a protein – for example its enzymatic activity – might not be sufficient to explain the disease symptoms. Instead, the mutation might affect a second function, such as interactions with another protein. Even if all the functions of a protein are known, the function (or both) that is affected by a disease-causing mutation might be unclear. Similarly, moonlighting proteins can complicate the understanding of mutant phenotypes of model organisms developed from experimental methods that alter the level of expression of a protein, such as gene knockouts, RNA interference, anti-sense RNA or protein over-expression.

Some specific proteins whose expression levels differ between healthy and diseased cells can serve as biomarkers for diagnosis or for following the progression of the disease. However, if a protein moonlights, the presence or absence of a disease state might be only one of many factors that affect its expression level. In that case, the lack of a direct correlation between expression levels and disease state would prevent a moonlighting protein from being a good biomarker for the disease.

The ability of a protein to moonlight can also complicate selection of potential drug targets and the development of novel therapeutics to treat disease. It is important that a drug that alters a protein activity alters the correct activity. Modifying other protein activities not involved in the disease can result in increased toxicity and side effects. A review by Searls describes in more detail the importance of considering many potential mechanisms in the evolution of protein functions, including moonlighting, gene redundancy, orthology, paralogy, and crosstalk, in selecting a suitable target for drug development [35].

4.8. Word of caution

Although the preceding discussion emphasises that moonlighting might be quite common, it is important to consider two points of caution.

Although the presence of a protein in multiple cellular locations, multiple cell types, or multiprotein complexes, or the observation of unexpected results, can suggest that a protein is moonlighting, it is not a guarantee that the protein has multiple functions. Some single-function proteins are found in multiple locations or cell types because a single function might be used in both places, for example a kinase activity. A protein with the same function in two different locations is not a moonlighting function. Evidence that the protein truly does two different things in the two locations is needed, and so it is important that the initial observations are complemented with further biochemical characterisation or other studies before a protein is determined to be moonlighting. Combining the results of multiple experimental methods, such as biochemical assays of catalytic activity, yeast two-hybrid data, and mass spectrometric analysis of cellular location, would provide more evidence of multiple functions.

Another area of caution is in regards to assigning functions to proteins based on amino acid sequence homology. If one protein is a moonlighting protein, its homologues or other isoforms might have one, the other, or both functions. For example, the *E. coli* aspartate receptor is also the receptor for maltose binding protein; however, the homologous aspartate receptor from a related bacterium, *Salmonella typhimurium*, does not bind to maltose binding protein [36, 37]. Similarly, within one organism, one isoform of a protein might have multiple functions, whereas other isoforms might each have only one function. For example, three isoforms of lysyl hydroxylase (in collagen synthesis) share approximately 60% overall amino acid sequence identity, and all three have lysyl hydroxylase catalytic activity. However, only isoform 3 (LH3) also contains galactosylhydroxylysyl glucosyltransferase catalytic activity.

4.9. Conclusions

A variety of different proteins has been found to moonlight, with different functions, mechanisms to switch between functions, ways in which they can benefit an organism, and methods by which they might have evolved, and this variety suggests that many more proteins might moonlight. In general, identifying one function of a protein is not always followed by a search for additional functions of a protein, and although there are several types of experiments that can suggest that a protein moonlights, there is no general method to identify which additional proteins moonlight. However, moonlighting proteins might be a common mechanism of communication and cooperation between the many different functions and pathways within a complex modern cell or between different cell types within an organism, and they might help explain complex disease symptoms or unexpected phenotypes from gene knockout experiments in model organisms. Perhaps the identification of more moonlighting proteins might also help explain why the human genome encodes only approximately twice as many proteins as *S. cerevisiae*, a single-celled yeast.

4.10. Acknowledgements

Research on moonlighting proteins in the Jeffery laboratory is supported by grants from the American Cancer Society.

REFERENCES

1. Jeffery C J. Moonlighting proteins. Trends Biochem Sci 1999, 24: 8–11.
2. Lay A J, Jiang X-M, Kisker O, Flynn E, Underwood A, Condron R and Hogg P J. Phosphoglycerase kinase acts in tumour angiogenesis as a disulphide reductase. Nature 2000, 408: 869–873.

3. Xu W, Seiter K, Feldman E, Ahmed T and Chiao J W. The differentiation and maturation mediator for human myeloid leukemia cells shares homology with neuroleukin or phosphoglucose isomerase. Blood 1996, 87: 4502–4506.

4. Watanabe H, Takehana K, Date M, Shinozaki T and Raz A. Tumor cell autocrine motility factor is the neuroleukin/phosphohexose isomerase polypeptide. Cancer Res 1996, 56: 2960–2963.

5. Gurney M E, Apatoff B R, Spear G T, Baumel M J, Antel J P, Bania M B and Reder A T. Neuroleukin: a lymphokine product of lectin-stimulated T cells. Science 1986, 234: 574–581.

6. Gurney M E, Heinrich S P, Lee M R and Yin H S. Molecular cloning and expression of neuroleukin, neurotrophic factor for spinal and sensory neurons. Science 1986, 234: 566–574.

7. Furukawa T, Yoshimura A, Sumizawa T, Haraguchi M and Akiyama S-I. Angiogenic factor. Nature 1992, 356: 668.

8. Young J D, Lawrence A J, Maclean A G, Leung B P, McInnes I B, Canas B, Pappin D J C and Stevenson R D. Thymosin beta 4 sulfoxide is an anti-inflammatory agent generated by monocytes in the presence of glucocorticoids. Nat Med 1999, 5: 1424–1427.

9. Wu R R and Couchman J R. cDNA cloning of the basement membrane chondroitin sulfate proteoglycan core protein, bamacan: a five domain structure including coiled-coil motifs. J Cell Biol 1997, 136: 433–444.

10. Darwiche N, Freeman L A and Strunnikov A. Characterization of the components of the putative mammalian sister chromatid cohesion complex. Gene 1999, 233: 39–47.

11. Brix K, Summa W, Lottspeich F and Herzog V. Extracellularly occuring histone H1 mediates the binding of thyroglobulin to the cell surface of mouse macrophages. J Clin Invest 1998, 102: 283–293.

12. Soker S, Takashim S, Miao H Q, Neufeld G and Klagsbrun M. Neuropilin-1 is expressed by endothelial and tumor cells as an isoform-specific receptor for vascular endothelial growth factor. Cell 1998, 92: 735–745.

13. Chu E, Koeller D M, Casey J L, Drake J C, Chabner B A, Elwood P C, Zinn S and Allegra C J. Autoregulation of human thymidylate synthase messenger RNA translation by thymidylate synthase. Proc Natl Acad Sci USA 1991, 88: 8977–8981.

14. Barker D F and Campbell A M. Genetic and biochemical characterization of the birA gene and its product: evidence for a direct role of biotin holoenzyme synthetase in repression of the biotin operon in *Escherichia coli*. J Mol Biol 1981, 146: 469–492.

15. Ostrovsky de Spicer P and Maloy S. PutA protein, a membrane-associated flavin dehydrogenase, acts as a redox-dependent transcriptional regulator. Proc Natl Acad Sci USA 1993, 90: 4295–4298.

16. Kennedy M C, Mende-Mueller L, Blondin G A and Beiner H. Purification and characterization of cytosolic aconitase from beef liver and its relationship to the iron-responsive element binding protein. Proc Natl Acad Sci USA 1992, 89: 11730–11734.

17. Crennell S, Takimoto T, Portner A and Taylor G. Crystal structure of the multi-functional paramyxovirus hemagglutinin-neuraminidase. Nat Struct Biol 2000, 7: 1068–1074.

18. Citron B A, Davis M D, Milstien S, Gutierrez J, Mendel D B, Crabtree G R and Kaufman S. Identity of 4α-carbinolamine dehydratase, a component of the

phenylalanine hydroxylation system, and DCoH, a transregulator of homeodomain proteins. Proc Natl Acad Sci USA 1992, 89: 11891–11894.

19. Guo G G, Gu M and Etlinger J D. 240-kDa proteasome inhibitor (CF-2) is identical to delta aminolevulinic acid dehydratase. J Biol Chem 1994, 269: 12399–12402.

20. Wool I G. Extraribosomal functions of ribosomal proteins. Trends Biochem Sci 1996, 21: 164–165.

21. Zhu W, Rainville I R, Ding M, Bolus M, Heintz N H and Pederson D S. Evidence that the pre-mRNA splicing factor Clf1p plays a role in DNA replication in *Saccharomyces cerevisiae*. Genetics 2002, 160: 1319–1333.

22. Russell C S, Ben-Yehuda S, Dix I, Kupiec M and Beggs J D. Functional analyses of interacting factors involved in both pre-mRNA splicing and cell cycle progression in *Saccharomyces cerevisiae*. RNA 2000, 6: 1565–1572.

23. Ben-Yehuda S, Dix I, Russell C S, McGarvey M, Beggs J D and Kupiec M. Genetic and physical interactions between factors involved in both cell cycle progression and pre-mRNA splicing in *Saccharomyces cerevisiae*. Genetics 2000, 156: 1503–1517.

24. Chung S, McLean M R and Rymond B C. Yeast ortholog of the *Drosophila* crooked neck protein promotes spliceosome assembly through stable U4/U6.U5 snRNP addition. RNA 1999, 5: 1042–1054.

25. Gonzalez F, Delahodde A, Kodadek T and Johnstom S A. Recruitment of a 19S proteasome subcomplex to an activated promoter. Science 2002, 296: 548–550.

26. Picot D, Loll P J and Garavito R M. The X-ray crystal structure of the membrane protein prostaglandin H2 synthase-1. Nature 1994, 367: 243–249.

27. Heikkinen J, Risteli M, Wang C, Latvala J, Rossi M, Valtavaara M and Myllyla R. Lysyl hydroxylase 3 is a multifunctional protein possessing collagen glucosyltransferase activity. J Biol Chem 2000, 275: 36158–36163.

28. Stutts M J, Canessa C M, Olsen J C, Hamrick M, Cohn J A, Rossier B C and Boucher R C. CFTR as a cAMP-dependent regulator of sodium channels. Science 1995, 269: 847–850.

29. Suzuki C K, Rep M, van Dijl J M, Suda K, Grivell L A and Schatz G. ATP-dependent proteases that also chaperone protein biogenesis. Trends Biochem Sci 1997, 22: 118–123.

30. Piatigorsky J. Multifunctional lens crystallins and corneal enzymes. More than meets the eye. Ann NY Acad Sci 1998, 842: 7–15.

31. Ursini F, Heim S, Kiess M, Maiorino M, Roveri A, Wissing J and Flohe L. Dual function of the selenoprotein PHGPx during sperm maturation. Science 1999, 285: 1393–1396.

32. Mark D F and Richardson C C. *Escherichia coli* thioredoxin: a subunit of bacteriophage T7 DNA polymerase. Proc Natl Acad Sci USA 1976, 73: 780–784.

33. Jeffery C J, Bahnson B J, Chien W, Ringe D and Petsko G A. Crystal structure of rabbit phosphoglucose isomerase, a glycolytic enzyme that moonlights as neuroleukin, autocrine motility factor, and differentiation mediator. Biochemistry 1999, 39: 955–964.

34. Jeffery C J. Multifunctional proteins: examples of gene sharing. Ann Med 2003, 35: 28–35.

35. Searls D B. Pharmacophylogenomics: Genes, evolution and drug targets. Nature Rev Drug Discovery 2003, 2: 613–623.

36. Wolff C and Parkinson J S. Aspartate taxis mutants of the *Escherichia coli* tar chemoreceptor. J Bacteriol 1988, 170: 4509–4515.

37. Mowbray S L and Koshland D E J. Mutations in the aspartate receptor of *Escherichia coli* which affect aspartate binding. J Biol Chem 1990, 265: 15638–15643.

38. Tabor S, Huber H E and Richardson C C. *Escherichia coli* thioredoxin confers processivity on the DNA polymerase activity of the gene 5 protein of Bacteriophage T7. J Biol Chem 1987, 262: 16212–16223.

39. Cascalho M, Wong J, Steinberg C and Wabl M. Mismatch repair co-opted by hypermutation. Science 1998, 279: 1207–1210.

40. Thunnissen M M G M, Nordlunch P and Heggstrom J Z. Crystal structure of human leukotriene A4 hydrolase, a bifunctional enzyme in inflammation. Nat Struct Biol 2001, 8: 131–135.

41. Chen J-W, Dodia C, Feinstein S I, Jain M K and Fisher A B. 1-Cys peroxiredoxin, a bifunctional enzyme with glutathione peroxidase and phospholipase A2 activities. J Biol Chem 2000, 275: 28421–28427.

42. Numata O. Multifunctional proteins in Tetrahymena: 14-nm filament protein/citrate synthase and translation elongation factor-1 alpha. Int Rev Cytol 1996, 164: 1–35.

43. Modun B, Morrissey J and Williams P. The staphylococcal transferrin receptor: a glycolytic enzyme with novel functions. Trends Microbiol 2000, 8: 231–237.

44. Brew K, Vanaman T C and Hill R L. The role of alpha-lactalbumin and the A protein in lactose synthetase: a unique mechanism for the control of a biological reaction. Proc Natl Acad Sci USA 1968, 59: 491–497.

45. Bolduc J M, Spiegel P C, Chatterjee P, Brady K L, Downing M E, Caprara M C, Waring R B and Stoddard B L. Structural and biochemical analysis of DNA and RNA binding by a bifunctional homing endonuclease and group I intron splicing factor. Genes Develop 2003, 17: 2875–2888.

46. Caprara M G, Mohr G and Lambowitz A M. A tyrosyl-tRNA synthetase protein induces tertiary folding of the group I intron catalytic core. J Mol Biol 1996, 257: 512–531.

47. Lim M L, Lum M G, Hansen T M, Roucou X and Nagley P. On the release of cytochrome c from mitochondria during cell death signaling. J Biomed Sci 2002, 9: 488–506.

5

Molecular Chaperones: The Unorthodox View

Brian Henderson and Alireza Shamaei-Tousi

5.1. Introduction

Like a Brian Rix farce, in which the characters' identities are continuously chang-ing, the functions of the class of protein known as molecular chaperones has been unfolding continuously over the past decade resulting in substantial con-fusion. However, like such farces, we are confident that the dénouement will be a complete surprise and will provide a new world picture of the processes with which molecular chaperones are involved. This short chapter aims to introduce the reader to the rapidly changing world of molecular chaperones as an aid to the reading of the rest of the chapters in this volume.

5.2. Molecular chaperones are protein folders

Our story starts with a huff and a puff with the study of the response of the polytene chromosomes of *Drosophila* to various stressors. This revealed novel patterns of specific chromosomal puffs, in response to heat, and a variety of other environmental stresses, representing the transcription of selected genes [1, 2]. The behaviour of cells exposed to various stresses became known as the heat shock response or the cell stress response and we now appreciate the very large number of environmental factors to which cells will respond in this stereotypical manner. The 'molecularisation' of the cell stress response occurred in the late 1980s with the pioneering work of Ellis and colleagues [3], who introduced both the concept of protein chaperoning and the term molecular chaperone. The enormous amount of work currently being carried out on the structural biology and molecular and cellular mechanisms of molecular chaperones has its genesis in this paper. The reader is referred to Chapter 1 in which John Ellis reviews the protein chaperoning function of molecular chaperones and warns of the pitfalls of incorrect definitions in relation to molecular chaperones and stress proteins.

5.3. Molecular chaperones are potent immunogens

Contemporaneously with the discovery of molecular chaperones as protein-folding 'machines' was the realisation that these proteins were potent immunogens involved in immune responses to infection [4] and also in autoimmunity [5]. The possibility of a connection between these two events was suggested by Irun Cohen [5], who continues this argument in Chapter 16. The discovery of the immune response to molecular chaperones was surprising because these proteins are highly conserved. Furthermore, human molecular chaperones such as chaperonin (Cpn) 10 and Cpn60 proteins can be considered to be bacterial molecules because the mitochondrion evolved from an α-proteobacterium [6]. The unexpected immunogenicity of molecular chaperones is presumably related to the capacity that these proteins have to activate myeloid cells. The interactions of molecular chaperones, ranging in mass from 8 kDa (ubiquitin) to 90 kDa (Hsp90), with myeloid and other cell types is detailed in many of the later chapters in this volume.

5.4. Molecular chaperones as moonlighting proteins

By the late 1980s and early 1990s the paradigm of molecular chaperones as intracellular proteins acting as 'catalysts' of protein folding was being forged [7]. Although this is clearly a major function for these proteins, evidence began to emerge that they have other intracellular and extracellular functions. The first line of evidence for this non-orthodox view of molecular chaperones was the finding of their presence on the surfaces of cells. For example, Cpn60 was identified on the cell surface of $\gamma\delta$ T cells [8]. The cellular disposition of molecular chaperones is detailed in Chapter 2. The binding of molecular chaperones to membranes is a continuing motif in the literature. For example, GroESL oligomers have been shown to stabilise artificial membranes [9]. A report suggests that type II chaperonins in *Archaea* function primarily to stabilise cellular membranes [10]. One experimental finding that has not been followed up was the report that Cpn60 induced pores in membranes [11]. Perhaps the most interesting association of molecular chaperones with cell membranes is the finding that the receptor for the potent pro-inflammatory Gram-negative bacterial component, lipopolysaccharide (LPS), contains the heat shock proteins Hsp70 and Hsp90 [12, 13].

LPS is a major issue for those working on the non-folding functions of molecular chaperones (e.g., [14]) because many of the proteins being used are recombinant proteins made in *Escherichia coli*. In the past few years a number of papers have appeared suggesting that all of the actions of molecular chaperones are due to LPS contamination [15–17]. This problem is dealt with by a number of the authors and there can be few fields of study in which more care is taken with LPS contamination of recombinant proteins. The recent finding that the *Helicobacter pylori* Cpn60 protein activates macrophages by a mechanism that does not involve the LPS (TLR4) or bacterial lipopeptide (TLR2) receptors reveals that non-proteinaceous bacterial contaminants are unlikely to account for the biological activity of molecular chaperones [18]. However, the watchword has to be vigilance.

These reports certainly begin to suggest that molecular chaperones may have functions in addition to their protein-folding actions. Thus it is obvious that molecular chaperones can also be grouped into the widening pool of proteins with multiple functions and now known as moonlighting proteins. The concept of moonlighting proteins has been reviewed in Chapter 4.

5.4.1. The unfolding moonlighting functions of molecular chaperones

In the past decade a surprisingly large number of apparent non-folding functions have been ascribed to one or another of the molecular chaperones (Table 5.1) and some of these proteins have a number of different biological actions. These results are still controversial and the molecular chaperone field is divided into those that believe that the non-folding actions of molecular chaperones are artefactual and those that hold that they are part of the systems biology of the cell stress response. The classic example of the former position is the belief that the cytokine-inducing actions of molecular chaperones are due to contamination of these proteins with LPS. Some of the criticism (in this case balanced criticism) about the extracurricular actions of molecular chaperones is voiced in Chapter 1. It is assumed that similar criticisms were levelled at the findings that most of the glycolytic enzymes have moonlighting functions. The protein currently holding the prize for most extracurricular activity is phosphoglucoisomerase (PGI). Over the past 20 years this protein, which has a CXXC motif identical to that found in the molecular chaperone thioredoxin and in certain chemokines, has been independently identified as three different cytokines and an implantation factor. Thus this protein is also neuroleukin [41], autocrine motility factor [42], differentiation and maturation mediator [43] and an implantation factor [44]. Any criticism of these findings seems to have dissipated

Table 5.1. Non-folding actions of molecular chaperones

Proteins	Molecular mass (kDa)	Additional functions
Ubiquitin	8	Antibacterial activity [19]
Thioredoxin	12	ADF – a T cell cytokine [20]
		A novel chemoattractant [21]
		Chemokine inhibitor [22]
		Modulates glucocorticoid action [23]
Chaperonin 10	10 (oligomer)	Early pregnancy factor [24]
		Osteolytic factor [25]
α-Crystallin	18–20	Activates microglia [26]
Cyclophilins	∼ 20	Secretory pro-inflammatory macrophage product [27]
		Chemotactic activity [28]
		Parasite inducer of IL-12 [29]
Hsp27	27	Induces IL-10; anti-inflammatory [30]
Hsp60/Cpn60	60	Modulates myeloid cell and vascular endothelial cell function
Hsp70	70	Cytokine inducer (various chapters in volume) or inhibitor [31]
		Receptor for LPS [11, 12]
Bip	70	Negative regulator of inflammation [32, 33]
Hsp90	90	Immunomodulator acting to present peptides to T lymphocytes
Grp94/Gp96	96	As above
		A cell surface receptor for Gram-negative bacteria [34, 35]
		A receptor for bacterial invasion [34, 35]
		A factor involved in surface TLR expression [36]
		A direct ligand for activating cells [37–40]

and they are now part of the mainstream of the biochemistry and cellular biology of glycolysis.

5.4.2. Moonlighting actions of individual molecular chaperones

The following discussion will briefly deal with the reported moonlighting actions of molecular chaperones and will deal with them in terms of increasing molecular mass (Table 5.1). Much of this information is dealt with more extensively elsewhere in this volume.

Ubiquitin: This is an 8.5-kDa intracellular protein involved in the controlled degradation of proteins and thus just comes under the remit of molecular chaperone. It has recently been discovered that this protein has antibacterial actions [19] and it therefore joins the multitude of proteins and peptides that function to defend us against bacteria.

Thioredoxin: This is a 12-kDa redox protein with a CXXC motif, which acts intracellularly as a hydrogen donor to ribonucleotide reductase. It has been known since the late 1980s that thioredoxin is a secreted cytokine [20], termed adult T cell-leukaemia-derived factor, with autocrine growth properties on T lymphocytes. Thioredoxin is found in the serum in normal individuals [45]. In addition to acting on T cells, thioredoxin is a unique chemoattractant with a different mechanism of action to the large family of chemotactic cytokines known as the chemokines [21]. Surprisingly, in spite of being identified as a chemoattractant, thioredoxin can also block cellular responses to LPS by suppressing the activity of known chemokines [22]. Indeed, circulating levels of thioredoxin appear to be important in AIDS, and it has been proposed that high levels of this molecular chaperone in the blood of HIV-infected individuals with low CD4$^+$ T cell counts directly impair survival by blocking pathogen-induced chemotaxis and thus prevent myeloid cell defences crucial for survival [46]. This chemotaxis suppressing activity of thioredoxin is likely to have a therapeutic effect and one study has shown that this molecular chaperone can block experimental inflammatory or fibrotic lung injury [47]. Finally, thioredoxin has also been shown to be involved in control of glucocorticoid action at the level of glucocorticoid-inducible gene expression. This interaction reveals a link between cellular and physiological stress responses [23]. The authors suggest that the homeostatic control of the multicellular organisms must require a link between the cellular stress responses and the physiological (organismal) stress response.

Chaperonin 10/Hsp10/early pregnancy factor: The fetus is equivalent to an allograft because the mother and fetus will generally express a different profile of major histocompatibility antigens. Thus an obvious question is how is fetal rejection controlled? Almost 30 years ago an immunosuppressive factor was identified in the sera of pregnant mothers and was termed early pregnancy factor (EPF) [48]. Significant efforts were made to identify EPF; however, it took until 1991 for the suggestion to be made that EPF was actually thioredoxin [49]. A second group identified EPF as chaperonin (Cpn) 10 (Hsp10) [50], a 10-kDa protein which forms a heptameric structure that interacts with Cpn60 to promote protein folding (Cpn10 is a co-chaperone). This finding raised an enormous amount of interest and criticism [51]. However, the recent cloning and expression of human EPF (Cpn10/Hsp10) in eukaryotic cells and in *E. coli* has revealed that the recombinant Cpn10 has EPF activity both *in vitro* and *in* *vivo* and that activity depended upon the presence of appropriate N-terminal modification [52]. The studies that suggested EPF was thioredoxin have not been repeated.

Another strand of this story comes from the work of Coates and colleagues, which is described to a limited extent in Chapter 6. Coates was the first to clone and express the *cpn10* gene of *Mycobacterium tuberculosis* [53] and had shown that this protein inhibited inflammation in both adjuvant arthritis in the rat [54] and experimental allergic asthma in the mouse [55]. In the former study, the activity of the whole molecule could be replicated by synthetic N-terminal peptides that are free of LPS.

Tuberculosis of the bone causes major damage and the *M. tuberculosis* Cpn10 was also found to be a potent inducer of bone resorption and the major osteolytic component of this organism [25]. Using synthetic peptides the active site in *M. tuberculosis* Cpn10 has been identified as the mobile loop [25]. The group that had identified EPF as Cpn10 has subsequently shown that human Cpn10 is able to inhibit inflammation in animals with experimental allergic encephalomyelitis, a much-used model of autoimmunity [56, 57]. Here is the first evidence of a molecular chaperone acting as a secreted hormone or cytokine and able to inhibit immune/inflammatory responses in a key process (pregnancy) [24]. Much more information is required before we can fully understand the role played by Cpn10 in the control of the early phase of pregnancy.

α-**Crystallin:** This is a member of the small heat shock protein family – proteins of approximate molecular mass of 20 kDa that form extremely large aggregates (see Chapter 1). It is reported that this protein activates microglial cells [26], which are the myeloid cell population in the brain with major roles in brain defences against infection.

Cyclophilins: A family of proteins (>30 genes in the human genome) with peptidyl-prolyl isomerase activity and a capacity to bind the cyclic peptide immunosuppressants such as cyclosporine. It has been reported that cyclophilin is secreted by LPS-activated macrophages and has chemotactic activity [27, 28]. Members of this protein family have been found in biological fluids including human milk [58, 59] and blood [59]. Elevated levels are also found in the synovial fluid of patients with rheumatoid arthritis [60] and in patients with sepsis [61]. *Toxoplasma gondii* is a protozoan responsible for toxoplasmosis in humans. It has been reported that this eukaryotic parasite releases a potent IL-12–stimulating protein which has recently been identified as C-18 cyclophilin. This protein activates dendritic cells (DCs) by binding to the CCR5 receptor [29]. These findings emphasise the diversity of the interactions that can occur between molecular chaperones and the chemokine system of cytokines.

Hsp27: The literature on this molecular chaperone is reviewed by Miller-Graziano in Chapter 13. The key observation is that exposure of human monocytes to Hsp27 induces the production of the anti-inflammatory cytokine, IL-10 [30]. This suggests that Hsp27 may have anti-inflammatory functions. The ability of extracellular molecular chaperones to act as inhibitors of immunity and inflammation appears to be a theme. Such anti-inflammatory actions of molecular chaperones rules out the possibility that the functions being described are due to LPS contamination.

Cpn60: There is now very good evidence that this molecule is a stimulator of a range of cells, including myeloid cells, vascular endothelial cells and epithelial cells, and this literature has been reviewed by Coates in Chapter 6 and the topic is also touched on in other chapters. Two issues will be addressed in this brief section. The first is the nature of the receptor for this protein. There appears to be a range to choose from, including CD14, TLR2 and TLR4 [62, 63]. However, a number of the Cpn60 proteins tested do not appear to bind to any of these receptors (e.g., [18]). The simplest explanation for this is that cells can discriminate between Cpn60 proteins from different species. The most striking demonstration of the ability to recognise differences in Cpn60 proteins is the finding that the salivary symbiont (*Enterobacter aerogenes*) of the insect predator known as the antlion produces a neurotoxin used by the insect in catching its prey. This neurotoxin turns out to be the Cpn60 protein of this bacterium. Strikingly, single-residue changes in the *E. coli* equivalent protein, GroEL, turn this best-studied of molecular chaperones into a potent insect neurotoxin [64]. The possible consequences of this will be discussed at the end of the chapter.

More recently it has been shown that eukaryotic Cpn60 interacts with Bax and Bak. These two are pro-apoptotic cytolysis proteins that stimulate the release of cytochrome c and apoptosis. Binding to Cpn60 may regulate the activity of these two pro-apoptotic proteins by preventing them from oligomerising and inserting into the mitochondrial membrane [65, 66]. This action resembles the role that Hsp90 has in the normal cell [67, 68]. The second issue is the presence of Cpn60 (Hsp60) in the blood of humans. The protein, which is N-terminally recognised and processed in mitochondria, can also be found on the surface of endothelial cells and macrophages [69]. The authors have established that the levels of Hsp60 in the blood of a population of normal individuals (healthy British civil servants) are stratified into three groups: (i) those below assay detection; (ii) those with measurable, but low levels and (iii) those with extremely high, biologically active, circulating levels. The latter can be in the hundreds of micrograms per millilitre range. There is no explanation for this stratification of plasma Hsp60 concentrations and these results imply a mechanism of production and/or release, or a mechanism of removal of Hsp60 that differs enormously within the normal human population. Preliminary data fail to support the hypothesis that the difference in levels is due to differences in transcriptional rates [70]. However, it has been shown that elevated levels of Hsp60 in the blood of healthy individuals could be associated with an unfavourable lipid profile, high TNF-α levels and low socioeconomic status. TNF-α plays an important role in atherogenesis and the development of acute coronary syndromes [71], and low socioeconomic status and social isolation have been related to chronic heart diseases [72]. This will be dealt with in more detail in Chapter 12. Some questions and speculations about human Cpn60 are detailed in Figure 5.1.

Hsp70: The human genome sequence has revealed that *Homo sapiens* is in possession of 13 hsp70 genes and thus one has to be careful when reviewing the literature on Hsp70 and its biological actions and receptors that one compares apples with apples and not with oranges. Other chapters in this volume (Chapters 7, 8, 9, 10 and 14) deal with aspects of a number of Hsp70 family proteins, including Bip. There is controversy in the literature about the receptor(s) used by exogenous Hsp70 to activate cells (see Chapter 10). There is also controversy about the nature of the signal induced by peptide-free Hsp70. Most studies of the human or mycobacterial Hsp70 show an activation of myeloid cell cytokine synthesis (see Chapters 7–10). However, in a recent report, *M. tuberculosis* Hsp70 has been shown to induce the production of the anti-inflammatory cytokine IL-10 and reduce the production of TNF-α [31].

Bip: This is another member of the Hsp70 protein family which is located in the lumen of the endoplasmic reticulum and was originally identified as an immunoglobulin heavy chain-binding protein. Transcription of Bip (also known as glucose-regulated protein (Grp) 78) is enhanced when the glucose concentration is lowered. The transcription of a number of molecular chaperones is regulated by environmental glucose levels and it is interesting to speculate why this evolved. For example, hypoglycaemia is one response to infection and to the key Gram-negative inflammogen, LPS [73]. As Corrigal and Panayi review in Chapter 14, Bip is now known to be a potent negative regulator of inflammation acting, like Hsp27, as an inducer of the anti-inflammatory cytokine IL-10 [32]. Bip also induces the production of soluble TNF receptor II and IL-1 receptor antagonist, inhibits the recall antigen response by peripheral blood mononuclear cells (PBMCs) to tuberculin purified protein derivative

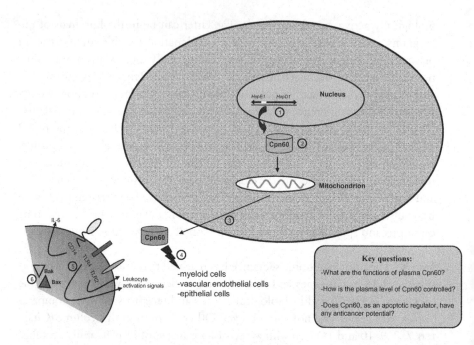

① The *Hsp60* gene, also called *HSPD1*, is linked head to head with the *Hsp10* (*HSPE1*) gene on chromosome 2. These genes are separated by a bidirectional promoter with two-fold greater transcriptional activity in the Hsp60 direction compared to the Hsp10 direction.

② Hsp60 is synthesized as a precursor with an N-terminal mitochondrial targeting sequence, or mitochondrial import peptide (MIP).

③ While Cpn60 is recognised to be mainly a mitochondrial protein, some proportion can also be found at discrete extramitochondrial sites such cell surfaces and circulation.

④ Cpn60 is known to activate myeloid cells, vascular endothelial cells and epithelial cells.

⑤ Soluble Cpn60 activates leukocytes, releasing TNF-α, nitric oxide and IL-6 via TLR2, TLR4 or CD14 receptors.

⑥ By binding to pro-apoptotic Bax and Bak, Cpn60 may have a regulatory role in apoptotic process in the normal cell.

Figure 5.1. Model for the actions of Cpn60 inside and outside of cells: some questions and speculations about mitochondrial Cpn60 (Hsp60), a protein that has been shown to have an increasing range of biological actions.

and down-regulates CD86 and HLA-DR expression of human PBMCs [33]. An obvious hypothesis is that inflammation results in lowered circulating glucose levels and that this triggers the transcription of a number of molecular chaperones, at least one of which, Bip, has anti-inflammatory properties. Is this a homeostatic regulatory network? Further work is required to determine if glucose-regulated molecular chaperones play a role in regulating immune responses.

Hsp90: This is a major cytoplasmic heat shock protein family with a growing range of key cellular functions – notably the ability to render cryptic protein gene polymorphisms that damage protein function [74]. Much of the current excitement about Hsp90 centres on its ability to bind peptides and present them to cytotoxic T cells, and it is believed that this activity will be the basis of a novel and very effective treatment for cancer [75]. In addition, Hsp90 is part of the LPS receptor complex and thus participates in the innate immune response to a key pathogen-associated molecular pattern [12, 18]. Related to this is the finding that the Hsp90 homologue, gp96 (also known as glucose-regulated protein (Grp) 94) is involved in the folding of Toll-like receptors (TLR).

Inactivation of the gene encoding gp96 results, not as one would expect – in cell death under stressed conditions – but in the inability to express Toll-like receptors on the cell surface. Thus the description of gp96 as a cell stress protein may be inaccurate [36]. While putatively a protein of the endoplasmic reticulum, gp96 is also found on cell surfaces throughout the vertebrate phylum and this surface expression is up-regulated by exposure of cells to bacteria and their constituents [76]. This cell surface targeting of gp96 is involved in the maturation of DCs [77]. Indeed, it turns out that gp96 is a receptor for one of the outer membrane proteins (OmpA) of *E. coli* and is involved in the ability of this bacterium to invade eukaryotic cells [34, 35]. In addition to being an infection-related receptor, soluble gp96 is also able to activate leukocytes through binding to receptors including the TLRs [37–39] and gp96 appears to act as a Th2-specific co-stimulatory molecule [40]. The binding of gp96 (with associated peptides) to cells utilises the cell surface 'receptor' CD91 [78]. However, there appears to be a price to pay for the surface expression of gp96 in terms of the induction of autoimmunity [79].

This quick trawl through the literature reveals that many molecular chaperones have a range of actions in addition to the proposed main function as protein-folding proteins. A good example of the changing paradigm in molecular chaperone biology is the finding that the intracellular pathogenic bacteria known as *Chlamydia* contain three *cpn60* genes. This is not unusual, as certain members of the Rhizobia can contain up to six *cpn60* genes. What is striking is that two of the Chlamydial genes encode proteins that have lost the ATPase domain required for the protein-folding ability of these molecules [80]. Confirmation of this loss of protein-folding activity is shown by the inability of the genes encoding these two aberrant proteins to complement an *E. coli* GroEL mutant [80]. These are not some form of pseudo-gene, because it has been shown that all three proteins are independently expressed by *Chlamydia* [81]. Thus, here is a situation in which cells have evolved proteins that look like Cpn60 molecules but which have no capacity to fold proteins. Can these be referred to as non-folding molecular chaperones? The key question is – what biological role do these non-folding chaperones have? Perhaps these proteins are not moonlighting and have completely different roles to play from the assumed protein-folding role.

5.5. A physiological role for molecular chaperones

As reviewed in Chapter 1, molecular chaperones, through their protein-folding and chaperoning actions, are vital for homeostatic cell function. The new paradigm that is the subject of most of this book is that some, if not all, molecular chaperones have additional non-folding actions and that these contribute to cell–cell signalling involved in homeostatic regulation of the organism. The necessary supporting evidence for this paradigm is that molecular chaperones must be found at the surface of cells and/or in the extracellular fluid in order to be able to transmit a signal from cell to cell. There is now substantial evidence that many of the molecular chaperones are found in the blood and there is emerging evidence that these proteins are found in other body fluids such as saliva [82] and seminal fluid [83]. Indeed, it must also be remembered that one molecular chaperone, clusterin (apolipoprotein J) is normally present in body fluids [84]. The extracellular disposition of molecular chaperones and the relationship that this might have to disease processes is described in detail in Chapter 12.

If molecular chaperones are found in the extracellular milieu, then this raises the thorny point about how they get out of the cell. Many have dismissed the idea that molecular chaperones could function as intercellular signalling molecules because there is no known mechanism to account for their secretion from cells. This is actually a poor argument, because for many years we had no idea how many key signalling proteins such as IL-1, IL-16, IL-18, fibroblast growth factor and annexin, to name but a few, were released from cells. As described in Chapter 3, a pathway that has been elucidated over the past decade or more – the leaderless secretory protein pathway – is responsible for the release of the aforementioned proteins and also of the molecular chaperone thioredoxin. It is now known that the glycolytic moonlighting protein, PGI, is secreted via this pathway. Other novel pathways of protein secretion also exist (e.g., [85]) and hence the absence of a mechanism for secreting molecular chaperones cannot be used as an argument that they cannot be released from cells. Discovering the mechanisms of molecular chaperone secretion should be a priority for those interested in the biology of these proteins.

So if we accept that molecular chaperones are released from cells, what do they do? It is now appreciated that we have no shortage of receptors for molecular chaperones such as Cpn60, Hsp70, Hsp90 and cyclophilin with experimental evidence identifying CD14, TLR2, TLR4, LOX1, CD40, CD91 and CCR5 as receptors for these various molecular chaperones. Of course, we do not have identifiable receptors for the other molecular chaperones. However, it is now clear that the various molecular chaperones described can bind to and activate a wide range of cells.

The authors propose that there are four major functions of extracellular molecular chaperones.

1. The first is as signals warning the multicellular organism that certain of its constituents are under stress and modulating the function of nearby cells in case the stress continues or expands. The biological response of cells of innate immunity to stress proteins may overlap with the concept of proteinaceous danger signals as postulated by Matzinger [86]. This warning signal function was termed stress broadcasting in a recent review [87]. It is not clear whether molecular chaperones in the circulation are part of this stress broadcasting mechanism or are involved in linking cellular stress to higher order systems control.

2. The second is as a physiological input signal to the immune system in the form of pro- and anti-inflammatory molecular chaperones which may better be called *stress* cytokines. The pro- and anti-inflammatory actions of molecular chaperones have been identified in earlier sections and will not be discussed further; they are also discussed elsewhere in this volume. This is a novel but testable hypothesis that we hope will be explored in the near future.

3. Increasing evidence exists that another function of molecular chaperones is to provide adjuvant-like signals through the ability of molecules like Hsp70 and Hsp90 to present peptides to antigen-presenting cells. This activity may be a key feature of Matzinger's danger model and is an obvious foundation for a novel therapy for cancer. This is reviewed in detail in Chapters 17 and 18.

4. This is extremely speculative and springs from the findings of the last decade or so that vertebrates live with a very large number of bacterial species. It is estimated that *H. sapiens* have 2–3,000 bacterial species as their constant companions. London Zoo only keeps about 6–700 species of animal. Contrast this with the 40 or 50 bacteria that cause human disease [88]. How do we discriminate between these friendly bacteria and the ones that mean us harm? We propose that one set of signals is the molecular chaperones. There is little evidence for this hypothesis as yet, other than the finding that many receptors recognise molecular chaperones and the work on *E. aerogenes*, which reveals that single nucleotide changes in GroEL can dramatically alter the biological actions of this protein [64].

As these speculations were being put onto paper, another idea emerged from the literature. It is now suggested that Cpn60 is involved in sperm capacitation, a key event required for fertilisation [89].

5.6. Conclusions

John Ellis, a pioneer in the study of molecular chaperones, has thrown down a gauntlet in Chapter 1 with the statement 'This view [that molecular chaperones have non-folding roles] has not found general acceptance, partly because it is novel [as was the concept of proteins folding proteins [90]] and partly because of the paucity of high-quality evidence compared with that available in support of the protein folding paradigm'. The authors would argue that there is now a large amount of 'high-quality' evidence in terms of papers in *Nature*, the *Journal of Experimental Medicine*, the *Journal of Clinical Investigation*, the *Journal of Immunology* and *International Immunology* to name but a few, which supports the hypothesis of the non-folding actions of molecular chaperones. As Sherlock Holmes was want to say 'when you have excluded the possible then the impossible must be true'.

Acknowledgements

The authors are grateful to the Arthritis Research Campaign (programme grant HO600) and to the British Heart Foundation (PG/03/029) for financial support.

REFERENCES

1. Ritossa F A. A new puffing pattern induced by temperature shock and DNP in *Drosophila*. Experientia 1962, 18: 571–573.
2. Ashburner M. Pattern of puffing activity in the salivary gland chromosomes of *Drosophila*. V. Response to environmental treatments. Chromosoma 1970, 31: 356–376.
3. Hemmingsen S M, Woolford C, van der Vies S M, Tilly K, Dennis D T, Georgopoulos G C, Hendrix R W and Ellis R J. Homologous plant and bacterial proteins chaperone oligomeric protein assembly. Nature 1988, 333: 330–334.
4. Young D B, Ivanyi J, Cox J H and Lamb J R. The 65kDa antigen of mycobacteria – a common bacterial protein? Immunol Today 1987, 8: 215–219.
5. Cohen I R and Young D B. Autoimmunity, microbial immunity and the immunological homunculus. Immunol Today 1991, 12: 105–109.
6. Horner D S, Hirt R P, Kilvington S, Lloyd D and Embley T M. Molecular data suggest an early acquisition of the mitochondrion endosymbiont. Proc Royal Soc London B Biol Sci 1996, 263: 1053–1059.
7. Hartl F U and Hayer-Hartl M. Molecular chaperones in the cytosol: from nascent chain to folded protein. Science 2002, 295: 1852–1858.

8. Fisch P, Malkovsky M, Kovats S, Sturm E, Braakman E, Klein B S, Voss S D, Morrissey L W, DeMars R, Welch W J, Bolhuis R L H and Sondel P M. Recognition by human Vγ9/Vδ2 T cells of a GroEL homolog on Daudi Burkitt's lymphoma cells. Science 1990, 250: 1269–1273.

9. Torok Z, Horvath I, Goloubinoff P, Kovacs E, Glatz A, Balogh G and Vigh L. Evidence for a lipochaperonin: association of active protein-folding GroESL oligomers with lipids can stabilize membranes under heat shock conditions. Proc Natl Acad Sci USA 1997, 94: 2192–2197.

10. Trent J D, Kagawa H K, Paavola C D, McMillan R A, Howard J, Jahnke L, Lavin C, Embaye T and Henze C E. Intracellular localization of a group II chaperonin indicates a membrane-related function. Proc Natl Acad Sci USA 2003, 100: 15589–15594.

11. Alder G M, Austen B M, Bashford C L, Mehkert A and Pasternak C A. Heat shock proteins induce pores in membranes. Biosci Reports 1990, 10: 509–518.

12. Triantafilou K, Triantafilou M and Dedrick R L. A CD14-independent LPS receptor cluster. Nat Immunol 2001, 2: 338–344.

13. Triantafilou K, Triantafilou M, Ladha S, Mackie A, Dedrick R L, Fernandez N and Cherry R. Fluorescence recovery after photobleaching reveals that LPS rapidly transfers from CD14 to Hsp70 and Hsp90 on the cell membrane. J Cell Sci 2001, 114: 2535–2545.

14. Bausinger H, Lipsker D, Ziylan U, Manie S, Briand J P, Cazenave J P, Muller S, Haeuw J F, Ravanat C, de la Salle H and Hanau D. Endotoxin-free heat-shock protein 70 fails to induce APC activation. Eur J Immunol 2002, 32: 3708–3713.

15. Gao B and Tsan M F. Induction of cytokines by heat shock proteins and endotoxin in murine macrophages. Biochem Biophys Res Commun 2004, 317: 1149–1154.

16. Gao B and Tsan M F. Endotoxin contamination in recombinant human Hsp70 preparation is responsible for the induction of TNFα release by murine macrophages. J Biol Chem 2003, 278: 174–179.

17. Gao B and Tsan M F. Recombinant human heat shock protein 60 does not induce the release of tumor necrosis factor α from murine macrophages. J Biol Chem 2003, 278: 22523–22529.

18. Gobert A P, Bambou J C, Werts C, Balloy V, Chignard M, Moran A P and Ferrero R L. *Helicobacter pylori* heat shock protein 60 mediates interleukin-6 production by macrophages via a Toll-like receptor (TLR)-2-, TLR-4-, and myeloid differentiation factor 88-independent mechanism. J Biol Chem 2004, 279: 245–250.

19. Metz-Boutigue M H, Kieffer A E, Goumon Y and Aunis D. Innate immunity: involvement of new neuropeptides. Trends Microbiol 2003, 11: 585–592.

20. Tagaya Y, Maeda Y, Mitsui A, Kondo N, Matsui H, Hamuro J, Brown N, Arai K, Yokota T and Wakasugi H. ATL-derived factor (ADF), an IL-2 receptor/Tac inducer homologous to thioredoxin; possible involvement of dithiol-reduction in the IL-2 receptor induction. EMBO J 1989, 8: 757–764.

21. Bertini R, Howard O M, Dong H F, Oppenheim J J, Bizzarri C, Sergi R, Caselli G, Pagliei S, Romines B, Wilshire J A, Mengozzi M, Nakamura H, Yodoi J, Pekkari K, Gurunath R, Holmgren A, Herzenberg L A, Herzenberg L A and Ghezzi P. Thioredoxin, a redox enzyme released in infection and inflammation, is a unique chemoattractant for neutrophils, monocytes, and T cells. J Exp Med 1999, 189: 1783–1789.

22. Nakamura H, Herzenberg L A, Bai J, Araya S, Kondo N, Nishinaka Y, Herzenberg L A and Yodoi J. Circulating thioredoxin suppresses lipopolysaccharide-induced neutrophil chemotaxis. Proc Natl Acad Sci USA 2001, 98: 15143–15148.
23. Makino Y, Okamoto K, Yoshikawa N, Aoshima M, Hirota K, Yodoi J, Umesono K, Makino I and Tanaka H. Thioredoxin: a redox-regulating cellular cofactor for glucocorticoid hormone action. Cross talk between endocrine control of stress response and cellular antioxidant defense system. J Clin Invest 1996, 98: 2469–2477.
24. Morton H. Early pregnancy factor: an extracellular chaperonin 10 homologue. Immunol Cell Biol 1998, 76: 483–496.
25. Meghji S, White P, Nair S P, Reddi K, Heron K, Henderson B, Zaliani A, Fossati G, Mascagni P, Hunt J F, Roberts M M and Coates A R. *Mycobacterium tuberculosis* chaperonin 10 stimulates bone resorption: a potential contributory factor in Pott's disease. J Exp Med 1997, 186: 1241–1246.
26. Bhat N R and Sharma K K. Microglial activation by the small heat shock protein, α-crystallin. Neuroreport 1999, 10: 2869–2873.
27. Sherry N, Yarlett A, Strupp A and Cerami A. Identification of cyclophilin as a proinflammatory secretory product of lipopolysaccharide-activated macrophages. Proc Natl Acad Sci USA 1992, 89: 3511–3515.
28. Xu Q, Lefeva M C, Fischkoff S A, Handschumacher R E and Lyttle C R. Leukocyte chemotactic activity of cyclophilin. J Biol Chem 1992, 267: 11968–11971.
29. Aliberti J, Valenzuela J G, Carruthers V B, Hieny S, Andersen J, Charest H, Reis e Sousa C, Fairlamb A, Ribeiro J M and Sher A. Molecular mimicry of a CCR5 binding-domain in the microbial activation of dendritic cells. Nat Immunol 2003, 4: 485–490.
30. De A K, Kodys K M, Yeh B S and Miller-Graziano C. Exaggerated human monocyte IL-10 concomitant to minimal TNF-α induction by heat-shock protein 27 (Hsp27) suggests Hsp27 is primarily an anti-inflammatory stimulus. J Immunol 2000, 165: 3951–3958.
31. Detanico T, Rodrigues L, Sabritto A C, Keisermann M, Bauer M E, Zwickey H and Bonorino C. Mycobacterial heat shock protein 70 induces interleukin-10 production: immunomodulation of synovial cell cytokine profile and dendritic cell maturation. Clin Exp Immunol 2004, 135: 336–342.
32. Corrigall V M, Bodman-Smith M D, Fife M S, Canas B, Myers L K, Wooley P, Soh C, Staines N A, Pappin D J, Berlo S E, van Eden W, van der Zee R, Lanchbury J S and Panayi G S. The human endoplasmic reticulum molecular chaperone BiP is an autoantigen for rheumatoid arthritis and prevents the induction of experimental arthritis. J Immunol 2001, 166: 1492–1498.
33. Corrigall V M, Bodman-Smith M D, Brunst M, Cornell H and Panayi G S. Inhibition of antigen-presenting cell function and stimulation of human peripheral blood mononuclear cells to express an anti-inflammatory cytokine profile by the stress protein BiP: Relevance to the treatment of inflammatory arthritis. Arthritis Rheum 2004, 50: 1164–1171.
34. Prasadarao N V, Srivastava P K, Rudrabhatla R S, Kim K S, Huang S H and Sukumaran S K. Cloning and expression of the *Escherichia coli* K1 outer membrane protein A receptor, a gp96 homologue. Infect Immun 2003, 71: 1680–1688.
35. Khan N A, Shin S, Chung J W, Kim K J, Elliott S, Wang Y and Kim K S. Outer membrane protein A and cytotoxic necrotizing factor-1 use diverse signaling mechanisms

for *Escherichia coli* K1 invasion of human brain microvascular endothelial cells. Microb Pathogenesis 2003, 35: 35–42.

36. Randow F and Seed B. Endoplasmic reticulum chaperone gp96 is required for innate immunity but not cell viability. Nat Cell Biol 2001, 3: 891–896.

37. Panjwani N N, Popova L and Srivastava P K. Heat shock proteins gp96 and hsp70 activate the release of nitric oxide by APCs. J Immunol 2002, 168: 2997–3003.

38. Vabulas R M, Braedel S, Hilf N, Singh-Jasuja H, Herter S, Ahmad-Nejad P, Kirschning C J, Da Costa C, Rammensee H G, Wagner H and Schild H. The endoplasmic reticulum-resident heat shock protein Gp96 activates dendritic cells via the Toll-like receptor 2/4 pathway. J Biol Chem 2002, 277: 20847–20853.

39. Radsak M P, Hilf N, Singh-Jasuja H, Braedel S, Brossart P, Rammensee H G and Schild H. The heat shock protein Gp96 binds to human neutrophils and monocytes and stimulates effector functions. Blood 2003, 101: 2810–2815.

40. Banerjee P P, Vinay D S, Mathew A, Raje M, Parekh V, Prasad D V, Kumar A, Mitra D and Mishra G C. Evidence that glycoprotein 96 (B2), a stress protein, functions as a Th2-specific costimulatory molecule. J Immunol 2002, 169: 3507–3518.

41. Chaput M, Claes V, Portetelle D, Cludts I, Cravador A, Burny A, Gras H and Tartar A. The neurotrophic factor neuroleukin is 90% homologous with phosphohexose isomerase. Nature 1988, 332: 454–455.

42. Watanabe H, Takehana K, Date M, Shinozaki T and Raz A. Tumor cell autocrine motility factor is the neuroleukin/phosphohexose isomerase polypeptide. Cancer Res 1996, 56: 2960–2963.

43. Xu W, Seiter K, Feldman E, Ahmed T and Chiao J W. The differentiation and maturation mediator for human myeloid leukemia cells shares homology with neuroleukin or phosphoglucose isomerase. Blood 1996, 87: 4502–4506.

44. Schulz L C and Bahr J M. Glucose-6-phosphate isomerase is necessary for embryo implantation in the domestic ferret. Proc Natl Acad Sci USA 2003, 100: 8561–8566.

45. Kogaki H, Fujiwara Y, Yoshiki A, Kitajima S, Tanimoto T, Mitsui A, Shimamura T, Hamuro J and Ashihara Y. Sensitive enzyme-linked immunosorbent assay for adult T-cell leukemia-derived factor and normal value measurement. J Clin Lab Anal 1996, 10: 257–261.

46. Nakamura H, De Rosa S C, Yodoi J, Holmgren A, Ghezzi P, Herzenberg L A and Herzenberg L A. Chronic elevation of plasma thioredoxin: inhibition of chemotaxis and curtailment of life expectancy in AIDS. Proc Natl Acad Sci USA 2001, 98: 2688–2693.

47. Hoshino T, Nakamura H, Okamoto M, Kato S, Araya S, Nomiyama K, Oizumi K, Young H A, Aizawa H and Yodoi J. Redox-active protein thioredoxin prevents proinflammatory cytokine- or bleomycin-induced lung injury. Am J Respir Crit Care Med 2003, 168: 1075–1083.

48. Morton H, Rolfe B and Clunie G J. An early pregnancy factor detected in human serum by the rosette inhibition test. Lancet 1977, 1(8008): 394–397.

49. Clarke F M, Orozco C, Perkins A V, Cock I, Tonissen K F, Robins A J and Wells J R. Identification of molecules involved in the 'early pregnancy factor' phenomenon. J Reprod Fertil 1991, 93: 525–539.

50. Cavanagh A C and Morton H. The purification of early-pregnancy factor to homogeneity from human platelets and identification as chaperonin 10. Eur J Biochem 1994, 222: 551–560.

51. Lash G E and Legge M. Early pregnancy factor: an unresolved molecule. J Assist Reprod Genet 1997, 14: 495–496.
52. Somodevilla-Torres M J, Morton H, Zhang B, Reid S and Cavanagh A C. Purification and characterisation of functional early pregnancy factor expressed in Sf9 insect cells and in Escherichia coli. Protein Expr Purif 2003, 32: 276–287.
53. Atkins D, al Ghusein H, Prehaud C and Coates A R M. Overproduction and purification of Mycobacterium tuberculosis chaperonin 10. Gene 1994, 150: 145–148.
54. Ragno S, Winrow V R, Mascagni P, Lucietto P, Di Pierro F, Morris C J and Blake D R. A synthetic 10-kD heat shock protein (hsp10) from Mycobacterium tuberculosis modulates adjuvant arthritis. Clin Exp Immunol 1996, 103: 384–390.
55. Riffo-Vasquez Y, Spina D, Page C, Tormay P, Singh M, Henderson B and Coates A R M. Effect of Mycobacterium tuberculosis chaperonins on bronchial eosinophilia and hyperresponsiveness in a murine model of allergic inflammation. Clin Exp Allergy 2004, 34: 712–719.
56. Zhang B, Walsh M D, Nguyen K B, Hillyard N C, Cavanagh A C, McCombe P A and Morton H. Early pregnancy factor treatment suppresses the inflammatory response and adhesion molecule expression in the spinal cord of SJL/J mice with experimental autoimmune encephalomyelitis and the delayed-type hypersensitivity reaction to trinitrochlorobenzene in normal BALB/c mice. J Neurol Sci 2003, 212: 37–46.
57. Athanasas-Platsis S, Zhang B, Hillyard N C, Cavanagh A C, Csurhes P A, Morton H and McCombe P A. Early pregnancy factor suppresses the infiltration of lymphocytes and macrophages in the spinal cord of rats during experimental autoimmune encephalomyelitis but has no effect on apoptosis. J Neurol Sci 2003, 214: 27–36.
58. Spik G, Haendler B, Delmas O, Mariller C, Chamoux M, Maes P, Tartar A, Montreuil J, Stedman K, Kocher H P, Roland Kellers R, Hiestand P C and Movva N R. A novel secreted cyclophilin-like protein (SCYLP). J Biol Chem 1991, 266: 10735–10738.
59. Allain F, Boutillon C, Mariller C and Spik G. Selective assay for CypA and CypB in human blood using highly specific anti-peptide antibodies. J Immunol Methods 1995, 178: 113–120.
60. Billich A, Winkler G, Aschauer H, Rot A and Peichl P. Presence of cyclophilin A in synovial fluids of patients with rheumatoid arthritis. J Exp Med 1997, 185: 975–980.
61. Tegeder I, Schumacher A, John S, Geiger H, Geisslinger G, Bang H and Brune K. Elevated serum cyclophilin levels in patients with severe sepsis. J Clin Immunol 1997, 17: 380–386.
62. Kol A, Lichtman A H, Finberg R W, Libby P and Kurt-Jones E A. Heat shock protein (HSP) 60 activates the innate immune response: CD14 is an essential receptor for HSP60 activation of mononuclear cells. J Immunol 2000, 164: 13–17.
63. Vabulas R M, Ahmad-Nejad P, da Costa C, Miethke T, Kirschning C J, Hacker H and Wagner H. Endocytosed HSP60s use toll-like receptor 2 (TLR2) and TLR4 to activate the toll/interleukin-1 receptor signaling pathway in innate immune cells. J Biol Chem 2001, 276: 31332–31339.
64. Yoshida N, Oeda K, Watanabe E, Mikami T, Fukita Y, Nishimura K, Komai K and Matsuda K. Protein function. Chaperonin turned insect toxin. Nature 2001, 411: 44.
65. Gupta S and Knowlton A A. Cytosolic heat shock protein 60, hypoxia, and apoptosis. Circulation 2002, 106: 2727–2733.

66. Kirchhoff S R, Gupta S and Knowlton A A. Cytosolic heat shock protein 60, apoptosis, and myocardial injury. Circulation 2002, 105: 2899–2904.

67. Knowlton A A and Sun L. Heat-shock factor-1, steroid hormones, and regulation of heat-shock protein expression in the heart. Am J Physiol Heart Circ Physiol 2001, 280: H455–464.

68. Pratt W B. The hsp90-based chaperone system: involvement in signal transduction from a variety of hormone and growth factor receptors. Proc Soc Exp Biol Med 1998, 217: 420–434.

69. Xu Q, Luef G, Weimann S, Gupta R S, Wolf H and Wick G. Staining of endothelial cells and macrophages in atherosclerotic lesions with human heat shock protein-reactive antisera. Arteriosclerosis Thrombosis 1993, 13: 1763–1769.

70. Shamaei-Tousi A, Steptoe A, Coates A and Henderson B. Circulating chaperonin 60 in the plasma of British civil servants. Biochem Soc Trans 2004, 'abstr 13'.

71. Libby P. Current concepts of the pathogenesis of the acute coronary syndromes. Circulation 2001, 104: 365–372.

72. Hemingway H and Marmot M. Evidence based cardiology: psychosocial factors in the aetiology and prognosis of coronary heart disease. Systematic review of prospective cohort studies. Brit Med J 1999, 318: 1460–1467.

73. Olson N C, Hellyer P W and Dodam J R. Mediators and vascular effects in response to endotoxin. Brit Vet J 1995, 151: 489–522.

74. Sangster T A, Lindquist S and Queitsch C. Under cover: causes, effects and implications of Hsp90-mediated genetic capacitance. Bioessays 2004, 26: 348–362.

75. Srivastava P. Roles of heat-shock proteins in innate and adaptive immunity. Nat Rev Immunol 2002, 2: 185–194.

76. Morales H, Muharemagic A, Gantress J, Cohen N and Robert J. Bacterial stimulation upregulates the surface expression of the stress protein gp96 on B cells in the frog Xenopus. Cell Stress Chaperones 2003, 8: 265–271.

77. Singh-Jasuja H, Scherer H U, Hilf N, Arnold-Schild D, Rammensee H-G, Toes R E M and Schild H. The heat shock protein gp96 induces maturation of dendritic cells and down-regulation of its receptor. Eur J Immunol 2000, 30: 2211–2215.

78. Binder R J and Srivastava P K. Essential role of CD91 in re-presentation of gp96-chaperoned peptides. Proc Natl Acad Sci USA 2004, 101: 6128–6133.

79. Liu B, Dai J, Zheng H, Stoilova D, Sun S and Li Z. Cell surface expression of an endoplasmic reticulum resident heat shock protein gp96 triggers MyD88-dependent systemic autoimmune diseases. Proc Nat Acad Sci USA 2003, 100: 15824–15829.

80. Karunakaran K P, Noguchi Y, Read T D, Cherkasov A, Kwee J, Shen C, Nelson C C and Brunham R C. Molecular analysis of the multiple GroEL proteins of Chlamydiae. J Bacteriol 2003, 185: 1958–1966.

81. Gerard H C, Whittum-Hudson J A, Schumacher H R and Hudson A P. Differential expression of three *Chlamydia trachomatis* hsp60-encoding genes in active vs persistent infections. Microb Pathogenesis 2004, 36: 35–39.

82. Fabian T K, Gaspar J, Fejerdy L, Kaan B, Balint M, Csermely P and Fejerdy P. Hsp70 is present in human saliva. Med Sci Monit 2003, 9: 62–65.

83. Utleg A G, Yi E C, Xie T, Shannon P, White J T, Goodlett D R, Hood L and Lin B. Proteomic analysis of human prostasomes. Prostate 2003, 56: 150–161.

84. Trougakos I P and Gonos E S. Clusterin/apolipoprotein J in human aging and cancer. Int J Biochem Cell Biol 2002, 34: 1430–1448.

85. Rammes A, Roth J, Goebeler M, Klempt M, Hartmann M and Sorg C. Myeloid-related protein (MRP) 8 and MRP14, calcium-binding proteins of the S100 family, are secreted by activated monocytes via a novel, tubulin-dependent pathway. J Biol Chem 1997, 272: 9496–9502.
86. Matzinger P. The Danger Model: A renewed sense of self. Science 2002, 296: 301–305.
87. Maguire M, Coates A R M and Henderson B. Chaperonin 60 unfolds its secrets of cellular communication. Cell Stress Chaperones 2002, 7: 317–329.
88. Wilson M, McNab R and Henderson B. *Bacterial Disease Mechanisms: An Introduction to Cellular Microbiology*. Cambridge University Press: 2002.
89. Asquith K L, Baleato R M, McLaughlin E A, Nixon B and Aitken R J. Tyrosine phosphorylation activates surface chaperones facilitating sperm-zona recognition. J Cell Sci 2004, 117: 3645–3657.
90. Ellis R J. Proteins as molecular chaperones. Nature 1987, 328: 378–379.

Extracellular Biology of Molecular Chaperones: Molecular Chaperones as Cell Regulators

6

Cell-Cell Signalling Properties of Chaperonins

Anthony Coates and Peter Tormay

6.1. Introduction

In the beginning, bacteria evolved chaperonins (Cpns) to help in the folding of other proteins. Then, about one and a half billion years ago, bacteria began to live with eukaryotic cells. The career of the chaperonin began to expand beyond the protein folding area, and they developed new functions in order to adapt to the eukaryotic evolutionary niche. As the eukaryotes became ever more complex, the chaperonins evolved into cell-cell signalling molecules. The sophistication of this new role has only recently begun to emerge.

The chaperonins that are the subject of this chapter belong to the 60- and 10-kDa classes and are called Cpn60 and Cpn10, respectively. The folding actions of these proteins have been described in detail in Chapter 1. If there is more than one Cpn60 or Cpn10 in any one species, they are called Cpn60.1, Cpn60.2 and so on [1]. Cpn60 proteins are also called heat shock protein (Hsp) 60 or 65, and Cpn10s are also named Hsp10. For example, the *Mycobacterium tuberculosis* genome contains two *cpn60* genes. One of these, termed *cpn60.1* [2], appears to form an operon with the *cpn10* gene. This is the usual relationship in most bacteria. The second *cpn60* gene encodes Cpn60.2, the well-known Hsp65 protein of *M. tuberculosis*, and is found elsewhere in the genome.

6.2. What is a cell-cell signalling molecule?

A cell-cell signalling molecule is one that directly communicates a message from one cell to another. The signalling molecule may be attached to the outside of the broadcasting cell or it may be released from it and may attach to audience cells. The broadcasting cells may be bacterial or eukaryotic, because chaperonins exist in both eukaryotes and prokaryotes. The audience cells respond in a variety of ways, such as cytokine release.

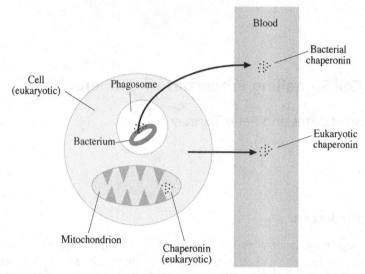

Figure 6.1. Chaperonins are produced by bacteria such as *M. tuberculosis* inside phagosomes. The chaperonin migrates into the phagosome and can be detected in tissues and in the blood. Eukaryotic chaperonins are found in the mitochondrion and can also be detected in the blood.

6.3. Are chaperonins released from cells?

Traditionally, chaperonins are regarded as intracellular, because they help to fold other proteins inside cells, and so, if this is their only function, why bother to wander into the extracellular space? The first suggestion that chaperonins might occur extracellularly was the observation that early pregnancy factor, subsequently reported to be Cpn10, could be detected in the serum of animals and of humans [3]. Since then, numerous reports have described chaperonins, both eukaryotic and prokaryotic, in extracellular situations [4–13]. For example, bacterial chaperonins have been detected on the cell surface [10], in the extracellular matrix [12] and in the serum of patients with tuberculosis [14] (Figure 6.1). Mammalian chaperonins are also detected on the cell surface [15], are secreted from glial cells and neuroblastoma cells [8] and are found in the serum of humans [5, 16]. It is not known for certain whether these chaperonins are secreted from live cells or whether they are released from dying ones. However, there is evidence that supports the idea that some chaperonins are secreted.

For instance, inside the macrophage phagosome, ingested *M. tuberculosis* Cpn10 accumulates in the wall of the bacterium and in the matrix of the phagosomes [17] (Figure 6.1). The chaperonins dissociate into partially helical monomers which interact with acidic lipids. This may represent two important steps in the mechanism of secretion of the protein into the external environ-ment. So it seems that some chaperonins, although lacking signal peptides, interact with membranes in a similar way to such peptides. The cell location and secretion of these proteins is dealt with in detail in Chapter 2 and to lesser extents in other chapters in this volume.

Interestingly, human Cpn10 behaves in a similar way to the mycobacterial protein and is found extracellularly; however, the Cpn10 of *Escherichia coli* is different in that it adopts a dimeric β-sheet which is not found outside the cell. These data suggest that Cpn10 is secreted by some species, but not by others.

An alternative, but not mutually exclusive, explanation for the presence of chaperonins outside the cell is that these proteins are released as the cell dies. This could be a kind of death message. In certain situations, it is possible that both secretion and release from dying cells might be responsible for extracellular chaperonins. That the broadcaster is dead does not diminish the power of the message, as readers of Shakespeare will no doubt agree.

6.4. Do all chaperonins transmit the same message?

The answer is a definite 'No'. The first indication that different chaperonins transmit different messages was the observation that *M. tuberculosis* Cpn10 in-duces lysis of bone, whereas Cpn60 from the same species is totally inactive [18]. Subsequently, it has been demonstrated *in vivo* [19] that *M. tuberculosis* Cpn60.1 actually blocks bone resorption. Even Cpn60s within the same species have different properties. For example, *Rhizobium leguminosarum* Cpn60.3 in-duces cytokine production by human monocytes, whereas Cpn60.1 is inactive [20]. In the case of *M. tuberculosis* Cpn60s, Cpn60.1 is a more powerful inducer of pro-inflammatory cytokines than Cpn60.2 [21]. These data indicate that dif-ferent chaperonins, even though they may share very high sequence identity, carry different messages.

6.5. Who is in the audience?

This is one of the most intriguing aspects of this area of research. Effects on human peripheral blood monocytes have been observed by several different laboratories [20–25]. However, the audience also contains macrophages and endothelial cells [26, 27], epithelial cells [28], vascular smooth muscle cells [29], dendritic cells (DCs) [24, 30], bone cells such as osteoclasts [31, 32] and osteoblasts [18], and cells of the central nervous system (CNS) [8, 33]. The available evidence indicates that the audience is primarily from a myeloid back-ground, although the inclusion of nerve cells suggests that this may be too narrow a definition. Chapter 15 describes in detail the interactions of chaperonins with nerve cells.

6.6. Is every chaperonin a molecular messenger?

Available evidence suggests that most species produce chaperonins which convey signalling messages to eukaryotic cells. For example, *M. tuberculosis* Cpn60.2 induces human monocytes to synthesise pro-inflammatory cytokines [20–25]. Many other Cpn60s from diverse bacterial species such as *E. coli* [23, 34], *Chlamydia spp.* [26, 34] and *Helicobacter pylori* [35] induce cells to synthesise pro-inflammatory cytokines. Even plant Cpn60s [20] can stimulate cytokine secretion. The significance of chaperonins became even more interesting when reports began to emerge that mammalian Cpn60s from rats, mice, hamsters and humans [24, 26, 34] also induce eukaryotic cells to synthesise pro-inflammatory cytokines. Such a ubiquitous property of chaperonins indicates that these molecules have a significant role to play in nature, one that we are only just beginning to understand.

6.7. What effect does the chaperonin signal have on the audience?

Chaperonins induce mammalian cells to produce pro-inflammatory cytokines [8, 18, 20–33]. For example, seven Cpn60s from different species all induce pro-inflammatory cytokine release from mouse macrophages [34]. However, data are available that suggest chaperonins have a greater significance than this in the world of non-folding biology: human Cpn10 has been shown to be early pregnancy factor [11, 36] and is present in red blood cells and in secretory granules [37]. Cpn60 from *Actinobacillus actinomycetemcomitans* [38], which is an oral pathogen in humans, induces proliferation of epithelial cells in 24 hours and increases the rate of epithelial cell death after prolonged incubation (144 hours). In cardiac muscle cells, Cpn10 and Cpn60 suppress ubiquitination of insulin-like growth factor-1 receptor and augment insulin-like growth factor-1 receptor signalling [39]. In the central nervous system, a chaperonin homologue is neuroprotective [33]. However, it is whole organ and animal studies that have revealed the truly startling biology of chaperonins.

In 1981 a major antigen of *M. tuberculosis* was identified by immunising mice with whole bacteria and generating the monoclonal antibody TB78 [40]. The antigen that bound to TB78 turned out to be Cpn60.2 (reviewed in [41]) which, curiously, can both induce and attenuate autoimmune arthritis and diabetes in animals [42, 43] (reviewed in [41]). Chaperonins are now being developed as vaccines for human diseases such as cancer [44] and diabetes [45]. Certain of the chapters in this volume deal with this in more detail (Chapters 16–18). It is thought that these properties are mediated by T lymphocytes. However, chaperonins have been called multiplex proteins [41], which means that they each have a number of parallel biological properties, of which immunogenicity and protein folding are but two.

The clearest example of a non-immunological, non-folding property of chaperonins is their ability to modulate bone formation. The first report of whole-organ bone modulation by chaperonins was that the Cpn60 proteins from *A. actinomycetemcomitans* and *E. coli* were potent stimulators of *in vitro* bone resorption [31]. Further work strengthened this observation, demonstrating that these proteins stimulated the proliferation and differentiation of myeloid precursor cells into mature bone-resorbing osteoclasts [32] and that the human Cpn60 protein was also a potent stimulator of bone resorption. In contrast, *M. tuberculosis* Cpn60s do not resorb bone [18], despite the fact that Cpn10 from this species is the primary bone resorbing molecule in the bacterium and may be responsible for the marked bone destruction that is seen in clinical cases of spinal tuberculosis, also known as Pott's disease.

However, in the rat adjuvant arthritis model, which is a T cell–driven disease with considerable osteoclastic bone remodeling [46], *M. tuberculosis* Cpn60.1 almost completely prevents bone destruction, whilst Cpn60.2 has no effect [19, 47]. The mechanism of action appears to be directly on bone formation (see the next section).

Another non-T cell–mediated effect is seen in the central nervous system. Rats that are given an intranasal dose of the Cpn60-like peptide activity-dependent neuroprotective protein (ADNP) are protected from neurodegeneration [33, 48]. An even stranger effect is a bacterial chaperonin in insect saliva which is used by the insect as a toxin to kill other insects [49].

This chapter will not review the T cell–mediated effects of chaperonins in autoimmune diabetes and arthritis because these are covered in the literature [41, 50, 51] and elsewhere in this volume. However, chaperonins seem to have effects in diseases that are not classically regarded as being T cell-mediated. For example, they are active in suppressing antigen-induced asthma in the mouse [52], despite the fact that asthma is an allergic condition. This is not to say that T cells play no part in asthma. The author and colleagues have found that *M. tuberculosis* Cpn60.1 blocks both the eosinophilia and the airway hyperresponsiveness found in this model and that it does so by a mechanism involving the priming of DCs. In contrast, Cpn60.2 was without effect ([52] and unpublished observations). These findings are supported by a recent publication which showed that *Mycobacterium leprae* Cpn60.2 is active in suppressing this asthma model, whereas *M. tuberculosis* Cpn60.2 is inactive [53].

Thus it seems that chaperonins remain full of biological surprises.

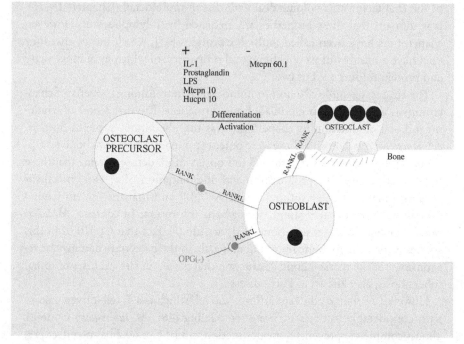

Figure 6.2. Osteoclast formation depends on the upregulation of the TNF family member RANKL (receptor activator of NF-κB ligand) on osteoblasts and the activation of osteoclast precursor cells through binding of RANKL to RANK. OPG binds RANKL and acts as a RANK antagonist, thereby inhibiting osteoclast formation (+ = molecules that enhance differentiation/activation; − = molecules that inhibit differentiation/activation).

6.8. Mechanism of action – cytokines and the cytokine ceiling

At the cellular level, there are two seemingly contradictory actions of chaperonins. Firstly, they stimulate the release of pro-inflammatory cytokines from eukaryotic cells. Secondly, they suppress cells that have been stimulated by either endogenous or exogenous agents. The effects of *M. tuberculosis* Cpn60.1 (MtCpn60.1) on bone are a case in point. This chaperonin stimulates cytokine release from cells [21], yet it suppresses bone resorption that has been triggered by exogenous bacterial lipopolysaccharide (LPS) or endogenous RANKL (receptor activator of NF-κB), a tumour necrosis factor (TNF) family member and key osteoclast-inducing cytokine. It is now recognised that the dynamic remodelling of bone, which is a consequence of the bone forming action of the osteoblast and the bone destroying activity of osteoclasts, is controlled by the regulation of RANKL expression on osteoblasts and its interaction with RANK on osteoclast precursor cells. Another TNF family member, osteoprotegerin (OPG), acts as a soluble RANK and antagonises RANKL/RANK interactions and inhibits osteoclast formation (Figure 6.2). In bone cell and explant culture studies it is now established that *M. tuberculosis* Cpn60.1 can block the activity of RANKL which interacts with RANK on osteoclast precursors [19]. Such interaction stimulates a complex signalling pathway involving TRAF-6, which is described in detail in Chapter 7. It has been established that inhibition is not due to (i) stimulation of synthesis of the negative regulator OPG, (ii) inhibition of the key MAP kinase c-JNK or (iii) binding of RANKL by Cpn60.1.

It is likely that DCs, the conductors of the immune system, are an important target for chaperonins and therein may hide secrets of their mechanism of action. Cpn60s can activate DCs to secrete pro-inflammatory cytokines and can induce these cells to mature [24]. Cpn60 induces the release of TNF-α, IL-12 and IL-1β from DCs [30], but only a small amount of IL-10, which suggests the presence of a bias towards Th1 (pro-inflammatory) responses. One interpretation of this is that Cpn60s may prime for destructive Th1 responses at sites of Cpn60 release [30]. An alternative hypothesis which has been suggested by the author (AC), called the 'cytokine ceiling hypothesis' suggests that whilst chaperonins do activate the innate immune system, they also suppress over-activation. This dampening down of, for example, a bacterial LPS-activated immune system, is an essential control mechanism which, if absent, would result in self-destruction by an innate immune system that is out of control. So, the 'cytokine ceiling' is an additional property of chaperonins. The innate immune system recognises bacterial pro-inflammatory signals (pattern-associated molecular patterns – PAMPs) as 'danger' molecules [54]; however, when the level of response reaches the 'cytokine ceiling', chaperonins induce suppression of an otherwise highly damaging immune response.

It is possible that some chaperonins are activators and others are suppressors (see Figure 6.3); however, it is also possible that one species of chaperonins can both activate and suppress the immune system. Furthermore, it likely that this paradigm applies to many other cell types such as bone and CNS cells, not just innate immune cells. For instance, *M. tuberculosis* chaperonins show three separate patterns of activity: (i) Cpn60.1 and Cpn60.2 both activate human monocytes [21], (ii) Cpn60.1 is a potent inhibitor of the differentiation of osteoclast precursor cells into mature bone resorbing osteoclasts [19] whilst Cpn60.2 is inactive in this regard and (iii) Cpn10 is a potent stimulator of osteoclast proliferation. Gram-negative bacterial and the human mitochondrial Cpn60 proteins also act as potent inducers of osteoclast formation. These data show that chaperonins can be very different from one another in terms of their interaction with selected target cells.

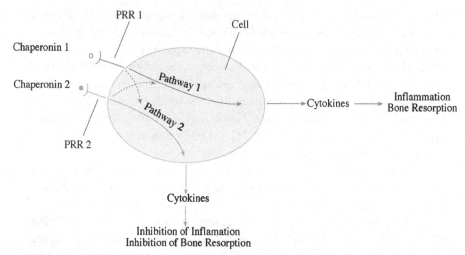

Figure 6.3. Chaperonins bind to PRRs. Shown here are hypothetical pathways 1 and 2 which lead to inflammation/bone resorption and inhibition, respectively.

6.9. Chaperonin receptors

Chaperonins activate cells by interacting with pattern-recognition receptors (PRRs) including CD14 and Toll-like receptors (TLRs) [26, 30, 55, 56] (Figure 6.3), and the interaction of Cpn60 proteins with the TLRs is discussed in detail in Chapter 7. However, chaperonins have such a diverse range of activities that it is likely that these receptors are probably the tip of the iceberg and it is anticipated that more receptors will be implicated soon. For example, although TLR4 is presumed by many workers to be the major (perhaps only) receptor for Cpn60 (see Chapter 7), early work on the Cpn60 protein of *A. actinomycetemcomitans* showed that it could activate myeloid cells from the TLR4 negative mutant, C3H/HeJ strain, of mouse [31]. Human and bacterial chaperonins seem to use the same receptors, and this suggests that bacteria may use this to modulate their host's immunity to their advantage.

Chaperonins can be considered PAMPs [57]. PAMPs are evolutionarily conserved components of pathogens recognised by non-clonal receptors of the host. A growing number of PRRs (recent examples being the TLRs) have been identified and are responsible for the ability of our innate immune system to rapidly recognise, and discriminate, infections by bacteria and fungi [58–60]. PAMPS include LPS, peptidoglycan, flagellae and CpG DNA [58–60] and are unique to pathogens. Interestingly, human chaperonins seem to use the same receptors as some of the bacterial PAMPs [55, 56]. This would be highly damaging to the host unless there was an upper cut-off point above which the chaperonins became suppressive, the so-called cytokine ceiling.

6.10. Is all the activity of chaperonins due to contaminating bacterial lipopolysaccharide?

Most recombinant chaperones are produced in *E. coli* and so are contaminated with LPS. In this and subsequent chapters the problems with such contamination will be discussed and evidence will be presented that the biological actions of molecular chaperones such as Cpn60 are not due to contaminating LPS. Many workers use commercially available columns with polymyxin B linked to agarose in order to remove LPS contaminating recombinant molecular chaperones. This has the disadvantage (as discussed in Chapter 10) of also resulting in the binding of the molecular chaperone to the column.

The author's (AC) group has solved this particular problem by washing affinity (e.g., Ni-NTA) columns containing recombinant chaperones with polymyxin B. This eliminates most of the contaminating LPS without removing the activity of the recombinant protein. Thus, powerful chaperonin activity can be detected when LPS is undetectable [21, 24]. If chaperonins are exposed to proteases or heat, their cell signalling activity is lost. Furthermore, when short peptides of chaperonins are made by chemical synthesis, chaperonin activity is seen [18]. In addition, *in vivo*, many activities of chaperonins cannot be reproduced by LPS. For example, *M. tuberculosis* Cpn60.1 blocks bone resorption [19]. This chaperonin also inhibits asthma in the mouse [52]. Single-amino-acid changes in another bacterial chaperonin render it toxic to insects [49] and a short mammalian chaperonin peptide protects rats from neurodegeneration [33, 48].

However, there are occasional reports of loss of activity of chaperonin which have been attributed to the absence of LPS [61]. In the author's experience, loss of activity of a chaperonin does occur from time to time; however, this is not due to absence of LPS, but rather from incorrect folding of the chaperonin or a technical problem associated with the biological assay. The involvement of LPS contamination in the observed biological activities of chaperonins has also been discussed elsewhere in this volume (Chapters 7, 8 and 10).

6.11. Conclusion

Chaperonins are multiplex molecules. Apart from helping cells to fold proteins, they have important non-folding activities, as well as being powerful antigens and cell-cell signalling molecules. Chaperonins are released from cells and can be detected in tissues and in the blood. They transmit messages from one cell to another, and messages transmitted by chaperonins from individual species can be different and distinct. The audience cells are primarily myeloid in origin; however, cells of the central nervous system are also involved.

The effects of the chaperonin signal are diverse. For example, chaperonins induce pro-inflammatory cytokine release, promote bone resorption, inhibit bone resorption, protect neural cells from degeneration, kill insects and protect mice from asthma. Chaperonins act through PRRs which include CD14 and Toll-like molecules. The mechanism of action has two main arms: firstly, stimulation of cytokine release, and secondly suppression. It is hypothesised that chaperonins suppress overactive cytokine responses at a certain high dangerous level, which is termed the cytokine ceiling. This protects the animal from damaging itself.

The extraordinary diversity of actions of chaperonins that are known today are almost certainly the tip of an iceberg.

REFERENCES

1. Coates A R M, Shinnick T M and Ellis J R. Chaperonin nomenclature. Mol Microb 1993, 8: 787.
2. Kong T H, Coates A R M, Butcher P D, Hickman C J and Shinnick T M. *Mycobacterium tuberculosis* expresses two chaperonin-60 homologs. Proc Natl Acad Sci USA 1993, 90: 2608–2612.
3. Cavanagh A C and Morton H. The purification of early-pregnancy factor to homogeneity from human platelets and identification as chaperonin 10. Eur J Biochem 1994, 222: 551–560.
4. Soltys B J and Gupta R S. Mitochondrial-matrix proteins at unexpected locations: are they exported? Trends Biochem Sci 1999, 24: 174–177.
5. Morton H. Early pregnancy factor: an extracellular chaperonin 10 homologue. Immunol Cell Biol 1998, 76: 483–496.
6. Goulhen F, Hafezi A, Uitto V J, Hinode D, Nakamura R, Grenier D and Mayrand D. Subcellular localization and cytotoxic activity of the GroEL-like protein isolated from *Actinobacillus actinomycetemcomitans*. Infect Immun 1998, 66: 5307–5313.
7. Kol A, Sukhova G K, Lichtman A H and Libby P. Chlamydial heat shock protein 60 localizes in human atheroma and regulates macrophage tumour necrosis factor-α and matrix metalloproteinase expression. Circulation 1998, 98: 300–307.
8. Bassan M, Zamostiano R, Giladi E, Davidson A, Wollman Y, Pitman J, Hauser J, Brenneman D E and Gozes I. The identification of secreted heat shock 60-like protein from rat glial cells and a human neuroblastoma cell line. Neurosci Lett 1998, 250: 37–40.
9. Coates A R M and Henderson B. Chaperonins in health and disease. Ann NY Acad Sci 1998, 851: 48–53.
10. Frisk A, Ison C A and Lagergard T. GroEL heat shock protein of *Haemophilus ducreyi*: association with cell surface and capacity to bind to eukaryotic cells. Infect Immun 1998, 66: 1252–1257.

11. Cavanagh A C. Identification of early pregnancy factor as chaperonin 10: implications for understanding its role. Rev Reprod 1996, 1: 28–32.
12. Esaguy N and Aguas A P. Subcellular localization of the 65-kDa heat shock protein in mycobacteria by immunoblotting and immunogold ultracytochemistry. J Submicrosc Cytol Pathol 1997, 29: 85–90.
13. Brenneman D E and Gozes I. A femtomolar-acting neuroprotective peptide. J Clin Invest 1996, 97: 2299–2307.
14. Sethna K B, Mistry N F, Dholakia Y, Antia N H and Harboe M. Longitudinal trends in serum levels of mycobacterial secretory (30 kD) and cytoplasmic (65 kD) antigens during chemotherapy of pulmonary tuberculosis patients. Scand J Infect Dis 1998, 30: 363–369.
15. Soltys B J and Gupta R S. Cell surface localization of the 60 kDa heat shock chaperonin protein (hsp60) in mammalian cells. Cell Biol Int 1997, 21: 315–320.
16. Lewthwaite J, Owen N, Coates A, Henderson B and Steptoe A. Circulating human heat shock protein 60 in the plasma of British civil servants. Circulation 2002, 106: 196–201.
17. Fossati G, Izzo G, Rizzi E, Gancia E, Niccolai N, Giannozzi E, Spiga O, Bono L, Marone P, Leone E, Mangili F, Harding S, Errington N, Walter C, Henderson B, Roberts M M, Coates A R M, Casetta B and Mascagni P. *Mycobacterium tuberculosis* chaperonin 10 is secreted in the macrophage phagolysosome: is secretion due to dissociation and the adoption of partially helical structure at the membrane? J Bact 2003, 185: 4256–4267.
18. Meghji S, White P, Nair S P, Reddi K, Heron K, Henderson B, Zaliani A, Fossati G, Mascagni P, Hunt J F, Roberts M M and Coates A R. *Mycobacterium tuberculosis* chaperonin 10 stimulates bone resorption: a potential contributory factor in Pott's disease. J Exp Med 1997, 186: 1241–1246.
19. Winrow V R, Meghji S, Mesher J, Coates A R M, Morris C J, Tormay P and Henderson B. *Mycobacterium tuberculosis* chaperonin 60.1, but not chaperonin 60.2, inhibits osteoclastic bone resorption by blocking RANKL activity.
20. Lewthwaite J, George R, Lund P A, Poole S, Tormay P, Sharp L, Coates A R M and Henderson B. *Rhizobium leguminosarum* chaperonin 60.3, but not chaperonin 60.1, induces cytokine production by human monocytes: activity is dependent on interaction with cell surface CD14. Cell Stress Chaperon 2002, 7: 130–136.
21. Lewthwaite J C, Coates A R M, Tormay P, Singh M, Mascagni P, Poole S, Roberts M, Sharp L and Henderson B. *Mycobacterium tuberculosis* chaperonin 60.1 is a more potent cytokine stimulator than chaperonin 60.2 (hsp 65) and contains a CD14-binding domain. Infect Immun 2001, 69: 7349–7355.
22. Friedland J S, Shattock R, Remick D G and Griffin G E. Mycobacterial 65-kD heat shock protein induces release of proinflammatory cytokines from human monocytic cells. Clin Exp Immunol 1993, 91: 58–62.
23. Tabona P, Reddi K, Khan S, Nair S P, Crean S J, Meghji S, Wilson M, Preuss M, Miller A D, Poole S, Carne S and Henderson B. Homogeneous *Escherichia coli* chaperonin 60 induces IL-1 and IL-6 gene expression in human monocytes by a mechanism independent of protein conformation. J Immunol 1998, 161: 1414–1421.
24. Bethke K, Staib F, Distler M, Schmitt U, Jonuleit H, Enk A H, Galle P R and Heike M. Different efficiency of heat shock proteins to activate human monocytes and dendritic cells: superiority of HSP60. J Immunol 2002, 169: 6141–6148.

25. Maguire M, Coates A R M and Henderson B. Cloning expression and purification of three chaperonin 60 homologues. J Chromatography 2003, 786: 117–125.
26. Kol A, Bourcier T, Lichtman A and Libby P. Chlamydial and human heat shock protein 60s activate human vascular endothelium, smooth muscle cells, and macrophages. J Clin Invest 1999, 103: 571–577.
27. Billack B, Heck D E, Mariano T M, Gardner C R, Sur R, Laskin D L and Laskin J D. Induction of cyclooxygenase-2 by heat shock protein 60 in macrophages and endothelial cells. Am J Physiol (Cell Physiol) 2002, 283: C1267–1277.
28. Zhang L, Pelech S L, Mayrand D, Grenier D, Heino J and Uitto V J. Bacterial heat shock protein-60 increases epithelial cell proliferation through the ERK1/2 MAP kinases. Exp Cell Res 2001, 266: 11–20.
29. Sasu S, LaVerda D, Qureshi N, Golenbock D T and Beasley D. *Chlamydia pneumoniae* and chlamydial heat shock protein 60 stimulate proliferation of human vascular smooth muscle cells via toll-like receptor 4 and p44/p42 mitogen-activated protein kinase activation. Circ Res 2001, 89: 244–250.
30. Flohé S B, Bruggemann J, Lendemans S, Nikulina M, Meierhoff G, Flohé S and Kolb H. Human heat shock protein 60 induces maturation of dendritic cells versus a Th1-promoting phenotype. J Immunol 2003, 170: 2340–2348.
31. Kirby A C, Meghji S, Nair S P, White P, Reddi K, Nishihara T, Nakashima K, Willis A C, Sim R, Wilson M and Henderson B. The potent bone-resorbing mediator of *Actinobacillus actinomycetemcomitans* is homologous to the molecular chaperone GroEL. J Clin Invest 1995, 96: 1185–1194.
32. Reddi K, Meghji S, Nair S P, Arnett T R, Miller A D, Preuss M, Wilson M, Henderson B and Hill P. The *Escherichia coli* chaperonin 60 (groEL) is a potent stimulator of osteoclast formation. J Bone Miner Res 1998, 13: 1260–1266.
33. Gozes I, Divinsky I, Pilzer I, Fridkin M, Brenneman D E and Spier A D. From vasoactive intestinal peptide (VIP) through activity-dependent neuroprotective protein (ADNP) to NAP: a view of neuroprotection and cell division. J Mol Neurosci 2003, 20: 315–322.
34. Habich C, Kempe K, van der Zee R, Burkart V and Kolb H. Different heat shock protein 60 species share pro-inflammatory activity but not binding sites on macrophages. FEBS Lett 2003, 533: 105–109.
35. Gobert A P, Bambou J C, Werts C, Balloy V, Chignard M, Moran A P and Ferrero R L. *Helicobacter pylori* heat shock protein 60 mediates interleukin-6 production by macrophages via a Toll-like receptor (TLR)-2-, TLR-4-, and myeloid differentiation factor 88–independent mechanism. J Biol Chem 2004, 279: 245–250.
36. Rolfe B, Cavanagh A, Forde C, Bastin F, Chen C and Morton H. Modified rosette inhibition test with mouse lymphocytes for detection of early pregnancy factor in human pregnancy serum. J Immunol Methods 1984, 70: 1–11.
37. Sadacharan S K, Cavanagh A C and Gupta R S. Immunoelectron microscopy provides evidence for the presence of mitochondrial heat shock 10-kDa protein (chaperonin 10) in red blood cells and a variety of secretory granules. Histochem Cell Biol 2001, 116: 507–517.
38. Zhang L, Pelech S and Uitto V J. Long-term effect of heat shock protein 60 from *Actinobacillus actinomycetemcomitans* on epithelial cell viability and mitogen-activated protein kinases. Infect Immun 2004, 72: 38–45.

39. Shan Y X, Yang T L, Mestril R and Wang P H. Hsp10 and Hsp60 suppress ubiq-uitination of insulin-like growth factor-1 receptor and augment insulin-like growth factor-1 receptor signaling in cardiac muscle: implications on decreased myocardial protection in diabetic cardiomyopathy. J Biol Chem 2003, 278: 45492–45498.

40. Coates A R M, Hewitt J, Allen B W, Ivanyi J and Mitchison D A. Antigenic diversity of *Mycobacterium tuberculosis* and *Mycobacterium bovis* detected by means of monoclonal antibodies. Lancet 1981, 2: 167–169.

41. Coates A R M. Immunological aspects of chaperonins. In Ellis, R. J. (Ed.) *The Chaperonins*. Academic Press, London 1996, pp 267–296.

42. van Eden W, Thole J E R, van der Zee R, Noordzij A, van Embden J D A, Hensen E J and Cohen I R. Cloning of the mycobacterial epitope recognized by T lymphocytes in adjuvant arthritis. Nature 1988, 331: 171–173.

43. Elias D, Markovits D, Reshef T, van der Zee R and Cohen I R. Induction and therapy of autoimmune diabetes in the non-obese diabetic mouse by a 65-kDa heat shock protein. Proc Natl Acad Sci USA 1990, 87: 1576–1580.

44. Srivastava P K. Immunotherapy of human cancer: lessons from mice. Nat Immunol 2000, 1: 363–366.

45. Raz I, Elias D, Avron A, Tamir M, Metzger M and Cohen I R. Beta-cell function in new-onset type 1 diabetes and immunomodulation with a heat-shock protein peptide (DiaPep277): a randomised, double-blind, phase II trial. Lancet 2001, 358: 1749–1753.

46. Billingham MEJ. Adjuvant arthritis: The first model. In Henderson, B., Edwards, J. C. W. and Pettipher, E. R. (Eds.) *Mechanisms and Models in Rheumatoid Arthritis*. Academic Press, London 1995, pp 389–409.

47. Winrow V R, Coates A R M, Tormay P, Henderson B, Singh M, Blake D R and Morris C J. Chaperonin 60.1 prevents bone destruction in Wistar rats with adjuvant-induced arthritis. Rheumatology 2002, 41 (abstr suppl 1): 47.

48. Gozes I and Brenneman D E. Activity-dependent neurotrophic factor (ADNF). An extracellular neuroprotective chaperonin? J Mol Neurosci 1996, 7: 235–244.

49. Yoshida N, Oeda K, Watanabe E, Mikami T, Fukita Y, Nishimura K, Komai K and Matsuda K. Protein function. Chaperonin turned insect toxin. Nature 2001, 411: 44.

50. Cohen I R. Peptide therapy for Type I diabetes: the immunological homunculus and the rationale for vaccination. Diabetologia 2002, 45: 1468–1474.

51. Cohen I R. The Th1/Th2 dichotomy, hsp60 autoimmunity, and type I diabetes. Clin Immunol Immunopathol 1997, 84: 103–106.

52. Riffo-Vasquez Y, Spina D, Page C, Tormay P, Singh M, Henderson B and Coates A R M. Effect of *Mycobacterium tuberculosis* chaperonins on bronchial eosinophilia and hyperresponsiveness in a murine model of allergic inflammation. Clin Exp Allergy 2004, 34: 712–719.

53. Rha Y-H, Taube C, Haczku A, Joeham A, Takeda K, Duez C, Siegel M, Ayditung M K, Born W K, Dakhama A and Gelfand E W. Effect of microbial heat shock proteins on airway inflammation and hyperresponsiveness. J Immunol 2002, 169: 5300–5307.

54. Matzinger P. An innate sense of danger. Semin Immunol 1998, 10: 399–415.

55. Asea A, Kraeft S-K, Kurt-Jones E A, Stevenson M A, Chen L B, Finberg R W, Koo G C and Calderwood S K. Hsp70 stimulates cytokine production through a

CD14-dependent pathway, demonstrating its dual role as a chaperone and cytokine. Nat Med 2000, 6: 435–442.

56. Vabulas R M, Ahmad-Nejad P, da Costa C, Miethke T, Kirschning C J, Hacker H and Wagner H. Endocytosed HSP60s use toll-like receptor 2 (TLR2) and TLR4 to activate the toll/interleukin-1 receptor signaling pathway in innate immune cells. J Biol Chem 2001, 276: 31332–31339.

57. Medzhitov R and Janeway C A J. Innate immunity: the virtues of a nonclonal system of recognition. Cell Stress Chaperon 1997, 91: 295–298.

58. Takeda K and Akira S. Toll receptors and pathogen resistance. Cell Microbiol 2003, 5: 143–153.

59. Girardin S E, Sansonetti P J and Philpott D J. Intracellular vs extracellular recognition of pathogens – common concepts in mammals and flies. Trends Microbiol 2002, 10: 193–199.

60. Colonna M. TREMS in the immune system and beyond. Nat Rev Immunol 2003, 3: 1–9.

61. Gao B and Tsan M F. Recombinant human heat shock protein 60 does not induce the release of tumor necrosis factor alpha from murine macrophages. J Biol Chem 2003, 278: 22523–22529.

7

Toll-Like Receptor-Dependent Activation of Antigen Presenting Cells by Hsp60, gp96 and Hsp70

Ramunas M. Vabulas and Hermann Wagner

7.1. Discovery of Toll-like receptors

The basic concept of the immune system postulates an ability to discriminate between self and non-self and to free the organism from the latter. Two major contributions advanced the comprehension of the cellular basis of self- versus non-self-discrimination. The first was the hypothesis regarding the expansion of antigen-recognising clones on encounter with a respective antigen, which allowed antigenic specificities of the resulting immune reactions to be explained. The co-stimulatory signal hypothesis represented another essential advancement. It postulated the necessity of a second, antigen-independent signal for lymphocyte activation. Its nature was put into an elegant metaphor of the 'immunologist's dirty little secret' [1], referring to substances of microbial origin that should be present concomitant with an antigen to prime an immune response to it.

Of a number of host receptors participating in detection of microbial constituents [2], Toll-like receptors (TLRs) currently represent the most interesting group. Their importance is assumed from the prominent cell activating capacity which they display after engagement with their cognate ligands. The name originates from the *Drosophila* homologue Toll, which was discovered as a part of the dorsoventral patterning cascade during the developmental larva stage of the fruit fly, and this seminal study established an additional, anti-microbial function for Toll in adult flies [3]. It demonstrated that mutants of the genes in the cassette between the Toll ligand Spätzle down to the IκB homologue Cactus showed a compromised inducibility of the anti-fungal peptide drosomycin upon fungal challenge and consequently succumbed to the infection. Moreover, there was an obvious specificity in discrimination of microbial classes, because the signalling cascade had no influence on the induction of anti-bacterial peptides.

Searches against the intracellular Toll domain were similarly performed and a human homologue hToll, now classified as the Toll-like receptor (TLR)4 was discovered [4]. Forced dimerisation of TLR4 induces a typical pro-inflammatory state, confirming not only structural but also functional similarity with the Toll of *Drosophila*. Ten human TLRs have now been described, and years of intense research have demonstrated the importance of this system for sensing a variety of microbial as well as endogenous products, including the main classes of human heat shock proteins.

7.2. Toll-like receptor structure

TLRs are type I membrane proteins. The extracellular (or lumenal) part of TLRs is composed of tandemly repeated modules enriched in leucines and, hence, called leucine-rich repeats (LRRs). LRRs are found in a variety of proteins with very different functions. The single LRR is usually 20–29 residues long and displays a characteristic leucine distribution pattern. By comparing sequences of LRRs from different proteins one can distinguish several classes which reflect the differences in the C-terminal part of the LRR module. On the other side, the leucine positions in the initial 10 amino acid stretches are well conserved and show an X-L-X-X-L-X-L-X-X-N pattern (X being any amino acid). The spatial visualisation of this conservation has been provided by crystal and nuclear magnetic resonance structures of several LRR proteins. They all indicate a horseshoe-like molecular shape in which parallel β-strands in perpendicular fashion line the concave horseshoe surface. β-strands fold out of the first, conserved LRR halves which explains the invariant appearance of the concave surface. The β-strand in each LRR module is followed by the less conserved stretch which forms more variable structures. The analysis of crystal structures of LRR proteins complexed with their ligands reveals that the concave surfaces and β–α loops provide the required interaction platform [5].

Unfortunately, no TLR structure has yet been solved, and the mechanistic aspects of activation therefore remain enigmatic. Because of the broad range of non-proteinaceous ligands implicated from functional studies, the structural determinants for TLR–ligand interactions might appear quite different to those already established. The recent bioinformatical analysis of the TLR family has implicated the insertions within the LRRs (typically at positions 10 and 15) to be important in controlling specificity of interaction [6]. Nevertheless, to test this and other hypotheses it will only be possible by solving X-ray structures of different TLR–ligand complexes.

The cytoplasmic domain of the TLR, the Toll/IL-1 receptor (TIR) domain, is responsible for all features of the receptor-mediated intracellular signalling of the TLR family. It was the TIR domain that provided the first evidence for the homology that spans such distant taxons as the fruit fly and the human and related proteins with such different functions as embryonic body polarisation and controlling the inflammatory response [7, 8]. Discovery of the TIR domain in proteins encoded by the plant disease-resistance genes underscore the conservation and, hence, the importance of this evolutionary 'invention' [9].

Together with the first insight into the heterogeneity of the TLR family, attempts were made to structurally analyse the TIR domain [10]. The $(\beta/\alpha)_5$ fold was predicted and the similarity to the bacterial chemotaxis regulator, CheY was implicated. The structural work proved the resemblance between human TIR domains and CheY and also revealed some specific properties [11]. Comparisons of TIR domain structures of different TLRs, including those that had been mutated or were non-functional, provided some insight into mechanistic aspects of the biology of these receptors. Two interaction interfaces on the domain surface appeared to be possible. The first of them, the R face, which varies in different receptors, appears to contribute to the specificity of the ligand recognition by specifying the oligomerisation of the receptors. Another one, the S face, which displays a conserved surface patch, contributes to the invariant aspect of the TLR family function by engaging the conserved intracellular machinery of cell activation.

7.3. Toll-like receptor signalling

After binding to their cognate extracellular ligands, many classes of membrane-bound receptors modulate or even re-programme cellular functions by recruiting and activating different sets of cytoplasmic mediators. The mediators, in turn, transmit inhibitory or activating signals to transcription factors, and these, consequently, alter the transcriptional profile and define the cellular response to the original external input.

The response to TLR engagement was first demonstrated by manipulating human TLR4 [4]. The extracellular part of the receptor was exchanged for the CD4 fragment, which drives spontaneous homodimerisation of the fusion protein. This forced ligand-independent signalling led to the paradigmatic state of immune alertness, namely, the activation of the transcription factor NF-κB and, consequently, the induction of pro-inflammatory cytokines and the upregulation of co-stimulatory molecules. Intense genetic and biochemical work ensued, and this has elucidated many aspects of the intracellular process. We will briefly discuss this work.

MyD88 is the first and essential adaptor molecule which becomes engaged after activation of every TLR, except for TLR3. MyD88-deficient mice show impaired responsiveness not only to most TLR ligands, but also to IL-1 and IL-18, thereby proving MyD88 to be a critical signalling component for any TIR domain-containing receptor [12, 13]. MyD88 possesses its own C-terminal TIR domain, which drives the heterodimerisation of the adaptor with the activated receptor. The N-terminal death domain of MyD88 then recruits IRAK1 and IRAK4 kinases. The IRAK4 phosphorylates and activates the IRAK1, which in turn initiates autophosphorylation and recruits TRAF6. Lack of IRAK4 reproduces the defects seen in MyD88 mutants [14]. The importance of IRAK1 is less clear, because its deficiency exhibits a less pronounced phenotype. In addition, there are two further IRAKs, neither of which have kinase activity – IRAK2 and IRAK-M. IRAK-M has been found to be a negative regulator of TLR signalling [15], whereas the exact role of IRAK2 remains unclear.

Of the family of six TRAF adaptor proteins, only TRAF6 is involved in TLR signalling. TRAF6, together with IRAK, dissociates from the activated receptor and binds to the preformed complex of TGF-β-activated kinase (TAK)1 and TAK1-binding proteins (TAB)1 and 2. TAK1 is a mitogen-activated protein kinase kinase kinase (MAP3K) involved in activation of IκB kinase (IKK). Activation of IKK appears to require atypical polyubiquitination [16, 17]. The TRAF6/TAK1/TAB1/TAB2 complex associates with the heterodimeric ubiquitin-conjugating enzyme Ubc13/Uev1A. This results in modification of TRAF6 with lysin63-linked polyubiquitin chain, which leads to IKK activation, IκB phosphorylation, ubiquitination and degradation. From the IκB released transcription factor NF-κB translocates to the nucleus and switches on the transcription of a large number of pro-inflammatory genes. TAK1 is also responsible for activation of MAP kinases (MAPKs) and c-Jun N-terminal kinases (JNKs), in this way broadening and diversifying transcriptional changes in response to receptor stimulation.

The specificity of the response to different ligands is assumed to arise from different adaptors associating with the respective receptors. In particular, and in contrast to other TLRs, TLR3 and TLR4 engagement is known to activate the transcription factor Interferon Regulatory Factor 3 (IRF3) and as a consequence to induce interferon-β. Furthermore, MyD88-deficient mice are not able to produce a number of cytokines upon challenge with lipopolysaccharide (LPS), whereas they still show NF-κB and JNK activation, albeit with slower kinetics [13]. This evidence prompted an intense search for additional adaptors.

The first candidate was a molecule named TIRAP/Mal; however, knock-out studies confirmed the function of the TIRAP/Mal in MyD88-dependent pathway [18, 19]. The specificity of another adaptor, the TRIF/Ticam1, was subsequently substantiated using classical and reverse genetic approaches [20, 21]. These studies proved the role of TRIF/Ticam in MyD88-independent effects upon TLR3 and TLR4 activation. The latest addition to the ever growing list of response specifiers is yet another TIR domain-containing molecule – TRAM. Analysis of TRAM-deficient mice located this adaptor in the exclusive position on the TLR4-triggered MyD88-independent pathway of cell activation [22].

There are a number of other receptor-proximal molecules implicated in one or another aspect of TLR function and further studies are awaited, for example, PI3-kinase, Tollip, Pellino or sterile alpml motif (SAM) and armadillo motif (ARM). The future will undoubtedly reveal additional complexities but, concomitantly, also a better understanding of TLR biology, thereby substantiating in molecular terms the hypothesis of the innate, and thus invariant system, of discrimination and the identification of 'foreign and dangerous'. The importance of this knowledge will become even more apparent during our further discussion showing that the TLR system has been accommodated to sense 'danger' by interacting with heat shock proteins, independently of their origin.

7.4. Heat shock protein signalling via Toll-like receptors

There are numerous circumstances in which the immune system becomes activated in the absence of infection. The most prominent examples could be aseptic necrosis or immunisation with syngeneic tumours. To explain these and other phenomena it has been postulated that the organism possesses its own (i.e., endogenous) danger signals that are normally concealed inside the cell. When released, these alert the organism that something is going wrong. Heat shock proteins, also called chaperones (see Chapter 1 for details on nomenclature) due to their assistance in protein folding and translocation, seem to be perfect candidates for endogenous danger signals. Firstly, most of them are essential and abundant. Secondly, some of them get strongly upregulated under stress conditions (hence, another name – stress proteins). Thirdly, they are normally located intracellularly. In addition, bacterial homologues have long been known to be immunodominant antigens in infections [23]. Pathogen heat shock protein–specific antibodies and T cell clones are often found in infected individuals, and their cross-reactivity with endogenous homologous molecules has been used to explain the pathogenesis of several autoimmune disorders. See Chapter 16 for more details of the immunology of chaperonins.

The essential role of heat shock proteins in a number of cellular processes [24] suggests that they should be evolutionarily conserved, and this is indeed the case. This sequence conservation also supports the cross-reactivity hypothesis and provides an additional basis for the proposition that heat shock proteins act as endogenous danger signals.

A strong thrust to the heat shock protein immunology field came with the discovery that some stress proteins are able to shuttle antigenic peptides into antigen presentation pathways and thereby prime adaptive immune responses [25–27]. Many details of this process have been elucidated including the characterisation of peptide association, the discovery of several receptors involved in the uptake of these proteins and some insights into the intracellular trafficking events. These aspects are discussed in Chapters 16–18. However, we will review here the stimulatory capacities of heat shock proteins and summarise what has been recently learned about their mechanisms.

7.4.1. Hsp60 signalling

In wishing to stress the point that heat shock proteins are more than an additional adjuvant for the immune system, it is salutary to start with Hsp60, the chaperone which does not have a documented ability to transport non-covalently bound antigens. In other chapters in this book the generic term chaperonin (Cpn)60 is used to describe this protein. The term Hsp60 tends to be used with eukaryotic (mitochondrial) proteins, but in this chapter it will be used to describe both human and bacterial proteins.

Hsp60 (also called GroEL – the *Escherichia coli* Hsp60) has the most impressive file on its role in immune responses among all chaperones. Immune reactions to bacterial Hsp60 become so prominent in some instances that they should be regarded as the characteristic sign of infection. Additionally, considerable interest in the inflammatory effects of Hsp60 arose from the field of atherosclerosis research. Chlamydial infection has been implicated in the pathogenesis of atherosclerosis, and chlamydial Hsp60 has been found in atherosclerotic lesions [28]. Interestingly, the human Hsp60 was co-localised with the chlamydial protein in the same study and it was shown to activate a number of different cell types. Subsequently, systematic analyses of human Hsp60 stimulatory features demonstrated a typical pro-inflammatory reaction pattern – specifically, the induction of TNF-α, IL-6, IL-12, IL-15 and nitric oxide (NO) production and synergy with IFN-γ – which was indistinguishable from stimulation with the classical bacterial inflammogen, LPS [29, 30]. These findings prompted the proposition that Hsp60 is an endogenous danger signal.

At the same time, evidence accumulated to establish the TLRs as the central axis of the infectious danger sensor. The question arose as to whether Hsp60 exploits the same receptor system. The first hint came with the analysis of CD14 involvement [31]. CD14 has been described as a high-affinity binding receptor for LPS [32] and for other bacterial products [33]. It was shown that cells, otherwise unresponsive to Hsp60, gained sensitivity to Hsp60 upon expression of CD14 [31]. Furthermore, anti-CD14 antibodies blocked Hsp60 activating potential on peripheral blood mononuclear cells. Attempts to analyse intracellular events following encounter with Hsp60 were undertaken and a transient activation of the transcription factor ATF2 was observed. ATF2 is a target of p38 MAPK, typically activated by a variety of bacterial products. CD14 is a glycosylphosphatidyl inositol (GPI)-anchored membrane protein without an intracellular domain and thus needs a partner to trigger signalling. CD14 can be found in complex with TLR4 on the cell surface and is thought to concentrate and pass ligands to it, thereby increasing TLR4 sensitivity [34]. Indeed, another study has provided evidence for the involvement of TLR4 in Hsp60-driven cell activation [35]. This study used C3H/HeJ mice, which are known to carry a mutation in the TLR4 TIR domain (P712H) rendering it unresponsive to TLR4 ligands [36]. C3H/HeJ mice failed to respond to Hsp60 stimulation in regard to each parameter tested, whereas C3H/HeN control mice showed normal reaction. Although there were no data on the direct intracellular events, the study gave clear indications for this kind of analysis, which followed soon afterwards.

The evidence provided by Vabulas et al. [37] remains the most direct data on Hsp60-triggered signalling. However, evidence for TLR4-independent signalling of Hsp60 proteins is presented in Chapter 6. Kinase activity measurements have shown the involvement of JNK and IKK in the macrophage activation process. Although the IKKβ subunit is typically responsible for the inducibility of IKK complex activity, IKKα activation was shown. Why this subunit is activated during Hsp60-induced signalling in macrophages and the consequences thereof require investigation.

As discussed earlier, IKK activation leads to IκB phosphorylation, ubiquitination and degradation allowing the NF-κB transcription factor to exit the cytoplasm and enter the nucleus. Analysis of IκB degradation in response to extracellular Hsp60 has shown good kinetic correlation to IKK activity. Together with NF-κB mobility shift assays [30], it has supported the involvement of the canonical pathway of NF-κB activation. Interestingly, analysis of different kinases measured either by kinase assays or by means of activated-state-specific antibodies have revealed delayed and sustained activation kinetics compared to most other bacterial stimuli ([37] and unpublished data). Only polyI:C, the surrogate double-stranded RNA which is the ligand of TLR3 [38], displays a similar kinetic pattern. Yet the involvement of TLR3 could be excluded by showing MyD88 dependence of Hsp60 signalling.

As discussed previously, the activity of all known TLRs, except TLR3, relies on the adaptor MyD88; hence, the demonstration of MyD88 dependence on JNK and IKK strongly supports the initial hypothesis that TLRs are involved in Hsp60 signalling. The definitive proof was supplied using the genetic reconstitution system. When Hsp60-unresponsive cells were transfected with cDNA encoding TLR2 or TLR4, the cells gained the sensitivity to Hsp60 as measured by an NF-κB–dependent luciferase reporter. Interestingly, TLR4-mediated signalling required co-transfection of MD2 co-receptor. MD2 is also known to be essential for LPS/TLR4 interaction [39]. The proof of specificity (i.e., the engagement of only selected receptors) provided the result from the reconstitution with TLR9. In this case Hsp60 failed to activate transfected cells. However, one should cautiously evaluate the latter data because of recent evidence on vesicular localisation of TLR9 [40] which possibly suggests that TLR9 is unavailable for interactions with extracellular ligands.

Another study, also aimed primarily at elucidation of heat shock protein signalling pathways, has supported and expanded previous findings [41]. The authors analysed the induction of cyclooxygenase-2 (COX2) and NO synthase-2 (NOS2) in macrophages and endothelial cells by Hsp60 by measuring the activity of respective luciferase reporters with intact and mutated transcription factor binding sites. They found that the inducibility of COX2 was dependent on the intact NF-κB binding site, cAMP-response element (CRE) and two NF-IL6 sites. One of the NF-IL6 elements was especially important for Hsp60 inducibility of the reporter. The mobility shift assays confirmed the earlier finding of NF-κB activation of others [30] and additionally validated the involvement of CRE binding protein. Furthermore, using phospho-specific antibodies and inhibitors of ERK and p38 MAPKs, the authors confirmed the involvement of ERK, JNK and p38 in cellular activation. Interestingly, IFN-γ was again shown to act synergistically with Hsp60 for the induction of NOS2, which is reminiscent of the synergy between IFN-γ and LPS.

The discussed studies and additional evidence [42–46] show that Hsp60 signalling, apart from the differences in the kinetics, is very similar to the signalling induced by other microbial ligands. This is in agreement with the involvement of TLRs. However, these similarities have prompted the proposition that the effects of heat shock proteins might result from contaminants in the heat shock protein preparations. This concern will be discussed later; however, at this point we would like to mention one aspect in advance, namely the requirement of Hsp60 uptake to initiate signalling which is, without exception, overlooked by protagonists of the contamination argument.

During the search for optimal conditions for Hsp60-driven macrophage activation, it was noted that serum exerts an inhibitory effect on the interaction [37]. Subsequent analysis demonstrated that the inhibition of Hsp60 signalling correlated with the blockade of Hsp60 uptake. Employing the clathrin-dependent endocytosis inhibitor monodansylcadaverin, the link between endocytosis of Hsp60 and its triggered signalling was established [37]. In contrast to the requirement for endocytosis in the signalling by Hsp60, LPS signalling is independent of uptake and can be initiated from the cell surface [40, 47].

7.4.2. Gp96 signalling

The connection between signalling and endocytosis makes sense in the case of heat shock proteins that are able to shuttle antigenic peptides. It is appealing to anticipate some economic rationale behind the evolutionary process having led to the situation in which endocytosis is a premise to evoke the signalling. Only those antigen-presenting cells that are able to internalise heat shock protein–peptide complexes become activated and thus acquire the immune response priming capacity. Gp96 and Hsp70 are chaperones that are capable of peptide shuttling. Analysis of their signalling capacity followed that of Hsp60 and soon allowed different aspects of the process, including the dependence on internalisation, to be compared.

Gp96 is an endoplasmic reticulum (ER)-resident chaperone, the function of which is less well understood than that of its cytoplasmic paralogue Hsp90. Nevertheless, gp96 has long been known to be able to confer tumour-specific immunity, a capacity that has been shown to be dependent on peptides with which it is associated [26, 27]. Further analysis revealed that gp96 can activate NF-κB and induce dendritic cell (DC) maturation [48, 49]. One study has claimed that this peptide-independent stimulatory capacity of gp96 is a major determinant of its anti-tumour activity [50].

Truncated forms of gp96 have been constructed, one lacking the ER-anchoring carboxy-terminal sequence Lys-Asp-Glu-Leu (KDEL) sequence and another without the complete C-terminal substrate binding domain. Both modifications render gp96 secretable, and the supernatants collected from the transfected cells were able to induce DC maturation in a similar manner to that induced by microbial products.

The interaction of human gp96 with TLRs has been investigated [51]. Resembling Hsp60-driven activation, it has been demonstrated that gp96 engages TLR2 and TLR4 to activate an NF-κB–dependent luciferase reporter. Again, MD2 appeared to be indispensable for TLR4 engagement. In regard to intracellular signalling, the activation of the typical inflammatory pattern of kinases was observed, namely, the activation of ERK, JNK and p38 MAP kinases. Because IκB degradation was detected, it was assumed that IKK also becomes activated. A clear dependency of the signalling on endocytosis was detected, as is the situation with Hsp60. However, given the capacity of gp96 to shuttle peptides, in this instance there was a more plausible basis to explain the requirement for cellular uptake in heat shock protein–mediated activation.

Comparison of Hsp60- and gp96-driven activation of immune cells suggested that the process is possibly more complicated than the simple ligand–receptor interaction. Bone marrow–derived DCs from TLR2-deficient and TLR4-mutant mice show partial defects in their response to Hsp60 as expected from genetic reconstitution data [37]. In contrast, similar experiments with gp96 unexpectedly showed that the response in DCs from mutant mice was completely controlled by TLR4 [51]. This finding could be explained by differences in the accessory proteins needed for response to ensue. Different uptake receptors could possibly act as critical specifiers of biological activity, and the requirement for these might differ between heat shock proteins and cell types. Thus, the cell type or even the developmental and maturational status of the same cell type seems to influence the type of reaction anticipated after stimulation with heat shock protein.

Additional evidence argues for interaction of gp96 and TLRs and comes from mutation studies of the 70Z/3 cell line [52]. The results from this study showed defects in the maturation of several cell surface receptors, notably, TLR1, TLR2 and TLR4, when gp96 was deleted or mutated. Interestingly, reconstitution of the missing gp96 or by the analogous parts or even the whole molecule of the cytoplasmic paralogue Hsp90 did not relieve the phenotype. However, distinct interactions between gp96 and TLRs in the ER as compared to the cell surface or endocytic compartments cannot be excluded and warrants further analysis.

7.4.3. Hsp70 signalling

Hsp70 is the second chaperone taking an exclusive place in immunology due to its role as a peptide shuttle [25]. The biology of this protein is also discussed in various other chapters in this volume. The analysis of Hsp70-triggered signalling events began with the proposal that CD14 is its receptor on the cell surface [53]. Besides demonstrating the inflammatory potential of human Hsp70 leading to NF-κB activation and the induction of pro-inflammatory cytokines IL-1β and IL-6, this study presented two additional, interesting and important findings. The first was the rapid induction of a transient cytoplasmic Ca^{2+} wave. This feature clearly distinguishes Hsp70 from LPS (described in more detail in Chapters 8 and 10). The second interesting observation was the unequal contribution of CD14 for the induction of different cytokines. Specifically, IL-1β and IL-6 were shown to depend on the presence of CD14, whereas TNF-α was produced independently of it. Unfortunately, the Ca^{2+} flux inhibitor turned out to be non-specific regarding CD14, in that CD14-dependent, IL-1β and IL-6 secretion, as well as CD14-independent TNF-α secretion upon Hsp70 stimulation, were compromised. Thus, the Ca^{2+} flux role in CD14-dependent effects remains an important, but still unresolved, issue. The study of human Hsp70 interactions with DCs followed and established this chaperone as a potent stimulus of innate immunity [54].

Finally, due to the established participation of CD14 and the insight gained with Hsp60 and gp96, the involvement of TLRs was investigated. Two studies simultaneously described TLR2 and TLR4 as receptors of Hsp70. Vabulas and colleagues [55] analysed the signalling pathway induced by Hsp70 in the macrophage cell line RAW264.7 by transfecting dominant negative forms of MyD88 and TRAF6 adaptors from the TLR signal pathway. Furthermore, a versatile tool for spatial dissection of signalling was used, namely, a cell line stably expressing functional MyD88-EGFP fusion protein. This allowed the visualisation of Hsp70-triggered signalling. The kinetics and pattern of MyD88 recruitment could be followed at the single-cell level. Once again, the results provided a strong argument against LPS contamination being responsible for the observed effects, since Hsp70 induced the intracellular MyD88 recruitment pattern, in contrast with LPS stimulation, which results in MyD88 recruitment to the macrophage surface [40, 47]. Furthermore, by employing transient transfections of TLR2 and TLR4 into an otherwise Hsp70-unresponsive fibroblast cell line, the authors demonstrated a specific involvement of those TLRs, because TLR9 transfection did not confer sensitivity to Hsp70.

Finally, to validate the results, MyD88-, TLR2- and TLR4-mutant mice were used. DCs from TLR4-mutant mice showed a complete defect in TNF-α and IL-12 production, which is reminiscent of the situation with gp96 and again implies the involvement of additional components in receptor complexes. A second study came to the same conclusion by using TLR stably transfected fibroblasts [56]. Apart from documenting the involvement of TLR2 and TLR4 in cellular activation by Hsp70, the authors confirmed their previous finding on CD14 involvement and showed TLR2/TLR4 synergism, which was established to be MyD88-independent.

Continuing work provides additional insights into Hsp70-induced responses and suggests broader implications. For example, an immediate release Hsp70 following heart surgery has been reported and might influence accompanying inflammatory effects [57]. Another study has demonstrated interactions of Hsp70 (and other heat shock proteins) with microglial cells, thereby suggesting an ability to evoke inflammation in the neural system [58]. The importance of understanding Hsp70-triggered cellular activation is therefore difficult to over-estimate.

7.4.4. Non-Toll-like receptor signalling

We have centred our discussion on heat shock protein–induced signalling around TLR-associated intracellular events because these have, to date, been more widely analysed and characterised. Nevertheless, TLR biology is still far from understood, especially with respect to the complexity of interactions with the diverse set of cell surface components [59]. In addition, given the documented versatility of heat shock protein uptake/scavenging mechanisms, it is tempting to speculate on additional non-TLR-driven signalling as a sequela to heat shock protein encounter. There has been some controversy regarding the route of gp96 uptake [60, 61]. Although gp96 uptake receptors under discussion (CD91, SR-A) are not primarily devoted to signalling, it is feasible to assume that they could introduce additional qualities into the primary reaction triggered by TLRs. The same holds true in the case of Hsp70, because CD91 has also been shown to participate in the uptake of Hsp70 [62] as has the scavenger receptor LOX1 [63]. Transient Ca^{2+} rises, induced by extracellular Hsp70 and discussed earlier, could represent one aspect of the TLR-independent process [53]. Interestingly, CD40 has been shown to interact with bacterial and human Hsp70 [64, 65]. The signalling potential of CD40 is well established and is highly relevant for the immune system. It would be interesting to explore this particular interaction in more detail. See Chapter 10 for a fuller discussion of CD40/Hsp70 interactions.

The peptides originating from microbial or endogenous heat shock proteins and their capacity to stimulate and expand peptide-specific T cell clones should also be highlighted [66]. Although such responses would involve different intracellular signalling cascades including the T cell receptor (TCR) signalling axis, the outcome could be as impressive as that following TLR engagement.

7.5. Contamination issue

Due to the extreme sensitivity of the immune system to some bacterial products, most notably LPS, concerns regarding reagent contamination inevitably accompany investigations of immune cell activation. This concern becomes of primary importance if the same receptor systems are implicated, as in the case of TLR2/TLR4 and heat shock proteins. From the beginning, much effort has been expended in order to exclude the possibility that contamination of heat shock protein preparations used for investigations accounts for the observed responses. Nevertheless, several reports proposing that some [67] or all of the stimulatory effects of heat shock proteins [68–70] originate from contaminating LPS have appeared. After examination of the available data, some major points of criticism of this 'contamination theory' should be mentioned:

1. Responsiveness to any ligand is dependent on the responsive cell type. Negative data obtained using a single cell line are not strong evidence on which to include or exclude an effect. The danger of ending up with a particular clone over short or long periods is well known and real. This unintended sub-cloning often results in loss of some and gain of additional properties, such as accessory receptors and defective signalling pathways to name but two. In some instances, the methods sections in publications acknowledge these problems. For example, in one study cell harvesting might be reported to require scraping, whereas in another the 'same' cell line is described as being only loosely attached or semi-adherent.

2. The possibility that different sub-units in receptor complexes are required for different heat shock proteins adds complexity to the analysis. If stimulation were to originate from LPS contamination, then the same DC would respond similarly. However, these cells respond to gp96 and Hsp70 irrespective of TLR2 expression and the response is fully dependent on TLR4. In contrast, Hsp60 uses both receptors in the same cells. These functional distinctions argue against the contamination theory.

3. Correct culture and, most importantly, activation conditions are of great concern. It is difficult to compare data collected from experiments in which stimulations are performed in serum-containing medium with those performed in the absence of serum because serum had been found to inhibit the process. At this point it is appropriate to mention the importance of generous upwards titrations in the event of negative results.

4. It is worth considering whether clarifying the source or even species of the protein would resolve some of these issues. Although homologous heat shock proteins are well conserved, one cannot exclude the possibility that small differences in primary sequence or in post-translational modification are critical for the activity.

5. Serum is known to enhance the sensitivity to LPS because of the presence of LPS binding protein (LBP), which interacts with LPS and passes it to CD14 on the myeloid cell surface [32]. If a considerable stimulatory capacity of heat shock protein preparations originates from contaminating LPS, how should one explain the strong inhibitory effect of serum on stimulation of macrophages with Hsp60 and gp96? Hsp70 has not been analysed in this respect by means of direct assays.

6. Last, but not least, is the endocytosis argument, an argument which is regrettably never discussed by proponents of the 'contamination theory'. LPS is known to signal from the cell surface and does not need endocytosis for its activity [40, 47]. In contrast, Hsp60 and gp96 clearly require uptake to initiate the signalling [37, 51]. Moreover, experiments with Hsp70 have shown that internalisation is required for TLR signalling, because the TLR-specific adaptor MyD88 is primarily recruited to the endocytic compartments, as visualised by means of MyD88-EGFP fusion protein [55]. This is in an obvious contrast to the cell surface recruitment of MyD88 after TLR4 engagement.

Even this incomplete list of concerns shows how important critical discussion and innovative experimental approaches are to advancing our understanding of the mechanisms of heat shock proteins and immune system interaction. Two recent publications demonstrate *in vivo* the adjuvant effects of Hsp70 and gp96, which simply can not be explained by contamination [71, 72]. Further evidence for the hypothesis that the biological effects of chaperones are not due to contaminants is provided in many of this book's chapters.

7.6. Conclusion

After presenting known facts on heat shock protein signalling we are now faced with two questions. The first question is – why TLRs? TLRs have evolved into a powerful system which is capable of sensing a broad range of invaders. TLRs not only mobilise cells for innate first line defence, they also condition the immune system for antigen-specific priming and thus provide a bridge between innate and adaptive immunity. Nothing else could be more suitable for alerting vertebrates to other kinds of danger. Endogenous heat shock proteins released as a consequence of tissue damage and signalling via TLR immediately attract all the resulting inflammatory power against the initiating insult. This scenario satisfies the original theoretical considerations [1, 73] and has received support from experimental data [74–77].

The second question, which asks how both pathogenic and host proteins can share the same receptor, is more difficult to answer. To gain some insight into this, one should remember that until recently it was believed that the only function of heat shock proteins was in the assistance in protein folding and prevention of aggregation. Binding and release of exposed or otherwise distorted regions on client polypeptides are features that are common to different classes of chaperones and constitute the mechanism of their biochemical action [78]. It is tempting to speculate that these common features account for the observed overlap in their interaction with TLRs. To test this and other assumptions at the molecular level is the challenge and the task for the future. Simultaneously, it is an opportunity to gain a better insight into the functions of TLRs and heat shock proteins, and this might reveal new avenues via which important physiological and pathological processes can be modulated.

REFERENCES

1. Janeway C A J. Approaching the asymptote? Evolution and revolution in immunology. Cold Spring Harb Symp Quant Biol 1989, 54 Pt 1: 1–13.
2. Gordon S. Pattern recognition receptors: doubling up for the innate immune response. Cell 2002, 111: 927–930.
3. Lemaitre B, Nicolas E, Michaut L, Reichhart J M and Hoffmann J A. The dorsoventral regulatory gene cassette spatzle/Toll/cactus controls the potent antifungal response in *Drosophila* adults. Cell 1996, 86: 973–983.
4. Medzhitov R, Preston-Hurlburt P and Janeway C A J. A human homologue of the *Drosophila* Toll protein signals activation of adaptive immunity. Nature 1997, 388: 394–397.
5. Kobe B and Deisenhofer J. A structural basis of the interactions between leucine-rich repeats and protein ligands. Nature 1995, 374: 183–186.
6. Bell J K, Mullen G E, Leifer C A, Mazzoni A, Davies D R and Segal D M. Leucine-rich repeats and pathogen recognition in Toll-like receptors. Trends Immunol 2003, 24: 528–533.
7. Gay N J and Keith F J. *Drosophila* Toll and IL-1 receptor. Nature 1991, 351: 355–356.
8. Schneider D S, Hudson K L, Lin T Y and Anderson K V. Dominant and recessive mutations define functional domains of Toll, a transmembrane protein required for dorsal-ventral polarity in the *Drosophila* embryo. Genes Dev 1991, 5: 797–807.
9. Whitham S, Dinesh-Kumar S P, Choi D, Hehl R, Corr C and Baker B. The product of the tobacco mosaic virus resistance gene N: similarity to toll and the interleukin-1 receptor. Cell 1994, 78: 1101–1115.
10. Rock F L, Hardiman G, Timans J C, Kastelein R A and Bazan J F. A family of human receptors structurally related to *Drosophila* Toll. Proc Natl Acad Sci USA 1998, 95: 588–593.
11. Xu Y, Tao X, Shen B, Horng T, Medzhitov R, Manley J L and Tong L. Structural basis for signal transduction by the Toll/interleukin-1 receptor domains. Nature 2000, 408: 111–115.
12. Adachi O, Kawai T, Takeda K, Matsumoto M, Tsutsui H, Sakagami M, Nakanishi K and Akira S. Targeted disruption of the MyD88 gene results in loss of IL-1- and IL-18-mediated function. Immunity 1998, 9: 143–150.
13. Kawai T, Adachi O, Ogawa T, Takeda K and Akira S. Unresponsiveness of MyD88-deficient mice to endotoxin. Immunity 1999, 11: 115–122.
14. Suzuki N, Suzuki S, Duncan G S, Millar D G, Wada T, Mirtsos C, Takada H, Wakeham A, Itie A, Li S, Penninger J M, Wesche H, Ohashi P S, Mak T W and Yeh W C. Severe impairment of interleukin-1 and Toll-like receptor signalling in mice lacking IRAK-4. Nature 2002, 416: 750–756.

15. Kobayashi K, Hernandez L D, Galan J E, Janeway C A J, Medzhitov R and Flavell R A. IRAK-M is a negative regulator of Toll-like receptor signaling. Cell 2002, 110: 191–202.
16. Deng L, Wang C, Spencer E, Yang L, Braun A, You J, Slaughter C, Pickart C and Chen Z J. Activation of the IkappaB kinase complex by TRAF6 requires a dimeric ubiquitin-conjugating enzyme complex and a unique polyubiquitin chain. Cell 2000, 103: 351–361.
17. Wang C, Deng L, Hong M, Akkaraju G R, Inoue J and Chen Z J. TAK1 is a ubiquitin-dependent kinase of MKK and IKK. Nature 2001, 412: 346–351.
18. Horng T, Barton G M, Flavell R A and Medzhitov R. The adaptor molecule TIRAP provides signalling specificity for Toll-like receptors. Nature 2002, 420: 329–333.
19. Yamamoto M, Sato S, Hemmi H, Sanjo H, Uematsu S, Kaisho T, Hoshino K, Takeuchi O, Kobayashi M, Fujita T, Takeda K and Akira S. Essential role for TIRAP in activation of the signalling cascade shared by TLR2 and TLR4. Nature 2002, 420: 324–329.
20. Hoebe K, Du X, Georgel P, Janssen E, Tabeta K, Kim S O, Goode J, Lin P, Mann N, Mudd S, Crozat K, Sovath S, Han J and Beutler B. Identification of Lps2 as a key transducer of MyD88-independent TIR signalling. Nature 2003, 424: 743–748.
21. Yamamoto M, Sato S, Hemmi H, Hoshino K, Kaisho T, Sanjo H, Takeuchi O, Sugiyama M, Okabe M, Takeda K and Akira S. Role of adaptor TRIF in the MyD88-independent toll-like receptor signaling pathway. Science 2003, 301: 640–643.
22. Yamamoto M, Sato S, Hemmi H, Uematsu S, Hoshino K, Kaisho T, Takeuchi O, Takeda K and Akira S. TRAM is specifically involved in the Toll-like receptor 4-mediated MyD88-independent signaling pathway. Nat Immunol 2003, 4: 1144–1150.
23. Kaufmann S H E. Heat shock proteins and the immune response. Immunol Today 1990, 11: 129–136.
24. Hartl F U and Hayer-Hartl M. Molecular chaperones in the cytosol: from nascent chain to folded protein. Science 2002, 295: 1852–1858.
25. Udono H and Srivastava P K. Heat shock protein 70-associated peptides elicit specific cancer immunity. J Exp Med 1993, 178: 1391–1396.
26. Arnold D, Faath S, Rammensee H-G and Schild H. Cross-priming of minor histo-compatibility antigen-specific cytotoxic T cells upon immunization with the heat shock protein gp96. J Exp Med 1995, 182: 885–889.
27. Suto R and Srivastava P K. A mechanism for the specific immunogenicity of heat shock protein-chaperoned peptides. Science 1995, 269: 1585–1588.
28. Kol A, Sukhova G K, Lichtman A H and Libby P. Chlamydial heat shock protein 60 localizes in human atheroma and regulates macrophage tumour necrosis factor-a and matrix metalloproteinase expression. Circulation 1998, 98: 300–307.
29. Chen W, Syldath U, Bellmann K, Burkart V and Kold H. Human 60-kDa heat-shock protein: a danger signal to the innate immune system. J Immunol 1999, 162: 3212–3219.
30. Kol A, Bourcier T, Lichtman A and Libby P. Chlamydial and human heat shock protein 60s activate human vascular endothelium, smooth muscle cells, and macrophages. J Clin Invest 1999, 103: 571–577.
31. Kol A, Lichtman A H, Finberg R W, Libby P and Kurt-Jones E A. Heat shock protein (HSP) 60 activates the innate immune response: CD14 is an essential receptor for HSP60 activation of mononuclear cells. J Immunol 2000, 164: 13–17.

32. Wright S D, Ramos R A, Tobias P S, Ulevitch R J and Mathison J C. CD14, a receptor for complexes of lipopolysaccharide (LPS) and LPS binding protein. Science 1990, 249: 1431–1433.
33. Pugin J, Heumann I D, Tomasz A, Kravchenko VV, Akamatsu Y, Nishijima M, Glauser M P, Tobias P S and Ulevitch R J. CD14 is a pattern recognition receptor. Immunity 1994, 1: 509–516.
34. da Silva Correia J, Soldau K, Christen U, Tobias P S and Ulevitch R J. Lipopolysaccharide is in close proximity to each of the proteins in its membrane receptor complex. Transfer from CD14 to TLR4 and MD-2. J Biol Chem 2001, 276: 21129–21135.
35. Ohashi K, Burkart V, Flohé S and Kolb H. Heat shock protein 60 is a putative endogenous ligand of the Toll-like receptor-4 complex. J Immunol 2000, 164: 558–561.
36. Poltorak A, He X, Smirnova I, Liu M Y, van Huffel C, Du X, Birdwell D, Alejos E, Silva M, Galanos C, Freudenberg M, Ricciardi-Castagnoli P, Layton B and Beutler B. Defective LPS signaling in C3H/HeJ and C57BL/10ScCr mice: mutations in Tlr4 gene. Science 1998, 282: 2085–2088.
37. Vabulas R M, Ahmad-Nejad P, da Costa C, Miethke T, Kirschning C J, Hacker H and Wagner H. Endocytosed HSP60s use Toll-like receptor 2 (TLR2) and TLR4 to activate the toll/interleukin-1 receptor signaling pathway in innate immune cells. J Biol Chem 2001, 276: 31332–31339.
38. Alexopoulou L, Holt A C, Medzhitov R and Flavell R A. Recognition of double-stranded RNA and activation of NF-kappaB by Toll-like receptor 3. Nature 2001, 413: 732–738.
39. Shimazu R, Akashi S, Ogata H, Nagai Y, Fukudome K, Miyake K and Kimoto M. MD-2, a molecule that confers lipopolysaccharide responsiveness on Toll-like receptor 4. J Exp Med 1999, 189: 17777–17782.
40. Ahmad-Nejad P, Hacker H, Rutz M, Bauer S, Vabulas R M and Wagner H. Bacterial CpG-DNA and lipopolysaccharides activate Toll-like receptors at distinct cellular compartments. Eur J Immunol 2002, 32: 1958–1968.
41. Billack B, Heck D E, Mariano T M, Gardner C R, Sur R, Laskin D L and Laskin J D. Induction of cyclooxygenase-2 by heat shock protein 60 in macrophages and endothelial cells. Am J Physiol (Cell Physiol) 2002, 283: C1267–1277.
42. Sasu S, LaVerda D, Qureshi N, Golenbock D T and Beasley D. *Chlamydia pneumoniae* and chlamydial heat shock protein 60 stimulate proliferation of human vascular smooth muscle cells via toll-like receptor 4 and p44/p42 mitogen-activated protein kinase activation. Circ Res 2001, 89: 244–250.
43. Zhang L, Pelech S L, Mayrand D, Grenier D, Heino J and Uitto V J. Bacterial heat shock protein-60 increases epithelial cell proliferation through the ERK1/2 MAP kinases. Exp Cell Res 2001, 266: 11–20.
44. Bulut Y, Faure E, Thomas L, Karahashi H, Michelsen K S, Equils O, Morrison S G, Morrison R P and Arditi M. Chlamydial heat shock protein 60 activates macrophages and endothelial cells through Toll-like receptor 4 and MD2 in a MyD88-dependent pathway. J Immunol 2002, 168: 1435–1440.
45. Flohé S B, Bruggemann J, Lendemans S, Nikulina M, Meierhoff G, Flohé S and Kolb H. Human heat shock protein 60 induces maturation of dendritic cells versus a Th1-promoting phenotype. J Immunol 2003, 170: 2340–2348.

46. Zanin-Zhorov A, Nussbaum G, Franitza S, Cohen I R and Lider O. T cells respond to heat shock protein 60 via TLR2: activation of adhesion and inhibition of chemokine receptors. FASEB J 2003, 17: 1567–1569.

47. Latz E, Visintin A, Lien E, Fitzgerald K A, Monks B G, Kurt-Jones E A, Golenbock D T and Espevik T. Lipopolysaccharide rapidly traffics to and from the Golgi apparatus with the toll-like receptor 4-MD-2-CD14 complex in a process that is distinct from the initiation of signal transduction. J Biol Chem 2002, 277: 47834–47843.

48. Basu S, Binder R J, Suto R, Anderson K M and Srivastava P K. Necrotic but not apoptotic cell death releases heat shock proteins, which deliver a partial maturation signal to dendritic cells and activates the NF-kB pathway. Int Immunol 2000, 12: 1539–1546.

49. Singh-Jasuja H, Scherer H U, Hilf N, Arnold-Schild D, Rammensee H-G, Toes R E M and Schild H. The heat shock protein gp96 induces maturation of dendritic cells and down-regulation of its receptor. Eur J Immunol 2000, 30: 2211–2215.

50. Baker-LePain J C, Sarzotti M, Fields T A, Li C Y and Nicchitta C V. GRP94 (gp96) and GRP94 N-terminal geldanamycin binding domain elicit tissue nonrestricted tumor suppression. J Exp Med 2002, 196: 1447–1459.

51. Vabulas R M, Braedel S, Hilf N, Singh-Jasuja H, Herter S, Ahmad-Nejad P, Kirschning C J, Da Costa C, Rammensee H G, Wagner H and Schild H. The endoplasmic reticulum-resident heat shock protein Gp96 activates dendritic cells via the Toll-like receptor 2/4 pathway. J Biol Chem 2002, 277: 20847–20853.

52. Randow F and Seed B. Endoplasmic reticulum chaperone gp96 is required for innate immunity but not cell viability. Nat Cell Biol 2001, 3: 891–896.

53. Asea A, Kraeft S-K, Kurt-Jones E A, Stevenson M A, Chen L B, Finberg R W, Koo G C and Calderwood S K. Hsp70 stimulates cytokine production through a CD14-dependent pathway, demonstrating its dual role as a chaperone and cytokine. Nat Med 2000, 6: 435–442.

54. Kuppner M C, Gastpar R, Gelwer S, Nossner E, Ochmann O, Scharner A and Issels R D. The role of heat shock protein (hsp70) in dendritic cell maturation: hsp70 induces the maturation of immature dendritic cells but reduces DC differentiation from monocyte precursors. Eur J Immunol 2001, 31: 1602–1609.

55. Vabulas R M, Ahmad-Nejad P, Ghose S, Kirschning C J, Issels R D and Wagner H. HSP70 as endogenous stimulus of the Toll/interleukin-1 receptor signal pathway. J Biol Chem 2002, 277: 15107–15112.

56. Asea A, Rehli M, Kabingu E, Boch J A, Baré O, Auron P E, Stevenson M A and Calderwood S K. Novel signal transduction pathway utilized by extracellular HSP70. Role of Toll-like receptor (TLR) 2 and TLR4. J Biol Chem 2002, 277: 15028–15034.

57. Dybdahl B, Wahba A, Lien E, Flo T H, Waage A, Qureshi N, Sellevold O F, Espevik T and Sundan A. Inflammatory response after open heart surgery: release of heat-shock protein 70 and signaling through Toll-like receptor-4. Circulation 2002, 105: 685–690.

58. Kakimura J, Kitamura Y, Takata K, Umeki M, Suzuki S, Shibagaki K, Taniguchi T, Nomura Y, Gebicke-Haerter P J, Smith M A, Perry G and Shimohama S. Microglial activation and amyloid-beta clearance induced by exogenous heat-shock proteins. FASEB J 2002: 601–603.

59. Underhill D M. Toll-like receptors: networking for success. Eur J Immunol 2003, 33: 1767–1775.

60. Binder R J, Han D K and Srivastava P K. CD91: a receptor for heat shock protein gp96. Nat Immunol 2000, 1: 151–155.
61. Berwin B, Hart J P, Rice S, Gass C, Pizzo S V, Post S R and Nicchitta C V. Scavenger receptor-A mediates gp96/GRP94 and calreticulin internalization by antigen-presenting cells. EMBO J 2003, 22: 6127–6136.
62. Basu S, Binder R J, Ramalingam T and Srivastava P K. CD91 is a common receptor for heat shock proteins gp96, hsp90, hsp70 and calreticulin. Immunity 2001, 14: 303–313.
63. Delneste Y, Magistrelli G, Gauchat J, Haeuw J, Aubry J, Nakamura K, Kawakami-Honda N, Goetsch L, Sawamura T, Bonnefoy J and Jeannin P. Involvement of LOX-1 in dendritic cell-mediated antigen cross-presentation. Immunity 2002, 17: 353–362.
64. Wang Y, Kelly C G, Karttunen T, Whittall T, Lehner P J, Duncan L, MacAry P, Younson J S, Singh M, Oehlmann W, Cheng G, Bergmeier L and Lehner T. CD40 is a cellular receptor mediating mycobacterial heat shock protein 70 stimulation of CC-chemokines. Immunity 2001, 15: 971–983.
65. Becker T, Hartl F U and Wieland F. CD40, an extracellular receptor for binding and uptake of Hsp70-peptide complexes. J Cell Biol 2002, 158: 1277–1285.
66. van Eden W. Stress proteins as targets for anti-inflammatory therapies. Drug Discov Today 2000, 5: 115–120.
67. Reed R C, Berwin B, Baker J P and Nicchitta C V. GRP94/gp96 elicits ERK activation in murine macrophages. A role for endotoxin contamination in NF-kappa B activation and nitric oxide production. J Biol Chem 2003, 278: 31853–31860.
68. Bausinger H, Lipsker D, Ziylan U, Manie S, Briand J P, Cazenave J P, Muller S, Haeuw J F, Ravanat C, de la Salle H and Hanau D. Endotoxin-free heat-shock protein 70 fails to induce APC activation. Eur J Immunol 2002, 32: 3708–3713.
69. Gao B and Tsan M F. Endotoxin contamination in recombinant human Hsp70 preparation is responsible for the induction of TNFα release by murine macrophages. J Biol Chem 2003, 278: 174–179.
70. Gao B and Tsan M F. Recombinant human heat shock protein 60 does not induce the release of tumor necrosis factor alpha from murine macrophages. J Biol Chem 2003, 278: 22523–22529.
71. Millar D G, Garza K M, Odermatt B, Elford A R, Ono N, Li Z and Ohashi P S. Hsp70 promotes antigen-presenting cell function and converts T-cell tolerance to autoimmunity *in vivo*. Nat Med 2003, 9: 1469–1476.
72. Liu B, Dai J, Zheng H, Stoilova D, Sun S and Li Z. Cell surface expression of an endoplasmic reticulum resident heat shock protein gp96 triggers MyD88-dependent systemic autoimmune diseases. Proc Nat Acad Sci USA 2003, 100: 15824–15829.
73. Matzinger P. The Danger Model: A renewed sense of self. Science 2002, 296: 301–305.
74. Gallucci S, Lolkema M and Matzinger P. Natural adjuvants: endogenous activators of dendritic cells. Nat Med 1999, 11: 1249–1255.
75. Sauter B, Albert M L, Francisco L, Larsson M, Somersan S and Bhardwaj N. Consequences of cell death: exposure to necrotic tumour cells, but not primary tissue cells or apoptotic cells, induces the maturation of immunostimulatory dendritic cells. J Exp Med 2000, 191: 423–433.
76. Li M, Carpio D F, Zheng Y, Bruzzo P, Singh V, Ouaaz F, Medzhitov R M and Beg A A. An essential role of the NF-kappa B/Toll-like receptor pathway in induction of

inflammatory and tissue-repair gene expression by necrotic cells. J Immunol 2001, 166: 7128–7135.

77. Somersan S, Larsson M, Fonteneau J F, Basu S, Srivastava P and Bhardwaj N. Primary tumor tissue lysates are enriched in heat shock proteins and induce the maturation of human dendritic cells. J Immunol 2001, 167: 4844–4852.

78. Bukau B and Horwich A L. The Hsp70 and Hsp60 chaperone machines. Cell 1998, 92: 351–366.

8

Regulation of Signal Transduction by Intracellular and Extracellular Hsp70

Alexzander Asea and Stuart K. Calderwood

8.1. Introduction

There is a clear dichotomy between the effects of the 70-kDa heat shock protein (Hsp70) when expressed intracellularly and when released into the extracellular space. Intracellular Hsp70 is primarily implicated as a protein chaperone that transports and folds naïve, aberrantly folded, or mutated proteins, resulting in cytoprotection when cells are exposed to stressful stimuli – most notably heat shock itself. Intracellular Hsp70 also functions as a regulatory molecule and has a largely inhibitory function in cellular metabolism. In contrast, Hsp70 is implicated in immune activation, given that exposure of immunocompetent cells to exogenous Hsp70 triggers acute inflammatory responses, activates innate immunity and enhances anti-tumour surveillance. This review focuses on recent advances in understanding the contrasting roles of Hsp70 as an intracellular molecular chaperone and extracellular signalling ligand and highlights its relevance to host defence against pathogens and malignant transformation.

The formative studies of Hsp70, dating back over 30 years, suggested a strictly intracellular function for Hsp70 with the properties of a molecular chaperone – a protein that modulates the tertiary structures of other proteins and protects cells from stress [1, 2]. However, in 1989 Hightower and colleagues showed that cells in culture contain a pool of Hsp70 which is loosely associated with the cell and which could be released into the extracellular medium after heat shock or even after mild washing with tissue culture medium [3]. Subsequent *in vitro* studies indicated that such Hsp70 released from cells could be taken up by neuronal cells under which circumstances it enhances the molecular chaperoning power of these cells [4]. Later work identified the presence of extracellular Hsp70 in human subjects, as indicated by free Hsp70 and Hsp70 antibodies circulating in the bloodstream [5]; the biology of circulating molecular chaperones is discussed in Chapter 12. However, the studies that accelerated the interest in extracellular

Hsp70 and its functions were those of Srivastava and colleagues who showed that tumour-associated antigens bind stably to molecular chaperones such as Hsp70 and that these complexes can form the basis of an effective anti-cancer vaccine [6]. This aspect is discussed in Chapter 18. All aspects of the extracellular functions of Hsp70 are currently under active investigation, and we aim here to provide an overview of our present understanding.

8.2. Intracellular signalling functions of Hsp70

As befits a molecular chaperone, intracellular Hsp70 plays a role in cellular signalling events that involves associations in high-molecular-weight complexes with other proteins [7, 8]. Hsp70 functions as a wide-spectrum negative regulator and plays a restraining role in a wide range of processes via its ability to inhibit the activities of protein kinases and transcription factors [7, 8]. Indeed, elevated Hsp70 inhibits a plethora of intracellular processes and, probably for this precise reason, Hsp70 levels are strictly regulated via negative control of its transcription factor heat shock factor 1 (HSF1) and its destabilisation at the mRNA level [9–12]. In addition, HSF1 itself can negatively regulate the promoters of cytokine genes and genes involved in cell proliferation [13, 14]. Particularly in the context of the inflammatory response and the innate immune response, therefore, intracellular Hsp70 and the heat shock system exhibit roles as negative regulators [13–16]. This property, as we shall discuss later, is in stark contrast to the pro-inflammatory and pro-immune functions of extracellular Hsp70.

8.3. Sources of extracellular Hsp70

It is evident that intracellular Hsp70 can escape into the interstitial fluid and is found at significant levels in the bloodstream [5, 17]. However, the cellular source of such Hsp70 and the mechanisms involved in the release of this intracellular protein are not well defined. See Chapter 2 for discussion of the cellular dispositions of mitochondrial Hsp70. As mentioned, cells in tissue culture can contain a pool of loosely associated Hsp70 which can be removed by gentle washing [3]. Such Hsp70 could be associated with the lipid rafts in the outer leaflet of the plasma membrane or could perhaps be loosely bound to other cell surface molecules [18–20]. This, of course, begs the question of how this intracellular protein is transported to the cell surface. One possible mechanism is via its release in association with proteins such as transferrin that are transported in exosomes bound to Hsp70 [21]. Details of this mechanism are provided in Chapter 3.

There is, in addition, considerable speculation that Hsp70 might be released into the extracellular space after destruction of the plasma membrane in cells

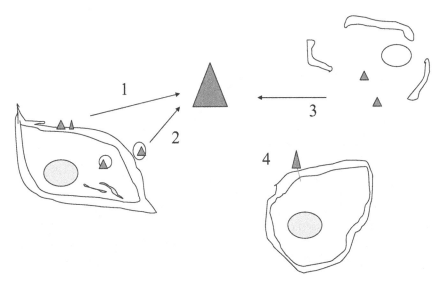

Figure 8.1. Origins of extracellular Hsp70. Extracellular Hsp70 (▲) may be derived from (1) a loosely bound pool on the plasma membrane, (2) through transport across the plasma membrane in exosomes, (3) from cells undergoing necrosis or (4) may be displayed in bound form on the cell surface.

undergoing necrotic death, and it might thereby act as a pro-inflammatory stimulus or a 'danger signal' for the immune system [22]. The extent to which this actually occurs *in vivo* is not clear, although, in the case of cancer cells, death is a continuous process and both apoptotic and necrotic cells are seen in the interior of tumours. It is also possible that Hsp70 is released after cancer therapy or as a consequence of ischaemic death following heart attacks or strokes [23].

A further source of extracellular Hsp70 appears to be brown fat from which Hsp70 is released in a quasi-endocrine manner, an effect that is exaggerated by exercise or behavioural stress [24, 25]. Such Hsp70 may be taken up by neuronal cells in the central nervous system that are chronically depleted of Hsp70 due to low endogenous HSF activity and the long distances involved in axonal transport of *de novo* synthesised proteins [4, 26, 27].

Yet another source of Hsp70 might be bound Hsp70 displayed on the surface of certain tumour cells [28]. Such bound Hsp70 is evidently able to interact with C-type lectin receptors on the surface of natural killer (NK) cells [29]. There is thus evidence for the existence of extracellular Hsp70 from a number of sources in the body which could act as a danger signal in the case of cancer or acute infection, as a target for tumour surveillance by NK cells, as a source of supplementary chaperones in the central nervous system or could play other as yet unknown functions [29–32] (Figure 8.1).

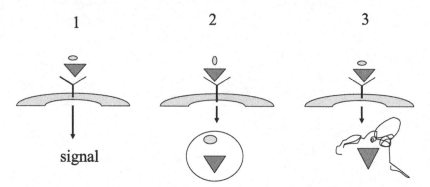

Figure 8.2. Consequences of Hsp70–cell surface binding. Hsp70 (▲) complexed with peptide cargo (●) (Hsp-PC) is shown binding to a cell surface receptor. Hsp70-PC is shown (1) activating intracellular signalling cascades, (2) entering the cell in endosomes through the process of receptor-mediated endocytosis and (3), after entering the cell as in (2), participating in molecular chaperone interactions with unfolded proteins.

8.4. Cell surface binding of Hsp70

Hsp70 and other chaperones released from cells appear able to bind with high affinity to cells such as B lymphocytes, peripheral blood monocytes, macrophages and dendritic cells (DCs), as well as to neuronal cells such as glial cells and astrocytes [29, 33–38]. The demonstration of high-affinity, saturable binding of Hsp70 to the cell surface has suggested the existence of specific 'Hsp70 receptors' (Figure 8.2). This hypothesis has been pursued, especially with regard to cells of the immune system, and a number of investigations have been carried out to determine specific binding structures. A number of candidate receptors have been proposed and their association with Hsp70 has been demonstrated. These include the α2-macroglobulin receptor CD91, the C-type lectin receptor and oxidised low-density lipoprotein (LDL) binding protein lectin-like oxidised low-density lipoprotein receptor-1 (LOX-1), and the tumour necrosis factor-α family protein CD40 [39–41]. Specific binding has been demonstrated in each case and Hsp70 binding is accompanied by endocytosis, which suggests a role for each ligand in both cell surface association and uptake of Hsp70 [39–41].

It is not as yet clear which receptor is the most important in the extracellular functions of Hsp70, although LOX-1 appears to be a promising candidate for an antigen-cross-presenting receptor [41]. At high concentrations, Hsp70 has been shown to associate with another member of the C-type lectin family, CD94 found on the surface of NK cells, and this interaction may underlie the enhanced ability of NK cells to kill tumour cells that express Hsp70 on the cell surface [29, 42]. In addition, Hsp70 has been found in lipid raft structures on the surface of cells concentrated together with important signalling receptors such as CD14, Toll-like receptors (TLRs) and chemokine receptors, and such interactions might be involved in the signalling events that are discussed in the next section [20]. At the present time, it is not clear whether cells possess a dedicated receptor for Hsp70, and Hsp70 thus remains an orphan ligand with, as yet, only hand-me-down receptors for its association with the cell surface. More details of Hsp70–receptor interactions are presented in Chapter 10.

8.5. Exogenous Hsp70-mediated signal transduction pathways

Much work on cell signalling has been carried out in macrophages, monocytes and DCs in which attempts have been made to determine signalling pathways upstream of the cytokine and antigen-presenting cell (APC) co-receptor gene expression which accompanies Hsp70 binding [34, 35, 43, 44]. Our group has evaluated the various steps involved in Hsp70-induced signal transduction and has demonstrated that Hsp70 binds with high affinity to the plasma membrane of APCs to elicit a rapid intracellular Ca^{2+} ($[Ca^{2+}]$) flux within 10 seconds [35]. This is an important signalling step which distinguishes Hsp70- from lipopolysaccharide (LPS)-induced signalling (discussed later), because the treatment of APCs with LPS does not result in an $[Ca^{2+}]_i$ flux [45].

The possibility that LPS contamination might confound our results was addressed by using Polymyxin B and Lipid IVa (LPS inhibitor) which abrogates LPS-induced, but not Hsp70-induced, cytokine expression. In addition, boiling for 1 hour abrogates Hsp70-induced, but not LPS-induced, cytokine expression. It is of note that Hsp70 also induces an increased Ca^{2+} flux in neuronal cells, suggesting that this is a general signalling response to Hsp70 binding [46].

We noted in studies on APCs that the rapid Hsp70-induced $[Ca^{2+}]_i$ flux is followed by the phosphorylation of I-κBα [35]. Activation of the transcription factor NF-κB is regulated by its cytoplasmic inhibitor I-κBα, via phosphorylation at Serine 32 (Ser-32) and 36 (Ser-36) which targets it for degradation by the proteosome and releases NF-κB to migrate to the nucleus and activate the promoter of target genes [47]. As early as 30 minutes after exposure to exogenous Hsp70, I-κBα is phosphorylated at Serine 32 (Ser-32) and 36 (Ser-36) in a calcium-dependent manner, and this results in the release and nuclear translocation of NF-κB [35]. Mechanistic studies using the HEK293 model system have revealed that Hsp70-induced NF-κB promoter activity is MyD88-dependent, CD14-dependent and is transduced via both TLR2 and TLR4 [43].

TLR2 and TLR4 are pattern-recognition receptors that recognise molecules associated with Gram-positive and Gram-negative bacteria, respectively [48, 49]. Hsp70 has the unique ability to interact with both of these signalling molecules

[43] and the presence of both TLR2 and TLR4 synergistically stimulates Hsp70-induced cytokine production [43]. Chapter 7 provides further discussion of these receptors. Interestingly, we have found that the synergistic activation of NF-κB promoter by co-expression of both TLR2 and TLR4 is MyD88-independent, which suggests an alterative pathway by which exogenous Hsp70 stimulates co-operation between TLR2 and TLR4 in cells of the immune system. As early as 2–4 hours after the exposure of APCs to exogenous Hsp70, there is significant release of TNF-α, IL-1β, IL-6 and IL-12 [35, 43]. By 3–5 days after exposure there is significant increase in proliferation of immature DCs and an augmentation in the expression of major histocompatibility complex (MHC) class II and the co-stimulatory molecule, CD86 [43]. These events might be highly significant to the pathways of antigen presentation, leading as they do to the maturation of DCs and antigen presentation to immune effector cells [31].

CD40 is a co-stimulatory molecule expressed on APCs that plays an important role in B lymphocyte function and autoimmunity [50] and CD40 binds Hsp70-peptide complexes via its exoplasmic domain [40]. The Hsp70–CD40 interaction is mediated by the NH_2-terminal ATPase domain of Hsp70 in its ADP-bound state and is augmented by the presence of substrate peptides in the COOH-terminal domain of Hsp70. The Hsp70–CD40 interaction is suppressed by Hip, a co-chaperone that is known to stabilise the Hsp70 ATPase domain in the ADP bound state [40].

Using the HEK293 cell model system, Hsp70–CD40 binding has been shown to stimulate signal transduction via the phosphorylation of p38 mitogen-activated protien kinase (previously shown to induce the release of TNF-α and secretion of IFN-γ [51]), and this results in the activation of NF-κB and uptake of peptide [40]. A detailed analysis of Hsp70–CD40 interactions is provided in Chapter 10. The oxidised LDL receptor LOX-1 on human DCs binds Hsp70, and incubation of cells with a neutralising anti-LOX-1 monoclonal antibody abrogates Hsp70 binding to DCs and suppresses Hsp70-induced antigen cross-presentation, although little is known of the signalling cascades that emanate from Hsp70–LOX-1 complexes [41]. However, LOX-1 is likely to exert a significant effect on Hsp70 uptake and antigen cross-presentation [41].

8.6. Conclusions

The data available in the literature therefore strongly support a role for extracellular Hsp70 as a danger signal for the immune response, as an emergency chaperone for the neuronal system and as playing some as yet undefined role in stress and exercise. We are, however, at an early stage in the exploration of this field and little is certain as yet. The uncertainties include the significance of the various sources of extracellular Hsp70, the receptor systems that permit cells to respond to this novel ligand, and the physiological role of free or bound extracellular Hsp70. What seems to be clear is that Hsp70, as a consequence of its capacity to carry processed peptides as cargo, permits the intracellular milieu of cells to be sampled by homeostatic cells such as APCs in a way similar to the MHC system [52]. This may represent a very primitive surveillance system which evolved many aeons prior to the development of the immune system [52]. In addition, cells may sense the structure of extracellular Hsp70 itself, regardless of peptide cargo, as a danger signal due to its capacity for massive induction by stress [35].

The relative physiological responses to heat shock proteins might depend on the amplitude and precise anatomical site of release. A gradual release of Hsp70 from a tissue in a regulated manner might be quite different from a massive Hsp70 release during toxic stress accompanied by other danger signals [25, 32]. Such signals might include other molecular chaperones such as Hsp60, Hsp90, gp96, Hsp110 and gp170, or molecules such as high mobility group box 1 protein (HMGB-1) and uric acid [36, 53–58]. Extracellular Hsp70 is potentially important in a number of diseases. In the case of cancer, Hsp70 offers an unique target for therapy with an agent that is already elevated in cancer and capable of capturing tumour-associated antigens, thus representing an Achilles heel that can be exploited by tumour immunotherapy approaches [6, 59, 60]. Extracellular Hsp70 might exert beneficial effects in neuronal and other cells in which extra chaperoning power can be transferred to vulnerable targets, permitting survival of acute stress [4, 27, 61].

Acknowledgements

We thank our colleagues Philip Auron, Rolph Issels, Hansjörg Schild, Chris Nichitta, Yue Xie, Betsy Repasky and John Subjeck for many helpful discussions and Dr. Auron in particular for invaluable materials. We also thank Olivia Bare, Maria Bausero and Edith Kabingu, for expert technical assistance. This work was supported in part by National Institutes of Health (NIH) Grants CA47407, CA31303, CA50642, CA77465 (to SKC) and the NIH grant RO1CA91889, Joint Center for Radiation Therapy Foundation Grant, Harvard Medical School and Institutional support from the Department of Medicine, Boston University School of Medicine (to AA).

REFERENCES

1. Li G C and Werb Z. Correlation between synthesis of heat shock proteins and development of thermotolerance in Chinese hamster fibroblasts. Proc Natl Acad Sci USA 1982, 79: 3218–3222.
2. Lindquist S and Craig E A. The heat-shock proteins. Ann Rev Genet 1988, 22: 631–677.
3. Hightower L E and Guidon P T. Selective release from cultured mammalian cells of heat-shock (stress) proteins that resemble glia-axon transfer proteins. J Cell Physiol 1989, 138: 257–266.
4. Tytell M, Greenberg S G and Lasek R J. Heat shock-like protein is transferred from glia to axon. Brain Res 1986, 363: 161–164.
5. Pockley A G, Shepherd J and Corton J. Detection of heat shock protein 70 (Hsp70) and anti-Hsp70 antibodies in the serum of normal individuals. Immunol Invest 1998, 27: 367–377.
6. Srivastava P K and Amato R J. Heat shock proteins: the 'Swiss Army Knife' vaccines against cancers and infectious agents. Vaccine 2001, 19: 2590–2597.
7. Nollen E A and Morimoto R I. Chaperoning signaling pathways: molecular chaperones as stress-sensing 'heat shock' proteins. J Cell Sci 2002, 115: 2809–2816.
8. Pratt W B and Toft D O. Regulation of signaling protein function and trafficking by the hsp90/hsp70-based chaperone machinery. Exp Biol Med 2003, 228: 111–133.
9. Chu B, Soncin F, Price B D, Stevenson M A and Calderwood S K. Sequential phosphorylation by mitogen-activated protein kinase and glycogen synthase kinase 3 represses transcriptional activation by heat shock factor-1. J Biol Chem 1996, 271: 30847–30857.
10. Feder J H, Rossi J M, Solomon J, Solomon N and Lindquist S. The consequences of expressing hsp70 in *Drosophila* cells at normal temperatures. Genes Dev 1992, 6: 1402–1413.
11. Wang X, Grammatikakis N, Siganou A and Calderwood S K. Regulation of molecular chaperone gene transcription involves the serine phosphorylation, 14-3-3 epsilon binding, and cytoplasmic sequestration of heat shock factor 1. Mol Cell Biol 2003, 23: 6013–6026.
12. Zhao M, Tang D, Lechpammer S, Hoffman A, Asea A, Stevenson M A and Calderwood S K. Double-stranded RNA-dependent protein kinase (pkr) is essential for thermotolerance, accumulation of HSP70, and stabilization of ARE-containing HSP70 mRNA during stress. J Biol Chem 2002, 277: 44539–44547.
13. Xie Y, Chen C, Stevenson M A, Hume D A, Auron P E and Calderwood S K. NF-IL6 and HSF1 have mutually antagonistic effects on transcription in monocytic cells. Biochem Biophys Res Commun 2002, 291: 1071–1080.
14. Xie Y, Zhong R, Chen C and Calderwood S K. Heat shock factor 1 contains two functional domains that mediate transcriptional repression of the c-fos and c-fms genes. J Biol Chem 2003, 278: 4687–4698.
15. Lau S S, Griffin T M and Mestril R. Protection against endotoxemia by HSP70 in rodent cardiomyocytes. Am J Physiol Heart Circ Physiol 2000, 278: H1439–1445.
16. McMillan D R, Xiao X, Shao L, Graves K and Benjamin I J. Targeted disruption of heat shock transcription factor 1 abolishes thermotolerance and protection against heat-inducible apoptosis. J Biol Chem 1998, 273: 7523–7528.

17. Wright B H, Corton J, El-Nahas A M, Wood R F M and Pockley A G. Elevated levels of circulating heat shock protein 70 (Hsp70) in peripheral and renal vascular disease. Heart Vessels 2000, 15: 18–22.

18. Broquet A H, Thomas G, Masliah J, Trugnan G and Bachelet M. Expression of the molecular chaperone Hsp70 in detergent-resistant microdomains correlates with its membrane delivery and release. J Biol Chem 2003, 278: 21601–21606.

19. Shin B K, Wang H, Yim A M, Le Naour F, Brichory F, Jang J H, Zhao R, Puravs E, Tra J, Michael C W, Misek D E and Hanash S M. Global profiling of the cell surface proteome of cancer cells uncovers an abundance of proteins with chaperone function. J Biol Chem 2003, 278: 7607–7616.

20. Triantafilou M, Miyake K, Golenbock D T and Triantafilou K. Mediators of innate immune recognition of bacteria concentrate in lipid rafts and facilitate lipopolysaccharide-induced cell activation. J Cell Sci 2002, 115: 2603–2611.

21. Mathew A, Bell A and Johnstone R M. Hsp-70 is closely associated with the transferrin receptor in exosomes from maturing reticulocytes. Biochem J 1995, 308: 823–830.

22. Shi Y and Rock K L. Cell death releases endogenous adjuvants that selectively enhance immune surveillance of particulate antigens. Eur J Immunol 2002, 32: 155–162.

23. Pockley A G. Heat shock proteins, inflammation and cardiovascular disease. Circulation 2002, 105: 1012–1017.

24. Campisi J and Fleshner M. Role of extracellular HSP72 in acute stress-induced potentiation of innate immunity in active rats. J Appl Physiol 2003, 94: 43–52.

25. Campisi J, Leem T H, Greenwood B N, Hansen M K, Moraska A, Higgins K, Smith T P and Fleshner M. Habitual physical activity facilitates stress-induced HSP72 induction in brain, peripheral, and immune tissues. Am J Physiol Regul Integr Comp Physiol 2003, 284: 1R520–R530.

26. Bechtold D A and Brown I R. Heat shock proteins Hsp27 and Hsp32 localize to synaptic sites in the rat cerebellum following hyperthermia. Brain Res Mol Brain Res 2000, 75: 309–320.

27. Guzhova I, Kislyakova K, Moskaliova O, Fridlanskaya I, Tytell M, Cheetham M and Margulis B. *In vitro* studies show that Hsp70 can be released by glia and that exogenous Hsp70 can enhance neuronal stress tolerance. Brain Res 2001, 914: 66–73.

28. Multhoff G and Hightower L E. Cell surface expression of heat shock proteins and the immune response. Cell Stress Chaperones 1996, 1: 167–176.

29. Multhoff G. Activation of natural killer cells by heat shock protein 70. Int J Hyperthermia 2002, 18: 576–585.

30. Bechtold D A, Rush S J and Brown I R. Localization of the heat-shock protein Hsp70 to the synapse following hyperthermic stress in the brain. J Neurochem 2000, 74: 641–646.

31. Noessner E, Gastpar R, Milani V, Brandl A, Hutzler P J, Kuppner M C, Roos M, Kremmer E, Asea A, Calderwood S K and Issels R D. Tumor-derived heat shock protein 70 peptide complexes are cross-presented by human dendritic cells. J Immunol 2002, 169: 5424–5432.

32. Todryk S M, Melcher A A, Dalgleish A G and Vile R G. Heat shock proteins refine the danger theory. Immunology 2000, 99: 334–337.

33. Arnold-Schild D, Hanau D, Spehner D, Schmid C, Rammensee H-G, de la Salle H and Schild H. Receptor-mediated endocytosis of heat shock proteins by professional antigen-presenting cells. J Immunol 1999, 162: 3757–3760.
34. Asea A, Kabingu E, Stevenson M A and Calderwood S K. Hsp70 peptide-bearing and peptide-negative preparations act as chaperokines. Cell Stress Chaperon 2000, 5: 425–431.
35. Asea A, Kraeft S-K, Kurt-Jones E A, Stevenson M A, Chen L B, Finberg R W, Koo G C and Calderwood S K. Hsp70 stimulates cytokine production through a CD14-dependent pathway, demonstrating its dual role as a chaperone and cytokine. Nat Med 2000, 6: 435–442.
36. Lipsker D, Ziylan U, Spehner D, Proamer F, Bausinger H, Jeannin P, Salamero J, Bohbot A, Cazenave J P, Drillien R, Delneste Y, Hanau D and de la Salle H. Heat shock proteins 70 and 60 share common receptors which are expressed on human monocyte-derived but not epidermal dendritic cells. Eur J Immunol 2002, 32: 322–332.
37. Reed R C and Nicchitta C V. Chaperone-mediated cross-priming: a hitchhiker's guide to vesicle transport. Int J Mol Med 2000, 6: 259–264.
38. Sondermann H, Becker T, Mayhew M, Wieland F and Hartl F U. Characterization of a receptor for heat shock protein 70 on macrophages and monocytes. Biol Chem 2000, 381: 1165–1174.
39. Basu S, Binder R J, Ramalingam T and Srivastava P K. CD91 is a common receptor for heat shock proteins gp96, Hsp90, Hsp70 and calreticulin. Immunity 2001, 14: 303–313.
40. Becker T, Hartl F U and Wieland F. CD40, an extracellular receptor for binding and uptake of Hsp70-peptide complexes. J Cell Biol 2002, 158: 1277–1285.
41. Delneste Y, Magistrelli G, Gauchat J, Haeuw J, Aubry J, Nakamura K, Kawakami-Honda N, Goetsch L, Sawamura T, Bonnefoy J and Jeannin P. Involvement of LOX-1 in dendritic cell-mediated antigen cross-presentation. Immunity 2002, 17: 353–362.
42. Gross C, Hansch D, Gastpar R and Multhoff G. Interaction of heat shock protein 70 peptide with NK cells involves the NK receptor CD94. Biol Chem 2003, 384: 267–279.
43. Asea A, Rehli M, Kabingu E, Boch J A, Baré O, Auron P E, Stevenson M A and Calderwood S K. Novel signal transduction pathway utilized by extracellular HSP70. Role of Toll-like receptor (TLR) 2 and TLR4. J Biol Chem 2002, 277: 15028–15034.
44. Vabulas R M, Braedel S, Hilf N, Singh-Jasuja H, Herter S, Ahmad-Nejad P, Kirschning C J, Da Costa C, Rammensee H G, Wagner H and Schild H. The endoplasmic reticulum-resident heat shock protein Gp96 activates dendritic cells via the Toll-like receptor 2/4 pathway. J Biol Chem 2002, 277: 20847–20853.
45. McLeish K R, Dean W L, Wellhausen S R and Stelzer G T. Role of intracellular calcium in priming of human peripheral blood monocytes by bacterial lipopolysaccharide. Inflammation 1989, 13: 681–692.
46. Smith P J, Hammar K and Tytell M. Effects of exogenous heat shock protein (Hsp70) on neuronal calcium flux. Biol Bull 1995, 189: 209–210.
47. Baeuerle P A and Baltimore D. I kB: a specific inhibitor of the NF-kB transcription factor. Science 1988, 242: 540–546.

48. Akira S and Sato S. Toll-like receptors and their signaling mechanisms. Scand J Infect Dis 2003, 35: 555–562.
49. Pulendran B, Palucka K and Banchereau J. Sensing pathogens and tuning immune responses. Science 2001, 293: 253–256.
50. Bodmer J L, Schneider P and Tschopp J. The molecular architecture of the TNF superfamily. Trends Biochem Sci 2002, 27: 19–26.
51. Pullen S S, Dang T T, Crute J J and Kehry M R. CD40 signaling through tumor necrosis factor receptor-associated factors (TRAFs). Binding site specificity and activation of downstream pathways by distinct TRAFs. J Biol Chem 1999, 274: 14246–14254.
52. Srivastava P. Interaction of heat shock proteins with peptides and antigen presenting cells: chaperoning of the innate and adaptive immune responses. Ann Rev Immunol 2002, 20: 395–425.
53. Manjili M H, Wang X Y, Chen X, Martin T, Repasky E A, Henderson R and Subjeck J R. HSP110-HER2/neu chaperone complex vaccine induces protective immunity against spontaneous mammary tumors in HER-2/neu transgenic mice. J Immunol 2003, 171: 4054–4061.
54. Park J S, Svetkauskaite D, He Q, Kim J Y, Strassheim D, Ishizaka A and Abraham E. Involvement of toll-like receptors 2 and 4 in cellular activation by high mobility group box 1 protein. J Biol Chem 2004, 279: 7370–7377.
55. Shi Y, Evans J E and Rock K L. Molecular identification of a danger signal that alerts the immune system to dying cells. Nature 2003, 425: 516–521.
56. Takata K, Kitamura Y, Tsuchiya D, Kawasaki T, Taniguchi T and Shimohama S. Heat shock protein-90-induced microglial clearance of exogenous amyloid-β1-42 in rat hippocampus *in vivo*. Neurosci Lett 2003, 344: 87–90.
57. Wang X Y, Kazim L, Repasky E A and Subjeck J R. Immunization with tumor-derived ER chaperone grp170 elicits tumor-specific CD8$^+$ T-cell responses and reduces pulmonary metastatic disease. Int J Cancer 2003, 105: 226–231.
58. Singh-Jasuja H, Toes R E M, Spee P, Münz C, Hilf N, Schoenberger S P, Ricciardi-Castagnoli P, Neefjes J, Rammensee H-G, Arnold-Schild D and Schild H. Cross-presentation of glycoprotein 96-associated antigens on major histocompatibility complex molecules requires receptor-mediated endocytosis. J Exp Med 2000, 191: 1965–1974.
59. Blagosklonny M V. Re: Role of the heat shock response and molecular chaperones in oncogenesis and cell death. J Natl Cancer Inst 2001, 93: 239–240.
60. Ciocca D R, Clark G M, Tandon A K, Fuqua S A, Welch W J and McGuire W L. Heat shock protein hsp70 in patients with axillary lymph node-negative breast cancer: prognostic implications. J Natl Cancer Inst 1993, 85: 570–574.
61. Yenari M A. Heat shock proteins and neuroprotection. Adv Exp Med Biol 2002, 513: 281–299.

9

Hsp72 and Cell Signalling

Vladimir L. Gabai and Michael Y. Sherman

9.1. Introduction

Many signalling molecules such as steroid hormone receptors and other receptors, protein kinases and phosphatases are found associated with various types of heat shock proteins, including Hsp90, Hsp70, Hsp40 and other cochaperones. The functional role of these associations appears to be multi-faceted and the association of signalling proteins with these chaperone cohorts plays a pivotal role in initial folding and maturation of steroid hormone receptors and many kinases (e.g., Src). In addition, association with Hsp90 and its cochaperones is critical for the stability of signalling proteins, because inhibition of Hsp90 by geldanamycin and other specific inhibitors leads to rapid ubiquitin-dependent degradation of Raf-1, Akt and other kinases that normally associate with Hsp90 [1, 2]. In fact, the anti-cancer activities of Hsp90 inhibitors could be related to the degradation and downregulation of signalling pathways that are controlled by these kinases [1, 3]. In contrast to Hsp90, which protects from degradation, an association with Hsp70 might target these proteins for rapid ubiquitination (usually via a ubiquitin ligase CHIP) followed by proteolysis [4].

In addition to their critical role in folding, maturation and stability of various signalling components, chaperones may be directly involved in regulation of their activities. In fact, it appears that Hsp70 and other chaperones play a regulatory role in the activation of many signalling pathways that are elicited by heat shock and other stresses.

There are multiple members of the Hsp70 protein family. Many of them, such as Hsc73, are expressed constitutively and serve a number of housekeeping functions, such as the folding of newly synthesised proteins and the transport of proteins to various organelles, and are involved in the ubiquitin-proteasome-dependent degradation pathway. Other members of this protein family, such as

Hsp72, are typically not expressed in normal, unstressed cells; however, their expression is rapidly induced to high levels upon exposure of the cells to heat shock and other stresses that cause protein damage. These proteins are believed to prevent aggregation of stress-damaged polypeptides and to promote their rapid refolding [5, 6].

The stress-induced Hsp70 has also been shown to play an important role in controlling multiple signalling pathways, and it appears that these interactions are directly related to a well-known cell protective function of Hsp70. Furthermore, these interactions can contribute to a phenomenon of acquired stress tolerance, in which cells that have been exposed to mild heat stress followed by recovery become tolerant to a wide range of stressful treatments.

This chapter will focus on the regulatory role of Hsp70 family members in activation of several kinase pathways and on the implications of Hsp70-mediated cell signalling in tumour development.

9.2. Hsp70 and MAP kinase signalling pathways

A role for Hsp70 in the regulation of mitogen-activated protein (MAP) kinase pathways has been studied for several years. Early work demonstrated that artificial expression of recombinant Hsp70 at high levels in human lymphoid cells dramatically diminished stress-induced activation of two MAP kinases, p38 and c-Jun N-terminal kinase (JNK) [7, 8]. The initial hypothesis arising from these findings was that abnormal proteins accumulating in cells under stressful treatments activate stress signalling cascades, and that Hsp70 can indirectly suppress the activity of these kinases by inhibiting the accumulation of abnormal proteins and facilitating protein refolding. Accordingly, it was shown that the accumulation of abnormal proteins induced by amino-acid analogues or proteasome inhibitors are powerful activators of p38 and JNK cascades [7, 9–11]. On the other hand, in addition to those stresses that can elicit protein damage (such as heat shock, ethanol, oxidative stress), a number of other stimuli, including tumour necrosis factor (TNF), IL-1, UV irradiation, or osmotic stress, can activate p38 and JNK without causing apparent proteotoxicity. Hsp70 expressed at high levels can also reduce the activation of the JNK and p38 kinase cascades induced by these stimuli [7, 12, 13]. Therefore, it appears that the effects of Hsp70 on these pathways may not be related to the handling of damaged proteins that accumulate after stressful treatment and may not even be related to the chaperone activities of Hsp70.

An understanding of the inhibitory effect of Hsp70 on JNK activation is complicated by the fact that there are at least two major pathways of activation of JNK. Of note, there are two similarly regulated major isoforms of JNK in non-neuronal cells, JNK1 and JNK2, and a neuron-specific isoform, JNK3 [14]. Activation of JNK by UV, osmotic stress and cytokines has been shown to proceed through a signal transduction pathway which involves a cascade of protein kinases, starting from mitogen activated protein kinase kinase kinase (MEKKs), followed by dual-specificity kinases, SAPK/ERK kinase/stress activated protein kinase (SEK1) mitogen activated protein kinase kinase (MKK4) and MKK7, which in turn phosphorylate JNK at tyrosine and threonine, thus activating it [15]. Several kinases other than MEKKs, including apoptosis signal regulating kinase 1 (ASK1) and mixed lineage kinase (MLKs), can also activate MKK4 and MKK7, and therefore there is a network of signalling pathways that contribute to JNK activation. p38 and extracellular signal-regulated kinases (ERKs), the third group of MAP kinase pathways, are activated via homologous kinase cascades [15, 16].

Interestingly, heat shock, oxidative stress, and other protein-damaging stresses activate MAP kinases via a novel pathway that involves an inhibition of their de-phosphorylation [17]. Multiple phosphatases with varying specificities, including dual-specificity phosphatases, PP2C, PP2A and others, are involved in the de-phosphorylation of MAP kinases [18, 19], the major contributors to which are dual-specificity phosphatases. For example, a stress-inducible phosphatase map kinase phosphatase (MKP-1) can de-phosphorylate all types of MAP kinases [20, 21]. VHI-recated phosphatase (VHR) and M3/6 are more specific to JNK [22–24], whereas MKP-3 and MKP-2 are specific to ERKs [22, 23, 25].

Many of these phosphatases are very sensitive to heat shock and other protein-damaging stresses. For example, upon exposure of cells to even mild heat shock, M3/6 becomes inactive and rapidly aggregates, leading to an inhibition of the de-phosphorylation of JNK and therefore an increase in the activity of this kinase [26]. Other, as yet unknown phosphatases that de-phosphorylate JNK in cells are probably also very heat-sensitive, because heat shock inactivates JNK de-phosphorylation almost entirely [17].

Interestingly, dual-specificity phosphatases have an essential cysteine in the active site, which potentially can be easily oxidised. The presence of this cysteine may explain the high sensitivity of these phosphatases to oxidative stress [27, 28]. Unexpectedly, JNK phosphatase M3/6 has also been found to aggregate, not only as a result of heat denaturation but also upon accumulation in cells of an abnormal polypeptide, a fragment of huntingtin with expanded polyglutamine domain (a cause of neurodegenerative Huntington's disease) [29]. The mechanism of M3/6 aggregation under these conditions is unclear. It might be that M3/6 phosphatase is so unstable that it requires molecular chaperones to maintain

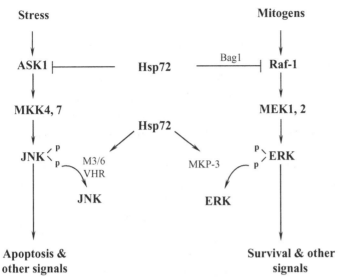

Figure 9.1. Effects of Hsp72 on MAP kinase pathways.

its normal conformation even at normal temperature. Then, accumulation of the mutant huntingtin fragment may bind chaperones and titrate them from a complex with the M3/6 phosphatase, leading to its aggregation and inactivation. Interestingly, endogenous levels of Hsp70 are strongly reduced in cells expressing the mutant huntingtin [29], which has been proposed to contribute to the M3/6 phosphatase aggregation and JNK activation.

In suppressing JNK activation by heat shock and other protein-damaging stresses, Hsp70 alleviates the inhibition of phosphatases (Figure 9.1). Similarly, over-production of Hsp70 protects the M3/6 phosphatase in cells that express the mutant huntingtin fragment with extended polyglutamine [29]. This function of Hsp70 is clearly dependent on its chaperone activity, because an Hsp70 mutant with a short C-terminal deletion, which abrogates the refolding activity, fails to preserve JNK dephosphorylation in heat-shocked cells [30]. Interestingly, protection of JNK phosphatases is not the only site of Hsp70 action in the JNK signalling pathway. In fact, in suppressing JNK activation induced by TNF and other non-protein-damaging stimuli, Hsp70 inhibits the upstream kinase cascade [13]. In contrast to protection of JNK phosphatases, for Hsp70-mediated inhibition of the kinase cascade, the refolding activity is dispensable, because the C-terminal deletion mutant of Hsp70 is as efficient in inhibiting the cascade as normal Hsp70 [13].

In line with these observations, it has been demonstrated that Hsp70 can directly interact and suppress activity of ASK1, an upstream component of p38 and JNK signalling cascades, which is activated by stimuli such as TNF or UV irradiation [31]. Hsp70 associated with ASK1 via an ATP-binding domain and deletion of this domain of Hsp70 abrogates interactions with ASK1 [31]. It is not clear whether interactions between Hsp70 and ASK1 are direct or are mediated by a distinct factor. It is possible that a co-chaperone such as Bag-1 may participate in these interactions, as it does with interactions between Hsp70 and a kinase Raf-1 (see discussion following).

The effect of proteotoxic stresses on MAP kinase de-phosphorylation seems to be quite general. Indeed, a dual-specificity phosphatase, MKP-3, which is involved in inactivation of ERK1/2, is also highly sensitive to heat shock and rapidly aggregates under these conditions [32]. Furthermore, Hsp70, but not the C-terminal mutant, was able to prevent heat shock–induced MKP-3 aggregation, thus suppressing ERK1/2 activation. As with the JNK pathway, Hsp70 can inhibit the ERK signalling pathway, also at a distinct site in the kinase cascade, and this activity of Hsp70 does not require its chaperone function [32]. Song and colleagues have reported that a co-chaperone, Bag-1, which can form a complex with a component of ERK-activating cascade, Raf-1, enhances Raf-1 activity [33]. Hsp70, when expressed at high levels, binds to Bag-1 and titrates it from the complex with Raf-1, leading to inactivation of the latter [33]. Interestingly, Bag-1 is known to interact with Hsp70 via its ATPase domain [34], which is intact in the C-terminal deletion mutant. This interaction may explain why the chaperone activity of Hsp70 is dispensable for the regulation of ERK kinase cascade. Further studies are needed to establish whether a similar Bag-1-dependent mechanism operates in Hsp70-mediated inhibition of the JNK signalling cascade.

Data from several laboratories indicate that this novel capacity of Hsp70 to control JNK signalling is important for the protection of cells from apoptosis induced by certain stimuli. For example, triggering apoptosis via specific activation of JNK via the expression of an active form of an upstream kinase MEKK1 can be efficiently blocked by Hsp70 [35]. In these cells, tolerance to UV-induced apoptosis caused by mild heat shock pre-treatment was clearly associated with Hsp70-mediated JNK inhibition [35]. In suppression of H_2O_2-induced apoptosis, or apoptosis induced by a constitutively active form of ASK1, JNK inhibition was probably associated with blocking of ASK1 by Hsp70 [31]. Furthermore, the ATPase domain of Hsp70, which is critical for the control of ASK1, is also essential for the inhibition of apoptosis under these conditions [31]. Although inhibition of JNK, which could be achieved by Hsp70 mutants lacking the chaperone function, is sufficient for suppressing apoptosis caused by UV irradiation [35], it does not block heat-induced apoptosis [13, 30]. Indeed, Hsp70 mutants

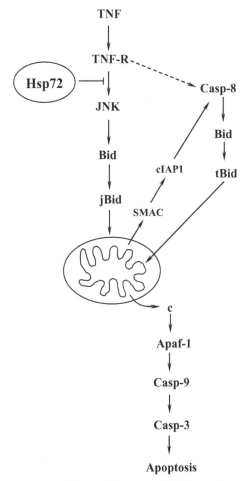

Figure 9.2. Effect of Hsp72 on TNF-induced apoptotic pathway.

that lack chaperone function efficiently suppressed activation of JNK by heat shock but were unable to protect cells from the consequences [13, 30].

The best-studied example of Hsp70-mediated inhibition of apoptosis is TNF-induced apoptosis of primary human fibroblasts, in which the suppression of JNK plays the major protective role [13]. The activation of JNK by TNF leads to caspase-8 independent cleavage of a BH3-domain protein Bid, which regulates mitochondrial integrity, and the release of a caspase regulator Smac/Diablo from mitochondria, with the subsequent activation of caspase-8 [36]. Hsp70-mediated inhibition of JNK in TNF-treated cells led to a suppression of Bid cleavage and an inhibition of subsequent apoptotic events [13] (Figure 9.2). Interestingly, neither JNK or Hsp70 could regulate the Bid-independent apoptotic pathway, which takes over at later time points [13].

Of note is that the suppression of JNK may not be the only target of Hsp70 in its inhibition of apoptotic signalling. For example, Hsp70 inhibits formation of the apoptosome [37, 38]. However, these effects of Hsp70 were studied in *in vitro* experiments, and the relevance of these data to the effects of Hsp70 *in vivo* have been challenged [74, 75].

An important development in recent years has been the finding that Hsp70-mediated JNK suppression is involved not only in apoptosis, but also in necrosis. In contrast to apoptosis, necrosis has always been considered to be a spontaneous, unregulated process. However, it now appears that some forms of necrosis may be tightly regulated and programmed [39]. For example, ischaemia-induced necrosis of cardiac cells is controlled by several signalling pathways, including JNK and p38 [40–42]. In the myogenic cell H9c2, simulated ischaemia (transient ATP depletion) leads to necrosis involving rapid JNK-dependent mitochondrial de-energisation. Expression of Hsp70 (or its C-terminal mutant) prevents JNK activation, mitochondrial de-energisation, and cell necrosis [32, 42]. This Hsp70-mediated suppression of stress kinases in ischaemic cells may be the basis for the protective effects of Hsp70 over-expression that have been demonstrated in a number of organs, including heart, brain, liver or kidney.

9.3. Control of the heme-activated kinase by Hsc73

The first example of a kinase that is directly regulated by an Hsp70 family member was HRI, a heme-regulated kinase of the α-subunit of eukaryotic translation initiation factor 2 (eIF-2α). HRI is present in reticulocytes (and possibly other cell types; R. Matts, personal communication). As with many other kinases, the initial folding and maturation of HRI requires an ensemble of chaperones, including Hsp90, Hsc73, Cdc37 and others [43, 44]. Upon maturation, HRI has low activity under normal conditions. However, upon heme limitation, HRI is phosphorylated and activated, and phosphorylates eIF-2α, leading to an inhibition of translation [45, 46]. Alternatively, HRI can be activated by heat shock, oxidative stress or other stresses that cause protein damage [45, 47] (Figure 9.3). Therefore, HRI shuts down synthesis of globin, the main protein produced in reticulocytes, under conditions that may cause improper production of haemoglobin (i.e., heme limitation or chaperone overload).

A breakthrough in understanding the mechanisms of HRI regulation was the finding that the activation of HRI by protein-damaging stresses is likely to be mediated by a build-up of damaged polypeptides. In fact, in reticulo-cyte lysates, the influence of heat shock on HRI activation can be mimicked by the addition of denatured, reduced carboxymethylated bovine serum albu-min (BSA), but not of normal BSA [48]. Elegant *in vitro* experiments have demonstrated that normally mature HRI associates with a constitutively ex-pressed member of the Hsp70 family, Hsc73. This association maintains HRI in a latent form in haemin-supplied lysates. Denatured proteins accumulated in cells after heat shock or denatured proteins added to lysates interact with Hsc73 and competitively inhibit association of Hsc73 with HRI. This interaction shifts the equilibrium, leading to dissociation of Hsc73 from HRI and activation of the latter [44]. Hsc73-mediated regulation of HRI is parallel with, and is inde-pendent of regulation of HRI by haemin, since neither association of Hsc73 with HRI nor its dissociation after heat shock are influenced by haemin [43] (Figure 9.3).

It is not clear whether Hsc73 interacts with HRI directly or whether there is an adaptor protein that mediates these interactions. Interestingly, there are two distinct protein kinases, PKR and PERK, that also regulate translation via phos-phorylation of eIF-2α. Although the main regulator of PKR is double-stranded RNA, this kinase, like HRI, can also be activated by heat shock and other protein-damaging stresses [49, 50] (Figure 9.3). The mechanism of this regulation is cur-rently unknown, although the involvement of Hsp70 has been suggested. In fact, an important regulator of PKR is the inhibitor p58[ipk], a protein with J-domain and multiple tetratricopeptide repeat (TPR) domains, which can bind Hsp70 pro-teins [51, 52]. Interactions between p58[ipk] and Hsp70 can be mediated by Hsp40 [53]. p58[ipk] can function as a co-chaperone of DnaJ type, because it stimulates ATPase activity of Hsp70 *in vitro* [52]. Interestingly, interactions between p58[ipk] and Hsp70 play an important role in regulation of PKR, because the J-domain of p58[ipk] is indispensable for this regulation [52]. It has been hypothesised that p58[ipk] recruits Hsp70 into the complex with PKR, which leads to inactivation of this kinase [54] (Figure 9.3).

In addition, p58[ipk] can also inhibit PERK, another eIF-2α kinase [55, 56] (Fig-ure 9.3). Therefore, it appears that all major kinases that regulate translation via the phosphorylation of eIF-2α could be controlled by Hsp70 family members.

9.4. Hsp70, cell signalling and cancer

Normal tissues usually express a constitutive member of the Hsp70 family, Hsc73, but not the inducible form Hsp70 (Hsp72). In contrast, tumours often express both Hsp70 and Hsc73, and a number of reports indicate that high expression of Hsp70 in human tumours, especially of epithelial origin, correlates

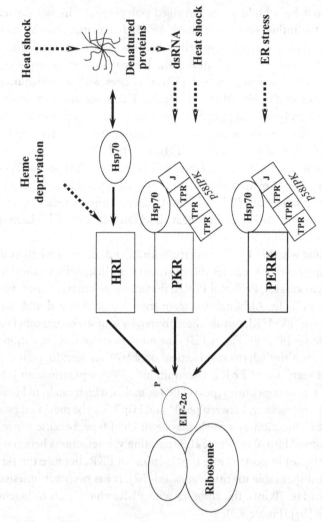

Figure 9.3. Regulation of HRI, PKR ad PERK kinases by Hsp70 and p58^{ipk}.

with high invasiveness, metastatic disease, poor prognosis of the disease and resistance to chemotherapy [57–60]. For example, in colorectal and lung cancers, expression of Hsp70 highly correlates with advanced clinical stages and positive lymph node involvement [61, 62]. These findings suggest that Hsp70 provides a selective advantage to tumour cells during cancer progression. One possible involvement of Hsp70 in tumour development could be its capacity to suppress apoptosis (see preceding discussion). Indeed, many oncogenes (e.g., myc) when activated can induce apoptosis in normal cells or enhance sensitivity of normal cells to various apoptotic stimuli [63, 64]. This response is considered to be an intrinsic cellular anti-cancer mechanism. Therefore, to avoid oncogene-activated apoptosis, upon cancer transformation cells develop special mechanisms for anti-apoptotic protection such as an over-expression of bcl-2, mutations in p53 or mutations in PTEN phosphatase.

In line with this observation, the expression of anti-apoptotic factors (e.g., bcl-2) significantly promotes cell transformation *in vitro* and tumour development *in vivo* following the expression of myc oncogene [64–66]. Hsp70 may serve as an alternative potent anti-apoptotic factor in the promotion of cancer transformation. In fact, the expression of Hsp70 in Rat-1 fibroblasts leads to the formation of foci, anchorage-independent growth, and formation of tumours in nude mice [67], and an over-production of Hsp70 in fibrosarcoma cells significantly enhances their tumourigenic potential in athymic mice [68]. Furthermore, transgenic mice that express human Hsp70 at high levels developed multiple lymphomas [69].

An important advance of recent years has been that, besides its possible role in cell transformation, Hsp70 may be critical for viability and proliferation of established cancer cell lines. Indeed, it has been demonstrated that the depletion of Hsp70 by adenovirus-encoded Hsp70 anti-sense RNA causes rapid death of various types of tumour cells, whereas non-transformed cells are resistant to such treatment [70]. Treatment with adenovirus expressing Hsp70 anti-sense RNA is also effective in human glioblastoma, breast and colon carcinoma grown as xenografts [71].

However, a caveat to these studies is that an adenoviral vector for anti-sense RNA delivery was employed, because adenovirus by itself causes subtle toxicity to tumour cells. This toxicity, combined with the effects of Hsp70 depletion, may lead to cancer cell killing. In fact, when we and others down-regulated Hsp70 in a milder way by retroviral expression of Hsp70 anti-sense RNA or siRNA, no significant tumour cell killing was observed; rather there was a specific sensitisation of cells to some drugs and stresses (Gabai et al, in press). Using retrovirus vector expressing Hsp70 in an anti-sense orientation, clones of PC-3 human prostate adenocarcinoma cells (PC-3/AS) which express 4-fold lower levels of Hsp70 than parental cells have been generated. These clones concomitantly expressed normal levels of Hsc73. As expected, PC-3/AS cells were significantly more sensitive to heat-induced apoptosis, as well as to apoptosis induced by TNF, cis-platinum, vinblastine or taxol. Surprisingly, these cells had a decreased etoposide-induced apoptosis, indicating that the effects of Hsp70 depletion are specific to certain types of apoptosis. Importantly, reduced levels of Hsp70 not only influence apoptotic cell death, but also overall cell survival. Indeed, short-term treatments with certain anti-cancer drugs or radiation did not induce apoptosis; rather it led to mitotic catastrophe and loss of colony-forming ability of cells. Down-regulation of Hsp70 markedly enhanced such a mitotic catastrophe.

In a variety of cells, activation of JNK and p38 has been shown to stimulate apoptosis, whereas stimulation of ERK and Akt kinases promotes survival. As discussed earlier, over-expression of Hsp70 suppresses stress-induced activation of JNK, p38, ASK1 and ERK kinases [7, 31, 33, 72] and hence we anticipated that Hsp70 downregulation in cancer cells should increase the activities of at least some of these kinases. However, no upregulation of p38 and JNK activities has been found. Furthermore, depletion of Hsp72 in PC-3 cell clones leads to a dramatic downregulation of endogenous and stress-induced ERK1/2 activity. This effect of Hsp72 depletion is associated with strong downregulation of the ERK-activation kinase cascade, probably at the level of Raf-1. It was previously demonstrated that treatment of cells with geldanamycin, an inhibitor of Hsp90, leads to ERK1/2 inactivation because of rapid ubiquitin-proteasome-dependent degradation of Raf-1 [1]. Although Hsp72 acts as a co-factor of Hsp90 in protein refolding, the levels of Raf-1 expression increased rather than decreased in PC-3/AS cells, indicating that Hsp72 downregulation did not reduce Raf-1 stability. At the same time, de-phosphorylation of the inhibitory site Ser-259 (which is required for Raf-1 activation [73]) was significantly reduced in PC-3/AS cells.

Therefore, it seems that a certain level of Hsp72 expression in tumour cells is necessary for maintaining activity of the major cell survival pathway, ERK, and downregulation of Hsp72 leads to a loss of ERK activity. These findings can potentially explain the increased sensitivity of PC-3/AS cells to stresses and drugs, because suppression of ERK can significantly affect viability of cells exposed to harsh conditions. Indeed, the specific inhibitor of ERK (U0126) made parental PC-3 cells as sensitive to heat shock as PC-3/AS cells. Taken together it would appear that the effects of Hsp72 on various signalling systems play a major role in tumour development and survival.

REFERENCES

1. Hostein I, Robertson D, DiStefano F, Workman P and Andrew Clarke P. Inhibition of signal transduction by the Hsp90 inhibitor 17-allylamino-17-demethoxygeldanamycin results in cytostasis and apoptosis. Cancer Res 2001, 61: 4003–4009.
2. Fujita N, Sato S, Ishida A and Tsuruo T. Involvement of Hsp90 in signaling and stability of 3-phosphoinositide-dependent kinase-1. J Biol Chem 2002, 277: 10346–10353.
3. Neckers L. Hsp90 inhibitors as novel cancer chemotherapeutic agents. Trends Mol Med 2002, 8: S55–61.
4. Wiederkehr T, Bukau B and Buchberger A. Protein turnover: a CHIP programmed for proteolysis. Cur Biol 2002, 12: R26–28.
5. Michels A A, Kanon B, Konings A W, Ohtsuka K, Bensaude O and Kampinga H H. Hsp70 and Hsp40 chaperone activities in the cytoplasm and the nucleus of mammalian cells. J Biol Chem 1997, 272: 33283–33289.
6. Nollen E A, Brunsting J F, Roelofsen H, Weber L A and Kampinga H H. *In vivo* chaperone activity of heat shock protein 70 and thermotolerance. Mol Cell Biol 1999, 19: 2069–2079.
7. Gabai V L, Meriin A B, Mosser D D, Caron A W, Rits S, Shifrin VI and Sherman M Y. HSP70 prevent activation of stress kinases: a novel pathway of cellular thermotolerance. J Biol Chem 1997, 272: 18033–18037.
8. Mosser D D, Caron A W, Bourget L, Denis-Larose C and Massie B. Role of the human heat shock protein hsp70 in protection against stress-induced apoptosis. Mol Cell Biol 1997, 17: 5317–5327.
9. Meriin A, Gabai V, Yaglom J, Shifrin V and Sherman M. Proteasome inhibitors activate stress-kinases and induce Hsp72: diverse effects on apoptosis. J Biol Chem 1998, 273: 6373–6379.
10. Hideshima T, Mitsiades C, Akiyama M, Hayashi T, Chauhan D, Richardson P, Schlossman R, Podar K, Munshi N C, Mitsiades N and Anderson K C. Molecular mechanisms mediating antimyeloma activity of proteasome inhibitor PS-341. Blood 2003, 101: 1530–1534.
11. Almond J B and Cohen G M. The proteasome: a novel target for cancer chemotherapy. Leukemia 2002, 16: 433–443.
12. Yaglom J A, Gabai V L, Meriin A B, Mosser D D and Sherman M Y. The function of HSP72 in suppression of c-Jun N-terminal kinase activation can be dissociated from its role in prevention of protein damage. J Biol Chem 1999, 274: 20223–20228.
13. Gabai V L, Mabuchi K, Mosser D D and Sherman M Y. Hsp72 and stress kinase c-jun N-terminal kinase regulate the bid-dependent pathway in tumor necrosis factor-induced apoptosis. Mol Cell Biol 2002, 22: 3415–3424.
14. Davis R J. Signal transduction by the JNK group of MAP kinases. Cell 2000, 103: 239–252.
15. Kyriakis J M a J A. Mammalian mitogen-activated protein kinase signal transduction pathways activated by stress and inflammation. Physiol Rev 2001, 81: 807–869.
16. Su B and Karin M. Mitogen-activated protein kinase cascades and regulation of gene expression. Cur Opin Immunol 1996, 8: 402–411.

17. Meriin A B, Yaglom J A, Gabai V L, Mosser D D, Zon L and Sherman M Y. Protein damaging stresses activate JNK via inhibition of its phosphatase: a novel pathway controlled by Hsp72. Mol Cell Biol 1999, 19: 2547–2555.
18. Millward T A, Zolnierowicz S and Hemmings B A. Regulation of protein kinase cascades by protein phosphatase 2A. Trends Biochem Sci 1999, 24: 186–191.
19. Saxena M and Mustelin T. Extracellular signals and scores of phosphatases: all roads lead to MAP kinase. Semin Immunol 2000, 12: 387–396.
20. Keyse S M. Protein phosphatases and the regulation of MAP kinase activity. Semin Cell Develop Biol 1998, 9: 143–152.
21. Keyse S M. An emerging family of dual specificity MAP kinase phosphatases. Biochim Biophys Acta 1995, 1265: 152–160.
22. Muda M, Theodosiou A, Gillieron C, Smith A, Chabert C, Camps M, Boschert U, Rodrigues N, Davies K, Ashworth A and Arkinstall S. The mitogen-activated protein kinase phosphatase-3 N-terminal noncatalytic region is responsible for tight substrate binding and enzymatic specificity. J Biol Chem 1998, 273: 9323–9329.
23. Muda M, Theodosiou A, Rodrigues N, Boschert U, Camps M, Gillieron C, Davies K, Ashworth A and Arkinstall S. The dual specificity phosphatases M3/6 and MKP-3 are highly selective for inactivation of distinct mitogen-activated protein kinases. J Biol Chem 1996, 271: 27205–27208.
24. Todd J L, Rigas J D, Rafty L A and Denu J M. Dual-specificity protein tyrosine phosphatase VHR down-regulates c-Jun N-terminal kinase (JNK). Oncogene 2002, 21: 2573–2583.
25. Chu Y, Solski P A, Khosravi-Far R, Der C J and Kelly K. The mitogen-activated protein kinase phosphatases PAC1, MKP-1, and MKP-2 have unique substrate specificities and reduced activity in vivo toward the ERK2 sevenmaker mutation. J Biol Chem 1996, 271: 6497–6501.
26. Palacios C, Collins M K and Perkins G R. The JNK phosphatase M3/6 is inhibited by protein-damaging stress. Cur Biol 2001, 11: 1439–1443.
27. Chen Y R, Shrivastava A and Tan T H. Down-regulation of the c-Jun N-terminal kinase (JNK) phosphatase M3/6 and activation of JNK by hydrogen peroxide and pyrrolidine dithiocarbamate. Oncogene 2001, 20: 367–374.
28. Theodosiou A and Ashworth A. Differential effects of stress stimuli on a JNK-inactivating phosphatase. Oncogene 2002, 21: 2387–2397.
29. Merienne K, Helmlinger D, Perkin G R, Devys D and Trottier Y. Polyglutamine expansion induces a protein-damaging stress connecting heat shock protein 70 to the JNK pathway. J Biol Chem 2003, 278: 16957–16967.
30. Mosser D D, Caron A W, Bourget L, Meriin A B, Sherman M Y, Morimoto R I and Massie B. The chaperone function of hsp70 is required for protection against stress-induced apoptosis. Mol Cell Biol 2000, 20: 7146–7159.
31. Park H-S, Cho S-G, Kim C K, Hwang H S, Noh K T, Kim M-S, Huh S-H, Kim M J, Ryoo K, Kim E K, Kang W J, Lee J-S, Seo J-S, Ko Y-G, Kim S and Choi E-J. Heat shock protein Hsp72 is a negative regulator of apoptosis signal-regulating kinase 1. Mol Cell Biol 2002, 22: 7721–7730.
32. Yaglom J A, Ekhterae D, Gabai V L and Sherman M Y. Regulation of necrosis of H9C2 myogenic cells upon transient energy deprivation: rapid de-energization of mitochondria precedes necrosis and is controlled by reactive oxygen species, stress-kinase JNK, HSP72 and ARC. J Biol Chem 2003, 278: 50483–50496.

33. Song J, Takeda M and Morimoto R I. Bag1-Hsp70 mediates a physiological stress signalling pathway that regulates Raf-1/ERK and cell growth. Nat Cell Biol 2001, 3: 276–282.

34. Bimston D, Song J, Winchester D, Takayama S, Reed J C and Morimoto R I. BAG-1, a negative regulator of Hsp70 chaperone activity, uncouples nucleotide hydrolysis from substrate release. EMBO J 1998, 17: 6871–6878.

35. Park H S, Lee J S, Huh S H, Seo J S and Choi E J. Hsp72 functions as a natural inhibitory protein of c-Jun N-terminal kinase. EMBO J 2001, 20: 446–456.

36. Deng Y, Ren X, Yang L, Lin Y and Wu X. A JNK-dependent pathway is required for TNFα-induced apoptosis. Cell 2003, 115: 61–70.

37. Beere H M, Wolf B B, Cain K, Mosser D D, Mahboubi A, Kuwana T, Tailor P, Morimoto R I, Cohen G M and Green D R. Heat-shock protein 70 inhibits apoptosis by preventing recruitment of procaspase-9 to the Apaf-1 apoptosome. Nat Cell Biol 2000, 2: 469–475.

38. Saleh A, Srinivasula S, Balkir L, Robbins P and Alnemri E. Negative regulation of the Apaf-1 apoptosome by Hsp70. Nat Cell Biol 2000, 2: 476–483.

39. Proskuryakov S Y, Konoplyannikov A G and Gabai V L. Necrosis: a specific form of programmed cell death? Exp Cell Res 2003, 283: 1–16.

40. Ma X L, Kumar S, Gao F, Louden C S, Lopez B L, Christopher T A, Wang C, Lee J C, Feuerstein G Z and Yue T L. Inhibition of p38 mitogen-activated protein kinase decreases cardiomyocyte apoptosis and improves cardiac function after myocardial ischemia and reperfusion. Circulation 1999, 99: 1685–1691.

41. Mackay K and Mochly-Rosen D. An inhibitor of p38 mitogen-activated protein kinase protects neonatal cardiac myocytes from ischemia. J Biol Chem 1999, 274: 6272–6279.

42. Gabai V L, Meriin A B, Yaglom J A, Wei J Y, Mosser D D and Sherman M Y. Suppression of stress kinase JNK is involved in Hsp72-mediated protection of myogenic cells from transient energy deprivation. Hsp72 alleviates the stress-induced inhibition of JNK dephosphorylation. J Biol Chem 2000, 275: 38088–38094.

43. Uma S, Thulasiraman V and Matts R L. Dual role for Hsc70 in the biogenesis and regulation of the heme-regulated kinase of the alpha subunit of eukaryotic translation initiation factor 2. Mol Cell Biol 1999, 19: 5861–5871.

44. Thulasiraman V, Xu Z, Uma S, Gu Y, Chen J J and Matts R L. Evidence that Hsc70 negatively modulates the activation of the heme-regulated eIF-2 alpha kinase in rabbit reticulocyte lysate. Eur J Biochem 1998, 255: 552–562.

45. Chen J J and London I M. Regulation of protein synthesis by heme-regulated eIF-2 alpha kinase. Trends Biochem Sci 1995, 20: 105–108.

46. Hinnebusch A G. The eIF-2 alpha kinases: regulators of protein synthesis in starvation and stress. Semin Cell Biol 1994, 5: 417–426.

47. Matts R L, Xu Z, Pal J K and Chen J J. Interactions of the heme-regulated eIF-2 alpha kinase with heat shock proteins in rabbit reticulocyte lysates. J Biol Chem 1992, 267: 18160–18167.

48. Matts R L, Hurst R and Xu Z. Denatured proteins inhibit translation in hemin-supplemented rabbit reticulocyte lysate by inducing the activation of the heme-regulated eIF-2 alpha kinase. Biochemistry 1993, 32: 7323–7328.

49. Williams B R. PKR; a sentinel kinase for cellular stress. Oncogene 1999, 18: 6112–6120.

50. Brostrom C O, Prostko C R, Kaufman R J and Brostrom M A. Inhibition of translational initiation by activators of the glucose-regulated stress protein and heat shock protein stress response systems. Role of the interferon-inducible double-stranded RNA-activated eukaryotic initiation factor 2 alpha kinase. J Biol Chem 1996, 271: 24995–25002.

51. Tang N M, Ho C Y and Katze M G. The 58-kDa cellular inhibitor of the double stranded RNA-dependent protein kinase requires the tetratricopeptide repeat 6 and DnaJ motifs to stimulate protein synthesis *in vivo*. J Biol Chem 1996, 271: 28660–28666.

52. Melville M W, Tan S L, Wambach M, Song J, Morimoto R I and Katze M G. The cellular inhibitor of the PKR protein kinase, P58(IPK), is an influenza virus-activated co-chaperone that modulates heat shock protein 70 activity. J Biol Chem 1999, 274: 3797–3803.

53. Melville M W, Hansen W J, Freeman B C, Welch W J and Katze M G. The molecular chaperone hsp40 regulates the activity of P58IPK, the cellular inhibitor of PKR. Proc Nat Acad Sci USA 1997, 94: 97–102.

54. Melville M W, Katze M G and Tan S L. P58IPK, a novel co-chaperone containing tetratricopeptide repeats and a J-domain with oncogenic potential. Cell Mol Life Sci 2000, 57: 311–322.

55. van Huizen R, Martindale J L, Gorospe M and Holbrook N J. P58IPK, a novel endoplasmic reticulum stress-inducible protein and potential negative regulator of eIF2 alpha signaling. J Biol Chem 2003, 278: 15558–15564.

56. Yan W, Frank C L, Korth M J, Sopher B L, Novoa I, Ron D and Katze M G. Control of PERK eIF2 alpha kinase activity by the endoplasmic reticulum stress-induced molecular chaperone P58IPK. Proc Nat Acad Sci USA 2002, 99: 15920–15925.

57. Ciocca D R, Clark G M, Tandon A K, Fuqua S A, Welch W J and McGuire W L. Heat shock protein hsp70 in patients with axillary lymph node-negative breast cancer: prognostic implications. J Natl Cancer Inst 1993, 85: 570–574.

58. Nanbu K, Konishi I, Mandai M, Kuroda H, Hamid A A, Komatsu T and Mori T. Prognostic significance of heat shock proteins Hsp70 and Hsp90 In endometrial carcinomas. Cancer Detect Prevent 1998, 22: 549–555.

59. Costa M J M, Rosas S L B, Chindano A, Lima P D S, Madi K and Carvalho M D D. Expression of heat shock protein 70 and P53 in human lung cancer. Oncol Rep 1997, 4: 1113–1116.

60. Vargasroig L M, Fanelli M A, Lopez L A, Gago F E, Tello O, Aznar J Z and Ciocca D R. Heat shock proteins and cell proliferation in human breast cancer biopsy samples. Cancer Detect Prevent 1997, 21: 441–451.

61. Hwang T S, Han H S, Choi H K, Lee Y J, Kim Y-J, Han M-Y and Park Y-M. Differential, stage-dependent expression of Hsp70, Hsp110 and Bcl-2 in colorectal cancer. J Gastroenterol Hepatol 2003, 18: 690–700.

62. Volm M, Koomagi R, Mattern J and Efferth T. Protein expression profile of primary human squamous cell lung carcinomas indicative of the incidence of metastases. Clin Exp Metastasis 2002, 19: 385–390.

63. Prendergast G C. Mechanisms of apoptosis by c-Myc. Oncogene 1999, 18: 2967–2987.

64. Nilsson J A and Cleveland J L. Myc pathways provoking cell suicide and cancer. Oncogene 2003, 22: 9007–9021.

65. Pelengaris S, Khan M and Evan G I. Suppression of Myc-induced apoptosis in beta cells exposes multiple oncogenic properties of Myc and triggers carcinogenic progression. Cell 2002, 109: 321–334.

66. Pelengaris S, Khan M and Evan G. c-MYC: more than just a matter of life and death. Nat Rev Cancer 2002, 269: 764–776.

67. Volloch V Z and Sherman M Y. Oncogenic potential of Hsp72. Oncogene 1999, 18: 3648–3651.

68. Jaattela M. Over-expression of hsp70 confers tumorigenicity to mouse fibrosarcoma cells. Int J Cancer 1995, 60: 689–693.

69. Seo J S, Park Y M, Kim J I, Shim E H, Kim C W, Jang J J, Kim S H and Lee W H. T cell lymphoma in transgenic mice expressing the human Hsp70 gene. Biochem Biophys Res Commun 1996, 218: 582–587.

70. Nylandsted J, Rohde M, Brand K, Bastholm L, Elling F and Jaattela M. Selective depletion of heat shock protein 70 (Hsp70) activates a tumor-specific death program that is independent of caspases and bypasses Bcl-2. Proc Natl Acad Sci USA 2000, 97: 7871–7876.

71. Nylandsted J, Wick W, Hirt U A, Brand K, Rohde M, Leist M, Weller M and Jaattela M. Eradication of glioblastoma, and breast and colon carcinoma xenografts by Hsp70 depletion. Cancer Res 2002, 62: 7139–7142.

72. Yaglom J, O'Callaghan-Sunol C, Gabai V and Sherman M Y. Inactivation of dual-specificity phosphatases is involved in the regulation of extracellular signal-regulated kinases by heat shock and Hsp72. Mol Cell Biol 2003, 23: 3813–3824.

73. Dhillon A S, Meikle S, Yazici Z, Eulitz M and Kolch W. Regulation of Raf-1 activation and signalling by dephosphorylation. EMBO J 2002, 21: 64–71.

74. Nylandsted J, Gyrd-Hansen M, Danielewicz A, Fehrenbacher N, Lademann U, Hoyer-Hansen M, Weber E, Multhoff G, Rohde M and Jaattela M. Heat shock protein 70 promotes cell survival by inhibiting lysosomal membrane permeabilization. J Exp Med 2004, 200:425-435.

75. Steel R, Doherty J P, Buzzard K, Clemons N, Hawkins C J and Anderson R L. Hsp72 inhibits apoptosis upstream of the mitochondria and not through interactions with Apaf-1. J Biol Chem 2004, 279:51490–51490.

10

Heat Shock Proteins, Their Cell Surface Receptors and Effects on the Immune System

Thomas Lehner, Yufei Wang, Trevor Whittall
and Lesley A. Bergmeier

10.1. Introduction

Heat shock or stress proteins are important intracellular protein chaperones that control their trafficking. The function of heat shock proteins in the immunopathology of infections, tumours and autoimmune diseases has been the subject of numerous experimental and clinical investigations over the past few decades and some of their properties are summarised in Table 10.1. Because there is extensive homology between mammalian and microbial heat shock proteins, immunological cross-reactions were considered to account for a number of autoimmune diseases. However, although the biological significance of lipopolysaccharide (LPS) found in Gram-negative bacteria has been well appreciated, heat shock proteins, which are found more widely in Gram-negative and -positive bacteria, especially those in the gut, have received more limited attention. A relatively new phase in this area of biology was initiated by the discovery of specific heat shock protein receptors and by rapid advances in our understanding of the signalling pathways. This chapter will deal with the receptors used by heat shock proteins, with particular reference to Hsp70, involvement of these proteins in the stimulation of chemokine production, maturation of dendritic cells (DCs), their intrinsic adjuvanticity and capacity to enhance immunogenicity.

10.2. Structural features of Hsp70

Although the overall three-dimensional structure of Hsp70 is not known, the structures of the two domains from various members of the family have been solved separately. The crystal structures of the ATPase domains of bovine Hsc70 (heat shock constitutive protein) and human Hsp70 have been determined [1, 2]. The domain consists of two approximately equal-sized lobes with a deep

Table 10.1. Properties of Heat Shock Proteins

1. Heat shock proteins are intracellular chaperones binding unfolded polypeptides to prevent misfolding and aggregation.
2. They bind peptides with a hydrophobic motif by non-covalent linkage.
3. Hsp70 and Hsp90 deliver exogenous antigen into the MHC class I as well as class II pathways, playing an important role in cross-priming free or antigen released from apoptotic cells.
4. Heat shock proteins stimulate production of the CC-chemokines, CCL3, CCL4 and CCL5 and cytokines, especially TNF-α, IL-12 and NO.
5. Heat shock proteins stimulate maturation of DCs in a similar way to that of CD40L.
6. They exert robust adjuvant function when linked to antigens and they are effective when administered systemically or by the mucosal route.
7. Generation of IL-12 and TNF-α by Hsp70 induces a Th1-α polarised adjuvant function.
8. Peptide epitopes within Hsp70 exert diverse immunomodulating functions.
9. Hsp70 can function as an alternative ligand to CD4$^+$ T cells, in activating the CD40–CD40L (CD154) co-stimulatory pathway, thereby enhancing immunity.
10. Tumour- or virus-specific peptides, non-covalently bound to Hsp70 or gp96, elicit CTL responses and exert protection against the specific tumours or viruses.
11. Heat shock proteins elicit innate immunity which may drive adaptive immune responses.

cleft between them. ATP binds at the base of the cleft. Two crystal forms of the human ATPase fragment which differ by a shift of 1–2 Å in one of the sub-domains have been obtained. This shift might be important in ATP binding and ADP release, and its presence indicates some degree of flexibility in this domain.

The crystal structure of the substrate-binding domain of DnaK from *Escherichia coli* with bound substrate (a seven-residue peptide with the sequence NRLLLTG) has been determined [3], and this consists of a β-sandwich sub-domain followed by an α-helical sub-domain. The β-sandwich sub-domain is formed by two stacked anti-parallel four-stranded β-sheets. The upper sheet forms the substrate binding site with loops L1,2 and L3,4 (between β-strands 1 and 2 and between β-strands 3 and 4, respectively), forming the sides of a channel that is the primary site of interaction with substrate. In the outer loop, L4,5 stabilises L1,2 by hydrogen bonds and hydrophobic interaction while L5,6 forms hydrogen bonds that stabilise L3,4. The α-helical sub-domain comprises five helices with the first and second helices (αA and αB) forming hydrophobic side-chain contacts with the β-sandwich. Helix αB extends over the entire substrate binding site and may stabilise substrate binding by interacting with all four loops that form this site; however, it does not interact directly with substrate.

The peptide substrate (NRLLLTG) complexed to DnaK adopts an extended conformation in which main-chain atoms form hydrogen bonds with DnaK, whereas side-chain contacts are predominantly hydrophobic [3]. The central

Table 10.2. Major receptors interacting with heat shock proteins and stimulating cytokine and chemokine production, and DC maturation

Receptor	Heat shock protein	Function	Reference
CD14	Human and chlamydial Hsp60 Human Hsp70	Stimulation of TNF-α, IL-12 and IL-6	[4, 5]
CD40	Microbial and human Hsp70	Stimulation of chemokines, TNF-α, IL-12 DC Maturation	[6–11]
CD91	Human Hsp70, Hsp90, gp96 and calreticulin (α2M)	Stimulation of TNF α, IL-12 and IL-1 β	[12, 13]
TLRs	Human Hsp60, Hsp70 and Hsp90	Stimulation of TNF-α and IL-12 DC maturation	[14–17]
LOX-1	Human Hsp70(LDL)	Scavenger receptor	[18, 19]

residue (Leu 4) is buried in a relatively large hydrophobic pocket of DnaK and, together with Leu 3, contributes most of the contacts with the protein. Binding is almost completely determined by a five-residue core (RLLLT) which is centred on Leu 4. The specificity of binding is principally determined by an interaction of the residue at the centre of the core sequence with the hydrophobic pocket of DnaK and it has been suggested that Ile, Met, Thr, Ser, and possibly Phe could be accommodated as alternatives to Leu. Hydrophobic residues are preferred at the positions adjacent to the central residues and, although there are fewer constraints on residues at the ends of the motif, negative charges are generally excluded. However, some differences in substrate specificity between members of the Hsp70 family have been described and these might reflect functional differences.

10.3. Heat shock protein receptors and co-receptors

A number of receptors that bind different heat shock protein family members have been identified and these are shown in Table 10.2. Despite their high degree of conservation, there is a diversity of receptor usage, in that different members of the heat shock protein family may use the same receptor whereas others may use different receptors. Interactions between heat shock proteins and their receptors might elicit two different, but related, functions: (a) non-specific stimulation of antigen-presenting cells generates production of chemokines [20] and cytokines [7, 21] and (b) internalisation of heat shock protein–peptide complexes by endocytosis and the translocation of heat shock proteins into the human leukocyte antigen (HLA) class I or II pathway [22, 23].

10.3.1. CD14

CD14 is a glycosylphosphatidylinositol-anchored protein expressed on the cell surface of monocytes and macrophages and, to a lesser extent, on other myeloid cells. It is a receptor for LPS and serum LPS-binding protein [24], in association with TLR molecules [25]. Because CD14 lacks a transmembrane domain, signalling is dependent on the associated molecules in multi-receptor complexes [26, 27].

CD14 was characterised as a receptor for human (hu) and chlamydial Hsp60 by Kol and colleagues in 2000 [4]. They demonstrated that Hsp60 treatment of peripheral blood mononuclear cells resulted in the production of IL-6 and activation of p38 mitogen-activated protein kinase (MAPK). This could be inhibited by antibody to CD14. However, transfection of CD14 into CHO cells resulted in these cells acquiring the ability to respond to LPS, but not to Hsp60. They suggested that CD14 is necessary, but not sufficient, for cellular responsiveness to Hsp60, and that other molecules are required, possibly for transduction of the activation signal. The possibility has been raised that inducible forms of human Hsp70 may also stimulate human monocytes to produce TNF-α, IL-1β and IL-6 [5]. This involves TLR2 and TLR4 and calcium-dependent activation of MyD88, IRAK and NF-κB [28]. In apparent contradiction, it has been reported that CD14$^-$ DCs are capable of internalising Hsp70 [29], although it is not clear that internalisation and stimulation are necessarily mediated by the same receptors. The reader is referred to Chapter 7 for more details on CD14–TLR–Hsp interactions.

10.3.2. CD40

CD40 is a 40–50-kDa glycoprotein, a member of the tumour necrosis factor (TNF) receptor superfamily, and is primarily expressed on B lymphocytes, monocytes and DCs [30]. CD40 can also be found on epithelial cells, some cancer cells and activated CD8$^+$ T cells [31, 32]. CD40 plays an important role in T cell–mediated immune responses. It is crucial for T cell–dependent B cell activation, differentiation and immunoglobulin class switching and germinal centre formation [30]. CD40 is also involved in the activation of antigen-presenting cells and mediates DC maturation. It induces CD8$^+$ cytotoxic T lymphocytes and the generation of memory CD8$^+$ T cells [32–34]. The natural ligand for CD40 is CD40 ligand (CD40L, CD154), and this is expressed by activated T cells.

We have reported that CD40 is a receptor for microbial Hsp70 (mHsp70) [6] and this was later confirmed and extended to include human Hsp70 [9].

We have put forward the hypothesis that the adjuvanticity of mHsp70 and mHsp65 is accounted for by their capacity to stimulate production of the CC-chemokines CCL3, CCL4 and CCL5, all of which attract the entire repertoire of immune cells [20]. Because both of the major co-stimulatory pathways, CD80/86–CD28 and CD40–CD40L, stimulate these CC-chemokines [35–37], we explored the possibility that heat shock proteins might interact with one of the co-stimulatory molecules [6]. Whereas antibodies to CD80 or CD86 had no effect, those to CD40 blocked Hsp70 stimulation of CC-chemokine production. Further in-depth investigations using HEK293 cells (human embryonic kidney cell lines) revealed that Hsp70 stimulated the production of CC-chemokines only if cells were transfected with human CD40, but not with control molecules. Immunoprecipitation studies revealed that Hsp70 physically associates with cell membrane CD40 when incubated with CD40-expressing cells, and surface plasmon resonance showed that Hsp70 can directly bind to CD40 molecules [6]. Hsp70–peptide complexes binding to CD40 deliver the peptide into the MHC class I pathway and this process is dependent on the ADP-loaded state of Hsp70 [9].

CD40 also mediates Hsp70 stimulation of monocytes and bone marrow-derived DCs, which are the principle antigen-presenting cells in priming $CD4^+$ and $CD8^+$ T cells responses. Treatment of human monocyte-derived immature DCs for two days with mHsp70 induces dramatic changes in the expression of phenotypes, including an increase in the expression of major histocompatibility complex (MHC) class II molecules, the co-stimulatory molecules CD80, CD86 and the CD83, CCR7 maturation markers [7]. The CC-chemokines and Th1-polarising cytokines – IL-12 and TNF-α – are also produced. The C-terminal portion of the molecule (Hsp70 359–610) is a more potent inducer of cytokine expression and DC maturation than the full-length Hsp70 molecule, whereas the N-terminal ATPase domain of Hsp70 exhibits no such biological effects. Indeed, there is evidence that the preceding functions can be induced by the peptide binding domain of mHsp70 (aa 359–494), and a stimulatory epitope (aa 407–426) has been identified [8]. Human and microbial Hsp70 bind the CD40 receptor; however, surprisingly, the human ATPase domain of Hsp70 binds to one site, whereas a microbial C-terminal domain binds another site of the CD40 molecule [6, 9]. It is not clear whether CD40L shares one of the two receptor sites on CD40 or whether it binds yet another site. Indeed, two other functional domains have been identified in the cytoplasmic tail of CD40; one is involved in the induction of extra-follicular B cells and another is required for germinal centre formation [38].

10.3.3. Toll-like receptors (TLRs)

Toll-like receptors (TLRs) are examples of pattern-recognition receptors, are expressed by the innate immune system and recognise specific pathogen-associated molecular patterns (PAMPs) expressed on microbial components [39]. To date, about 10 TLRs have been described within the TLR family and each receptor appears to recognise different microbial pathogenic elements. TLRs are primarily expressed in those cell types that are involved in the first line of defence, such as DCs, monocytes, neutrophils, epithelial cells and endothelial cells. Activation of TLRs leads to production of inflammatory cytokines, chemokines, nitric oxide (NO), complement proteins, enzymes (such as cyclooxygenase-2), adhesion molecules and immune receptors [40]. These innate immune responses are essential for the elimination of pathogens and the regulation of adaptive immunity.

TLRs are primary candidates as receptors for heat shock proteins, because these proteins are highly conserved among microbial organisms and may serve as PAMPs to activate the innate immune system by interacting with pattern recognition receptors. However, only human Hsp60 and Hsp70, but not microbial Hsp, have so far been found to stimulate TLRs [14, 15]. Bone marrow–derived macrophages from the mouse strain C3H/HeJ, which carry a mutant TLR4, do not respond to Hsp60 [14].

It is, however, not clear whether TLRs act as receptors for heat shock proteins or are involved in signalling cellular activation. Some studies suggest that cell surface TLR4 is essential for human Hsp60 activation of monocytes [14], whereas others indicate that endocytosis of Hsp60 or Hsp70 is a prerequisite for activation of the intracellular TLR2 and TLR4 signalling pathways and that the endocytosis process is independent of these receptors [16, 17]. There is no evidence to indicate that there is a direct interaction between Hsp and TLRs. However, the possibility that human heat shock proteins can activate TLRs suggests that these not only serve as receptors for PAMPs derived from pathogenic microbes but that they also recognise endogenous ligands. Endogenous Hsp are present predominantly in the cell cytoplasm and can be induced and released in pathological conditions, such as injury, necrosis and stress. The observation that intracellular TLR2 and TLR4 may interact with internalised bacteria [41] is consistent with interaction of internalised heat shock proteins with TLR or other receptors in phagosomes. Indeed, small Hsp70 fragments or peptides have potent functional activity [7, 8]. Induction of local inflammatory responses by TLRs and endogenous heat shock protein interactions in the milieu around damaged tissues is probably helpful for tissue repair and wound healing, but it can also play an important part in chronic inflammatory diseases. Chapter 7 should be read in the context of this paragraph because Vabulas and Wagner take a different line with regard to the importance of TLRs in the response of cells to chaperones.

10.3.4. CD91

CD91 is a receptor for $\alpha2$ macroglobulin (α2M) which is a protease inhibitor that binds to microbial pathogens and mediates phagocytosis by monocytes [42] but is poorly expressed on DCs. Several members of the heat shock protein families, including gp96, Hsp90, Hsp70 and calreticulin, despite being structurally distinct, interact with CD91 and are internalised by the receptor and the heat shock protein–bound peptides. This interaction is critical to the induction of CD8$^+$ cytotoxic T lymphocytes by heat shock protein–mediated cross-priming mechanisms [13, 43]. The heat shock protein gp96 binds directly to CD91, and this can be inhibited either by antibodies to CD91 or α2M [13]. It is not clear whether CD91 engagement by heat shock proteins activates antigen-presenting cells to produce cytokines, express co-stimulatory molecules or induces DC maturation [44].

10.3.5. LOX-1

LOX-1 is a scavenger receptor which may bind human heat shock proteins, especially Hsp70 [19]. It is found on endothelial cells, monocytes, immature DCs and smooth muscle cells [18, 19]. LOX-1 is a cell surface glycoprotein which binds modified lipoproteins and modified LDL (Ox-LDL), apoptotic cells and bacterially derived cell wall components [45, 46]. LOX-1 induces Hsp70-mediated cross-priming of CD8$^+$ cytotoxic T lymphocytes [19]. However, it is not clear whether LOX-1 engagement by heat shock proteins activates antigen-presenting cells to produce cytokines, express co-stimulatory molecules or induce DC maturation.

10.4. LPS contamination of heat shock protein preparations

Heat shock proteins stimulate innate immune cells to produce inflammatory cytokines, including TNF-α, IL-1β, IL-12, GM-CSF, NO and chemokines, such as MIP-1α, MIP-1β, RANTES, MCP-1 and MCP-2. These activities of heat shock proteins are similar to those of LPS which can contaminate heat shock protein preparations, and it is essential to exclude the possibility that the observed effects are elicited by contaminating LPS. Currently it is difficult to prepare LPS-free heat shock protein preparations, especially those expressed in *E. coli*. LPS activity is abrogated or greatly reduced by polymixin B treatment [5, 6], whereas heat shock protein stimulation is reduced by heat denaturation [5, 47]. Other reagents, such as RSLP derived from *Rhodopseudomonas spheroids* and lipid IVa, also inhibit LPS activity and have no effect on heat shock protein stimulation [5, 47]. As the stimulating activity of Hsp70 is calcium-dependent, the intracellular calcium chelator 1,2-bis(2-aminopheroxy)ethane-N,N,N, 'N'-tetraacetic acid-acetoxymethylester (BAPTA-AM) has been used to differentiate between the functions of LPS and Hsp70 [5, 6, 48]. Furthermore, treatment with proteinase K also inhibits Hsp70, but not LPS-stimulating activity [48]. The C3H/HeJ and C57BL/10ScCr inbred mouse strains are homozygous for a mutant *Lps* allele ($Lps^{d/d}$) which confers hyporesponsiveness to LPS challenge [49] and provides a model to study immunological functions of heat shock proteins. Genetic analysis revealed that C3H/HeJ mice have a point mutation within the coding region of the TLR4 gene, whereas C57BL/10ScCr mice exhibit a deletion of TLR4 [50]. In studies with monocytes and DCs using microbial Hsp70, inhibition with antibodies to CD40, but not to CD14, should discriminate between heat shock protein and LPS [6, 7].

Most of the LPS contamination (95–99%) in heat shock protein preparations can be removed using polymixin B immobilised on agarose affinity column. Polymixin B is a cationic cyclopeptide that can neutralise the biological activity of LPS by binding to its lipid A portion. It is noteworthy that heat shock proteins contain a hydrophobic domain that can interact with the affinity column and result in protein loss. This may be enhanced by heat shock protein binding to LPS on the column [27, 51]. Our experience with this method indicates that 60–70% of proteins can be recovered after one treatment with polymixin B and there is further protein loss if a second treatment is required. The residual LPS concentration in recovered heat shock protein preparations is typically less than 5 U per mg protein, as determined by the Limulus amoebocyte lysate assay [7]. Although the significance of this low level of LPS in Hsp70 preparations depends on the function under investigation, such low levels of LPS do not stimulate production of cytokines, CC-chemokines or maturation of DCs [6, 7]. The removal of LPS by washing columns of recombinant proteins with polymyxin B is described in Chapter 6. This overcomes the losses of protein described.

10.5. Signalling pathways

Engagement of heat shock proteins with receptors, such as CD40 and TLRs, might induce intracellular signalling which varies with the level of cell surface expression of the receptors on DCs and monocytes [52]. CD40 engagement with its natural ligand CD154 forms trimeric clusters and recruits adaptor proteins known as TNF receptor-associated factors (TRAFs) to the cytoplasmic tail [53]. CD40 has two cytoplasmic domains for binding TRAF. The TRAF6 binding site is within the membrane proximal cytoplasmic (Cmp) region, whereas TRAF2/3/5 is in the membrane distal cytoplasmic region (Cmd). See Chapter 7 for a more detailed discussion of the role of TRAFs in Hsp70-induced cell signalling cascades.

Binding of TRAF2, 3 and 5 results in formation of a signalling complex which includes multiple kinases, such as the MAPK family, NF-κB inducing kinase or receptor interacting protein (reviewed in [54]). Human Hsp70 uses the CD40 receptor and activates p38 MAPK [9]. However, another report has suggested that huHsp70 activates CD14 [28], as does huHsp60, with TLR2 and TLR4 receptors signalling via myeloid differentiation protein 88 and IL-1 receptor associated kinase (IRAK) [14, 15, 17, 55]. The activated IRAK associates with TRAF6 and activates p38 MAPK or NF-κB [28].

10.6. Interface between immunity and tolerance

The finding that CD40 is a receptor for mycobacterial Hsp70 (mHsp70) [6] and huHsp70 [9] might have profound implications for the development of acquired immune responses at the interface between adaptive immunity and tolerance. The interaction of CD40 with its ligand CD154 (or Hsp70) plays an important role in the development of the quality and magnitude of humoral and cellular immunity. Interference in the interaction between CD40 and CD154 can induce allogeneic and xenogeneic graft tolerance (reviewed in [54]). Indeed, there is evidence that CD154 blockade of immature DCs with specific antibodies might induce a state of antigen-specific tolerance [56].

Exposure to necrotic cells can induce maturation of DCs and immune stimulation via the release of antigens and heat shock proteins which are preferentially taken up by DCs [12, 57–60]. However, a state of tolerance might be induced by DCs that have captured apoptotic cells [61]. These findings and the cytokine stimulating activity of heat shock proteins suggest that heat shock proteins might play a role in the interface between immunity and tolerance. The state of maturation of DCs might be an important determining factor in the induction of immunity or tolerance. However, DCs within lymphoid tissue are able to form MHC–peptide complexes in the absence of maturation signals. On recognition of their ligands, naïve T cells divide and might undergo clonal deletion, resulting in a state of tolerance [62, 63]. However, if maturation signals are co-administrated with antigen, an immune response develops. DCs in the absence of maturation signals are also able to down-modulate established immune responses [64–67], probably by inducing regulatory T cells [62, 63].

The capacity of CD40 signalling to induce DC maturation has been demonstrated in a transgenic CD40$^{-/-}$ DC model in which TNF-α fails to rescue the immune deficiency, despite its ability to induce maturation of DCs [68]. CD40 ligation of lymphoid DCs might abrogate their tolerogenic activity to a normally tolerogenic peptide [69]. Indeed, Hsp70 has been shown to stimulate CD40 and convert T cell tolerance, in the presence of lymphocytic choriomeningitis virus (LCMV) peptide, into autoimmune diabetes [10]. Another study has also implicated an inducible form of Hsp70 in the presentation of a major autoantigen in multiple sclerosis [70]. Steinmann and Nussenzweig [71] suggested that chronic infection might result from tolerance being exploited by a pathogen, as when persistent micro-organisms are taken up by DCs that fail to mature. The DC might then induce tolerance either by deleting T cells or by inducing T regulatory cells. They suggested that this might occur in HIV infection, when a large amount of virus is produced and steady-state DCs express receptors for the virus, such as dendritic cell-specific ICAM-3 grabbing non-integrin (DC-SIGN).

10.7. Interface between innate and adaptive immunity

Utilisation of the co-stimulatory CD40 molecule as a receptor for Hsp70 raises the paradigm that Hsp70 might function at the interface between innate and adaptive immunity. The B7 (CD80/CD86) and CD28 co-stimulatory interaction plays a central role in providing the second signal necessary for the adaptive immune function between HLA peptide and TCR [72]. The HLA-bound peptide and co-stimulatory molecule CD40 on an antigen-presenting cell interact with the corresponding TCR and CD40L on T cells, respectively, to elicit an effective immune response. It is of interest to note that ligation of CD40 by CD40L or Hsp70 and of CD80/86 by CD28 elicits CC-chemokines [35–37] and this suggests the presence of a non-cognate immune response which is responsible for attracting the immunological repertoire of cells (monocytes, immature DC, T and B cells). The interaction between CD40 and CD40L or Hsp70 also elicits production of some cytokines (such as IL-12 and TNF-α) and may induce Th1 polarisation of the immune response [7]. Hsp70 may thus function as an alternative ligand to CD40L, stimulating the major co-stimulatory pathway CD40–CD40L. Hsp70 functions as a multi-purpose molecule by acting as a carrier of antigens, inducing immune cells to the Hsp70 antigen site and by eliciting the maturation of DCs. There is also the possibility that Hsp70-bound antigen is processed by antigen-presenting cells and chaperoned by Hsp70 into the MHC presentation pathway for recognition by T cells [73].

10.8. The role of heat shock proteins in innate and adaptive immunity and vaccination

An essential component of vaccines against infections and tumours is an adjuvant activity which can elicit innate immune responses that can drive the development of specific adaptive immunity. Microbial Hsp70 and Hsp65 have been used as carrier molecules or adjuvants to enhance systemic immune responses when covalently linked to synthetic peptides [74–76]. Indeed, these and gp96 can be fused, covalently linked or loaded with peptides to elicit specific immunity to tumours or viruses [77–80]. The adjuvanticity of microbial Hsp70 and Hsp65 has been demonstrated not only by systemic but also by mucosal immunisation in non-human primates [20]. Both systemic and mucosal adjuvanticity are dependent on stimulating the production of three CC-chemokines – CCL3, 4 and 5 (or MIP-1α and MIP-1β and RANTES). CCL5 is a potent chemoattractant for monocytes, CD4$^+$ T cells and activated CD8$^+$ T cells [81–84]. CCL3 and CCL4 attract CD4$^+$ T and B cells [85] and all three chemokines attract immature DCs [86]. Monocytes and DCs internalise antigens that are processed and presented on the cell surface. DCs then undergo maturation and migrate to the regional lymph nodes, in which they present the processed antigen to T and B cells and elicit cell-mediated and humoral immune responses.

In addition to the CC-chemokines, IL-12, TNF-α and NO are also elicited by Hsp70 or, more efficiently, by its C-terminal portion [7]. Because IL-12 is one of the most potent cytokines for inducing Th1 polarisation [87] this might be responsible for the Th1-polarised adjuvanticity. The C-terminal portion of Hsp70-linked peptide elicits higher serum IgG$_{2a}$ and IgG$_3$ subclasses of antibodies than the native Hsp70-bound peptide, which is consistent with a Th1-polarising activity [7]. Furthermore, the Th2-type cytokine (IL-4) was not produced in immunised macaques. Thus, the C-terminal portion might be used as a microbial adjuvant that attracts the entire immunological repertoire of cells by virtue of stimulating the production of CC-chemokines and eliciting a Th1 response by generating IL-12.

The presence of Hsp70 and Hsp65 in most microorganisms [88, 89] and their capacity to generate CC-chemokines raises the possibility that the well-recognised immunogenicity of whole organisms, as compared with a subunit antigen, is mediated by CC-chemokines generated by heat shock proteins [20]. This is consistent with the principle that the innate immune system might drive adaptive immunity [90, 91]. Heat shock proteins in micro-organisms might function as a natural adjuvant generating CC-chemokines and cytokines. This concept is also consistent with the 'danger hypothesis' of infection [92], with the heat shock protein inducing the innate system to secrete CC-chemokines and mobilise the repertoire of cells required to generate specific immune responses against the invading organism.

The significance of Hsp70 as an alternative ligand to CD40L [6] stimulating the major co-stimulatory pathway CD40–CD40L has been highlighted. In mice lacking CD40 (CD40 knockout mice, CD40$^{-/-}$) the production of IL-12 by bone marrow–derived DCs was substantially reduced following Hsp70 stimulation [9, 10]. In these CD40$^{-/-}$ mice, Hsp70 failed to enhance DC function or to prime CD4$^+$ and CD8$^+$ T cell antigen-specific responses, and protection from *Mycobacterium tuberculosis* infection depended on the alternative Hsp70–CD40 co-stimulatory pathway [11]. An intriguing report that over-expression of Hsp70 in *M. tuberculosis* reduces the level of infection observed during the chronic phase [93] might also be interpreted as reflecting an enhanced immunity to the organism resulting from the interaction between increased amounts of Hsp70 with CD40 expressed on macrophages and DCs. In another report, co-administration of the tolerogenic LCMV peptide with human Hsp70 might reverse tolerance and promote the induction of autoimmune diabetes by DCs [10].

The application of Hsp70 as a carrier of HIV gp120 and peptides derived from CCR5 in mucosal vaccination has been recently demonstrated in rhesus macaques [94]. Significant protection against SHIV 89.6P has been associated with the induction of specific serum and secretory antibodies, IL-2 and IFN-γ stimulated by the vaccine components, and a raised concentration of CC-chemokines which was inversely correlated with the proportion of CCR5$^+$ cells [94].

CD8$^+$ cytotoxic T lymphocytes (CTLs) can be generated by loading LCMV peptides onto human Hsp70, and this has been shown to elicit protective antiviral immunity in mice [80]. Human anti-influenza CTLs have been generated by pulsing DCs with mHsp70 loaded with peptides from influenza virus; the resulting CTL response is significantly greater than that induced by pulsing DCs with peptides alone [48]. There is also evidence that natural killer (NK) cells can be stimulated to proliferate by human Hsp70 and that this function resides in the C-terminal portion of Hsp70 [95]. Cell-surface-bound Hsp70 found on some tumour cells may induce migration of, and cytolysis by, CD56$^+$CD94$^+$ NK cells [96].

These investigators identified a peptide (aa 450–463) within the sequence of huHsp70 which enhances NK cell activity. A signal peptide derived from Hsp60 which binds HLA-E and interferes with CD94/NKG2A recognition, and enables NK cells to detect stressed cells, has also been identified [97]. Hsp70 and Hsp65 upregulate $\gamma\delta^+$ T cells, both *in vitro* and *in vivo* in non-human primates, and induce CD8-suppressor factors and CC-chemokines [98]. Indeed, a significant increase in $\gamma\delta^+$ T cells was found in rectal mucosal tissue and the draining lymph nodes in macaques immunised with SIVgp120 and p27 and protected from rectal mucosal challenge by SIV [98].

Acknowledgments

We wish to acknowledge the support received from the European Union (Grant No: LSHP-CT-2003-503240)

REFERENCES

1. Flaherty K M, Deluca-Flaherty C and McKay D B. Three-dimensional structure of the ATPse fragment of a 70K heat-shock cognate protein. Nature 1990, 346: 623–628.
2. Osipiuk J, Walsh M A, Freeman B C, Morimoto R I and Joachimiak A. Structure of a new crystal form of human hsp70 ATPse domain. D. Biol. Crystallogr. 1999, D55: 1105–1107.
3. Zhu X, Zhao X, Burkholder W F, Gragerov A, Ogata C M, Gottesman M E and Hendrickson W A. Structural analysis of substrate binding by the molecular chaperone DnaK. Science 1996, 272: 1606–1614.
4. Kol A, Lichtman A H, Finberg R W, Libby P and Kurt-Jones E A. Heat shock protein (HSP) 60 activates the innate immune response: CD14 is an essential receptor for HSP60 activation of mononuclear cells. J Immunol 2000, 164: 13–17.
5. Asea A, Kraeft S-K, Kurt-Jones E A, Stevenson M A, Chen L B, Finberg R W, Koo G C and Calderwood S K. Hsp70 stimulates cytokine production through a CD14-dependent pathway, demonstrating its dual role as a chaperone and cytokine. Nat Med 2000, 6: 435–442.
6. Wang Y, Kelly C G, Karttunen J T, Whittall T, Lehner P J, Duncan L, MacAry P, Younson J S, Singh M, Oehlmann W, Cheng G, Bergmeier L and Lehner T. CD40 is a cellular receptor mediating mycobacterial heat shock protein 70 stimulation of CC-chemokines. Immunity 2001, 15: 971–983.
7. Wang Y, Kelly C G, Singh M, McGowan E G, Carrara A S, Bergmeier L A and Lehner T. Stimulation of Th1-polarizing cytokines, C-C chemokines, maturation of dendritic cells, and adjuvant function by the peptide binding fragment of heat shock protein 70. J Immunol 2002, 169: 2422–2429.
8. Wang Y, Whittal T, McGowan E, Younson J, Kelly C, Bergmeier L A, Singh M, Lehner T. Identification of stimulating and inhibitory epitopes within the Hsp70 molecule which modulate cytokine production and maturation of dendritic cells. J Immunology 2005, 174: 3306–3316.
9. Becker T, Hartl F U and Wieland F. CD40, an extracellular receptor for binding and uptake of Hsp70-peptide complexes. J Cell Biol 2002, 158: 1277–1285.
10. Millar D G, Garza K M, Odermatt B, Elford A R, Ono N, Li Z and Ohashi P S. Hsp70 promotes antigen-presenting cell function and converts T-cell tolerance to autoimmunity in vivo. Nat Med 2003, 9: 1469–1476.

11. Lazarevic V, Myers A J, Scanga C A and Flynn J L. CD40, but not CD40L, is required for the optimal priming of T cells and control of aerosol M. tuberculosis infection. Immunity 2003, 19: 823–835.
12. Basu S, Binder R J, Suto R, Anderson K M and Srivastava P K. Necrotic but not apoptotic cell death releases heat shock proteins, which deliver a partial maturation signal to dendritic cells and activates the NF-κB pathway. Int Immunol 2000, 12: 1539–1546.
13. Binder R J, Han D K and Srivastava P K. CD91: a receptor for heat shock protein gp96. Nat Immunol 2000, 1: 151–155.
14. Ohashi K, Burkart V, Flohé S and Kolb H. Heat shock protein 60 is a putative endogenous ligand of the Toll-like receptor-4 complex. J Immunol 2000, 164: 558–561.
15. Vabulas R M, Ahmad-Nejad P, da Costa C, Miethke T, Kirschning C J, Hacker H and Wagner H. Endocytosed HSP60s use toll-like receptor 2 (TLR2) and TLR4 to activate the toll/interleukin-1 receptor signaling pathway in innate immune cells. J Biol Chem 2001, 276: 31332–31339.
16. Vabulas R M, Ahmad-Nejad P, Ghose S, Kirschning C J, Issels R D and Wagner H. HSP70 as endogenous stimulus of the Toll/interleukin-1 receptor signal pathway. J Biol Chem 2002, 277: 15107–15112.
17. Vabulas R M, Braedel S, Hilf N, Singh-Jasuja H, Herter S, Ahmad-Nejad P, Kirschning C J, Da Costa C, Rammensee H G, Wagner H and Schild H. The endoplasmic reticulum-resident heat shock protein Gp96 activates dendritic cells via the Toll-like receptor 2/4 pathway. J Biol Chem 2002, 277: 20847–20853.
18. Draude G, Hrboticky N and Lorenz R L. The expression of the lectin-like oxidized low-density lipoprotein receptor (LOX-1) on human vascular smooth muscle cells and monocytes and its down-regulation by lovastatin. Biochem Pharmacol 1999, 57: 383–386.
19. Delneste Y, Magistrelli G, Gauchat J, Haeuw J, Aubry J, Nakamura K, Kawakami-Honda N, Goetsch L, Sawamura T, Bonnefoy J and Jeannin P. Involvement of LOX-1 in dendritic cell-mediated antigen cross-presentation. Immunity 2002, 17: 353–362.
20. Lehner T, Bergmeier L A, Wang Y, Tao L, Singh M, Spallek R and van der Zee R. Heat shock proteins generate β-chemokines which function as innate adjuvants enhancing adaptive immunity. Eur J Immunol 2000, 30: 594–603.
21. Retzlaff C, Yamamoto Y, Hoffman P S, Friedman H and Klein T W. Bacterial heat shock proteins directly induce cytokine mRNA and interleukin-1 secretion in macrophage cultures. Infect Immun 1994, 62: 5689–5693.
22. Castellino F, Boucher P E, Eichelberg K, Mayhew M, Rothman J E, Houghton A N and Germain R N. Receptor-mediated uptake of antigen/heat shock protein complexes results in major histocompatibility complex class I antigen presentation via two distinct pathways. J Exp Med 2000, 191: 1957–1964.
23. Fujihara S M and Nadler S G. Intranuclear targeted delivery of functional NF-κB by 70kDa heat shock protein. EMBO J 1999, 18: 411–419.
24. Wright S D, Ramos R A, Tobias P S, Ulevitch R J and Mathison J C. CD14, a receptor for complexes of lipopolysaccharide (LPS) and LPS binding protein. Science 1990, 249: 1431–1433.

25. Yang R B, Mark M R, Gurney A L and Godwski P J. Signalling events induced by lipopolysaccharide-activated toll-like receptor 2. J Immunol 1999, 163: 639–643.

26. da Silva Correia J, Soldau K, Christen U, Tobias P S and Ulevitch R J. Lipopolysaccharide is in close proximity to each of the proteins in its membrane receptor complex. Transfer from CD14 to TLR4 and MD-2. J Biol Chem 2001, 276: 21129–21135.

27. Triantafilou K, Triantafilou M and Dedrick R L. A CD14-independent LPS receptor cluster. Nat Immunol 2001, 2: 338–344.

28. Asea A, Rehli M, Kabingu E, Boch J A, Baré O, Auron P E, Stevenson M A and Calderwood S K. Novel signal transduction pathway utilized by extracellular HSP70. Role of Toll-like receptor (TLR) 2 and TLR4. J Biol Chem 2002, 277: 15028–15034.

29. Lipsker D, Ziylan U, Spehner D, Proamer F, Bausinger H, Jeannin P, Salamero J, Bohbot A, Cazenave J P, Drillien R, Delneste Y, Hanau D and de la Salle H. Heat shock proteins 70 and 60 share common receptors which are expressed on human monocyte-derived but not epidermal dendritic cells. Eur J Immunol 2002, 32: 322–332.

30. van Kooten C and Banchereau J. Functions of CD40 on B cells, dendritic cells and other cells. Cur Opin Immunol 1997, 9: 330–337.

31. Young L S, Eliopoulos A G, Gallagher N J and Dawson C W. CD40 and epithelial cells: across the great divide. Immunol Today 1998, 19: 502–506.

32. Bourgeois C, Rocha B and Tanchot C. A role for CD40 expression on CD8[+] T cells in the generation of CD8[+] T cell memory. Science 2002, 297: 2060–2063.

33. Bennett S R, Carbone F R, Karamalis F, Flavell R A, Miller J F and Heath W R. Help for cytotoxic-T cell responses is mediated by CD40 signalling. Nature 1998, 393: 478–480.

34. Schoenberger S P, Toes R E, van der Voort E I, Offringa R and Melief C J. T cell help for cytotoxic T lymphocytes is mediated by CD40-CD40L interactions. Nature 1998, 393: 480–483.

35. Herold K C, Lu J, Rulifson I, Vezys V, Taub D, Grusby M J and Bluestone J A. Regulation of C-C chemokine production by murine T cells by CD28/B7 costimulation. J Immunol 1997, 159: 4150–4153.

36. Kornbluth R S, Kee K and Richman D D. CD40 ligand (CD154) stimulation of macrophages to produce HIV-1 suppressive chemokines. Proc Natl Acad Sci USA 1998, 99: 5205–5210.

37. McDyer J F, Dybul M, Goletz T J, Kinter A L, Thomas E K, Berzofsky J A, Fauci A S and Seder R A. Differential effects of CD40 ligand/trimer stimulation on the ability of dendritic cells to replicate and transmit HIV infection: evidence for CC-chemokine-dependent and -independent mechanisms. J Immunol 1999, 162: 3711–3717.

38. Yasui T, Muraoka M, Takaoka-Shichijo Y, Ishida I, Takegahara N, Uchida J, Kumanogoh A, Suematsu S, Suzuki M and Kikutani H. Dissection of B cell differentiation during primary immune responses in mice with altered CD40 signals. Int Immunol 2002, 14: 319–329.

39. Janeway C A J and Medzhitov R. Innate immune recognition. Ann Rev Immunol 2002, 20: 197–216.

40. Medzhitov R. Toll-like receptors and innate immunity. Nat Rev Immunol 2001, 1: 135–145.

41. Uronen-Hansson H, Allen J, Osman M, Squires G, Klein N and Callard R E. Toll-like receptor 2 (TLR2) and TLR4 are present inside human dendritic cells, associated with microtubules and the Golgi apparatus but are not detectable on the cell surface: integrity of microtubules is required for interleukin-12 production in response to internalized bacteria. Immunology 2004, 111: 173–178.

42. Armstrong P B and Quigley J P. α2 Macroglobulin: an evolutionary conserved arm of the innate immune system. Dev Comp Immunol 1999, 23: 375–390.

43. Basu S, Binder R J, Ramalingam T and Srivastava P K. CD91 is a common receptor for heat shock proteins gp96, hsp90, hsp70 and calreticulin. Immunity 2001, 14: 303–313.

44. Srivastava P. Roles of heat-shock proteins in innate and adaptive immunity. Nat Rev Immunol 2002, 2: 185–194.

45. Krieger M. The other side of scavenger receptors: pattern recognition for host defense. Cur Opin Lipidol 1997, 8: 275–280.

46. Gough P J and Gordon S. The role of scavenger receptors in the innate immune system. Microbes Infect 2000, 2: 305–311.

47. Panjwani N N, Popova L and Srivastava P K. Heat shock proteins gp96 and hsp70 activate the release of nitric oxide by APCs. J Immunol 2002, 168: 2997–3003.

48. MacAry P A, Javid B, Floto R A, Smith K G C, Singh M and Lehner P J. HSP70 peptide binding mutants separate antigen delivery from dendritic cell stimulation. Immunity 2004, 20: 95–106.

49. Coutinho A and Meo T. Genetic basis for unresponsiveness to lipopolysaccharide in C57BL/10Cr mice. Immunogenetics 1978, 7: 17–24.

50. Quershi S T, Larivière L, Leveque G, Clermont S, Moore K J, Gros P and Malo D. Endotoxin-tolerant mice have mutations in Toll-like receptor 4 (TLR4). J Exp Med 1999, 189: 615–625.

51. Randow F and Seed B. Endoplasmic reticulum chaperone gp96 is required for innate immunity but not cell viability. Nat Cell Biol 2001, 3: 891–896.

52. Kadowaki N, Ho S, Antonenko S, Malefyt R W, Kastelein R A and Bazan F. Subsets of human dendritic cell precursors express different toll-like receptor and respond to different microbial antigens. J Exp Med 2001, 163: 5786–5795.

53. Lee H H, Dempsey P W, Parks T P, Zhu X, Baltimore D and Cheng G. Specificities of CD40 signaling: involvement of TRAF2 in CD40-induced NF-κB activation and intercellular adhesion molecule-1 up-regulation. Proc Natl Acad Sci USA 1999, 96: 1421–1426.

54. Quezada S A, Jarvinen L Z, Lind E F and Noelle R J. CD40/CD154 interactions at the interface of tolerance and immunity. Ann Rev Immunol 2004, 22: 307–328.

55. Vabulas R M, Wagner H and Schild H. Heat shock proteins as ligands of Toll-like receptors. Cur Topics Microbiol Immunol 2002, 270: 169–184.

56. Markees T G, Phillips N E, Noelle R J, Shultz L D, Mordes J P, Greiner D L and Rossini A A. Prolonged survival of mouse skin allografts in recipients treated with donor splenocytes and antibody to CD40 ligand. Transplantation 1997, 64: 329–335.

57. Gallucci S, Lolkema M and Matzinger P. Natural adjuvants: endogenous activators of dendritic cells. Nat Med 1999, 11: 1249–1255.

58. Sauter B, Albert M L, Francisco L, Larsson M, Somersan S and Bhardwaj N. Consequences of cell death: exposure to necrotic tumour cells, but not primary tissue cells

or apoptotic cells, induces the maturation of immunostimulatory dendritic cells. J Exp Med 2000, 191: 423–433.

59. Shi Y and Rock K L. Cell death releases endogenous adjuvants that selectively enhance immune surveillance of particulate antigens. Eur J Immunol 2002, 32: 155–162.

60. Shi Y, Zhang W and Rock K L. Cell injury releases endogenous adjuvants that stimulate cytotoxic T cell responses. Proc Natl Acad Sci USA 2000, 97: 14590–14595.

61. Steinman R M, Turley S, Mellman I and Inaba K. The induction of tolerance by dendritic cells that have captured apoptotic cells. J Exp Med 2000, 191: 411–416.

62. Steinman R M, Hawiger D and Nussenzweig M C. Tolerogenic dendritic cells. Ann Rev Immunol 2003, 21: 685–711.

63. Lutz M B and Schuler G. Immature, semi-mature and fully mature dendritic cells: which signals induce tolerance or immunity. Trends Immunol 2002, 23: 445–449.

64. Menges M, Rößner S, Voigtländer C, Schindler H, Kukutsch N A, Bogdan C, Erb K, Schuler G and Lutz M B. Repetitive injections of dendritic cells matured with tumor necrosis factor alpha induce antigen-specific protection of mice from autoimmunity. J Exp Med 2002, 195: 15–21.

65. Dhodapkar M V, Steinman R M, Krasovsky J, Munz C and Bhardwaj N. Antigen-specific inhibition of effector T cell function in humans after injection of immature dendritic cells. J Exp Med 2001, 193: 233–238.

66. Legge K L, Gregg R K, Maldonado-Lopez R, Li L, Caprio J C, Moser M and Zaghouani H. On the role of dendritic cells in peripheral T cell tolerance and modulation of autoimmunity. J Exp Med 2002, 196: 217–227.

67. Ferguson T A, Herndon J, Elzey B, Griffith T S, Schoenberger S and Green D R. Uptake of apoptotic cells by lymphoid dendritic cells and cross-priming of CD8$^+$ T cells produce active immune unresponsiveness. J Immunol 2002, 168: 5589–5595.

68. Miga A J, Masters S R, Durell B G, Gonzalez M, Jenkins M K, Maliszewski C, Kikutani H, Wade W F and Noelle R J. Dendritic cell longevity and T cell persistence is controlled by CD154-CD40 interactions. Eur J Immunol 2001, 31: 959–965.

69. Grohmann U, Fallarino F, Silla S, Biachi R, Belladonna M L, Vacca C, Micheletti A, Fioretti M C and Puccetti P. CD40 ligation abrogates the tolerogenic potential of lymphoid dendritic cells. J Immunol 2001, 166: 277–283.

70. Mycko M P, Cwiklinska H, Szymanski J, Szymanska B, Kudla G, Kilianek L, Odyniec A, Brosnan C F and Selmaj K W. Inducible heat shock protein 70 promotes myelin autoantigen presentation by the HLA class II. J Immunol 2004, 172: 202–213.

71. Steinman R M and Nussenzweig M C. Avoiding horror autotoxicus: the importance of dendritic cells in peripheral T cell tolerance. Proc Natl Acad Sci USA 2002, 99: 351–358.

72. Bretscher P A. A two-step, two-signal model for the primary activation of precursor helper T cells. Proc Natl Acad Sci USA 1999, 96: 185–190.

73. Roth S, Willcox N, Rzepka R, Mayer M P and Melchers I. Major differences in antigen-processing correlate with a single Arg71↔Lys substitution in HLA-DR molecules predisposing to rheumatoid arthritis and with their selective interactions with 70-kDa heat shock protein chaperones. J Immunol 2002, 169: 3015–3020.

74. Lussow A R, Barrios C, van Embden J, van der Zee R, Verdini A S, Pessi A, Louis J A, Lambert P-H and Del Giudice G. Mycobacterial heat-shock proteins as carrier molecules. Eur J Immunol 1991, 21: 2297–2302.

75. Barrios C, Lussow J A, van Embden J, van der Zee R, Rappouli R, Costantino P, Louis J A, Lambert P-H and Del Giudice G. Mycobacterial heat-shock proteins as carrier molecules. II. The use of the 70kDa mycobacterial carrier for conjugated vaccines can circumvent the need for adjuvants and Bacillus Calmette Guerin priming. Eur J Immunol 1992, 22: 1365–1372.

76. Perraut R, Lussow A R, Gavoille S, Garraud O, Matile H, Tougne C, van Embden J, van der Zee R, Lambert P-H, Gysin J and Del Giudice G. Successful primate immunization with peptide conjugated to purified protein derrivative or mycobacterial heat shock proteins in the absence of adjuvants. Clin Exp Immunol 1993, 93: 382–386.

77. Suzue K and Young R A. Adjuvant-free hsp70 fusion protein system elicits humoral and cellular immune responses to HIV-1 p24. J Immunol 1996, 156: 873–879.

78. Udono H and Srivastava P K. Heat shock protein 70-associated peptides elicit specific cancer immunity. J Exp Med 1993, 178: 1391–1396.

79. Nieland T J F, Tan M C A A, Monee-van Muijen M, Koning F, Kruisbeek A M and van Bleek G M. Isolation of an immunodominant viral peptide that is endogenously bound to the stress protein GP96/GRP94. Proc Natl Acad Sci USA 1996, 93: 6135–6139.

80. Ciupitu A T, Petersson M, O'Donnell C L, Williams K, Jindal S, Kiessling R and Welsh R M. Immunization with a lymphocytic choriomeningitis virus peptide mixed with heat shock protein 70 results in protective antiviral immunity and specific cytotoxic T lymphocytes. J Exp Med 1998, 187: 685–691.

81. Schall T J, Bacon K, Toy K J and Goedell D V. Selective attraction of monocytes and T lymphocytes of the memory phenotype by cytokine RANTES. Nature 1990, 347: 669–671.

82. Murphy W J, Taub D D, Anver M, Conlon K, Oppenheim J J, Kelvin D J and Longo D L. Human RANTES induces the migration of human T lymphocytes into the peripheral tissues of mice with severe combined immune deficiency. Eur J Immunol 1994, 24: 1823–1827.

83. Meurer R, van Riper G, Feeney W, Cunningham P, Hora D J, Springer M S, MacIntyre D E and Rosen H. Formation of eosinophilic and monocytic intradermal inflammatory sites in the dog by injection of human RANTES but not human monocyte chemoattractant protein 1, human macrophage inflammatory protein 1 α, or human interleukin 8. J Exp Med 1993, 178: 1913–1921.

84. Kim J J, Nottingham L K, Sin J I, Tsai A, Morrison L, Oh J, Dang K, Hu Y, Kazahaya K, Bennett M, Dentchev T, Wilson D M, Chalian A A, Boyer J D, Agadjanyan M G and Weiner D B. CD8 positive influence antigen-specific immune responses through the expression of chemokines. J Clin Invest 1998, 102: 1112–1124.

85. Schall T J, Bacon K, Camp R D, Kaspari J W and Goeddel D V. Human macrophage inflammatory protein alpha (MIP-1α) and MIP-1β chemokines attract distinct populations of lymphocytes. J Exp Med 1993, 177: 1821–1826.

86. Dieu M C, Vanbervliet B, Vicari A, Bridon J M, Oldham E, Ait-Yahia S, Briere F, Zlotnik A, Lebecque S and Caux C. Selective recruitment of immature and mature dendritic cells by distinct chemokines expressed in different anatomic sites. J Exp Med 1998, 188: 373–386.

87. Trinchieri G. Interleukin-12: a cytokine produced by antigen presenting cells with immunoregulatory functions in the generation of T-helper cells type 1 and cytotoxic lymphocytes. Blood 1994, 84: 4008–4027.

88. Thole J E, van Schooten W C, Keulen W J, Hermans P W, Janson A A, de Vries R R, Kolk A H and van Embden J D. Use of recombinant antigens expressed in *Escherichia coli* K-12 to map B-cell and T-cell epitopes on the immunodominant 65-kilodalton protein of *Mycobacterium bovis* BCG. Infect Immun 1988, 56: 1633–1640.
89. Ivanyi J, Sharp K, Jackett P and Bothamley G. Immunological study of defined constituents of mycobacteria. Springer Semin Immunopathol 1988, 10: 279–300.
90. Medzhitov R M and Janeway C A J. Innate immunity: impact on the adaptive immune response. Cur Opin Immunol 1997, 9: 4–9.
91. Fearon D T and Locksley R M. The instructive role of innate immunity in the acquired immune response. Science 1996, 272: 50–53.
92. Matzinger P. Tolerance, danger, and the extended family. Ann Rev Immunol 1994, 12: 991–1045.
93. Stewart G R, Snewin V A, Walzl G, Hussell T, Tormay P, O'Gaora P, Goyal M, Betts J, Brown I N and Young D B. Overexpression of heat-shock proteins reduces survival of *Mycobacterium tuberculosis* in the chronic phase of infection. Nat Med 2001, 7: 732–737.
94. Bogers W M, Bergmeier L A, Ma J, Oostermeijer H, Wang Y, Kelly C G, Ten Haaft P, Singh M, Heeney J L and Lehner T. A novel HIV-CCR5 receptor vaccine strategy in the control of mucosal SIV/HIV infection. AIDS 2004, 18: 25–36.
95. Multhoff G, Mizzen L, Winchester C C, Milner C M, Wenk S, Eissner G, Kampinga H H, Laumbacher B and Johnson J. Heat shock protein 70 (Hsp70) stimulates proliferation and cytolytic activity of natural killer cells. Exp Hematol 1999, 27: 1627–1636.
96. Gastpar R, Gross C, Rossbacher L, Ellwart J, Riegger J and Multhoff G. The cell surface-localized heat shock protein 70 epitope TKD induces migration and cytolytic activity selectively in human NK cells. J Immunol 2004, 172: 972–980.
97. Michaëlsson J, Teixeira de Matos C T, Achour A, Lanier L L, Kärre K and Söderström K. A signal peptide derived from hsp60 binds HLA-E and interferes with CD94/NKG2A recognition. J Exp Med 2002, 196: 1403–1414.
98. Lehner T, Mitchell E, Bergmeier L, Singh M, Spallek R, Cranage M, Hall G, Dennis M, Villinger F and Wang Y. The role of γδ T cells in generating antiviral factors and β-chemokines in protection against mucosal simian immunodeficiency virus infection. Eur J Immunol 2000, 30: 2245–2256.

11

Molecular Chaperone–Cytokine Interactions at the Transcriptional Level

Anastasis Stephanou and David S. Latchman

11.1. Introduction

The heat shock proteins (Hsps) are a group of highly conserved proteins that have major physiological roles in protein homeostasis [1, 2]. In most cell types, 1–2% of total proteins consist of heat shock proteins even prior to stress, which suggests important roles for these proteins in the biology and physiology of the unstressed cell. These roles particularly concern regulating the folding and unfolding of other proteins. The term 'heat shock proteins' was coined because these proteins were first identified on the basis of their increased synthesis following exposure to elevated temperatures [3]. Subsequently, it has been clearly shown that they can be induced following a variety of stressful stimuli. Some heat shock proteins, such as Hsp90 (each heat shock protein is named according to its mass in kilodaltons – see Chapter 1 for more details), are detectable at significant levels in unstressed cells and increase in abundance following a suitable stimulus, whereas others such as Hsp70 exist in both constitutively expressed and inducible forms [4, 5].

The dual role of heat shock proteins in both normal and stressed cells evidently requires the existence of complex regulatory processes which ensure that the correct expression pattern is produced. Indeed, such processes must be operative at the very earliest stages of embryonic development, since the genes encoding Hsp70 and Hsp90 are amongst the first embryonic genes to be transcribed [6, 7].

The induction of heat shock proteins in response to various stresses is dependent on the activation of specific members of a family of transcription factors, the heat shock factors that bind to the heat shock element in the promoters of the genes encoding heat shock proteins [8]. Four heat shock factors (HSF-1 to -4) have been cloned from a number of organisms, and their roles have now been characterised (Table 11.1). HSF-1 and HSF-3 have been shown to be involved

Table 11.1. Functional role of different heat shock factors in regulating heat shock proteins

	Intracellular function	Knock-out phenotype
HSF-1	Stress-induced heat shock protein gene expression	Defective heat shock response
HSF-2	Non-stress-induced heat shock protein gene expression	NA
HSF-2α	HSF-2α expressed predominantly in adult tissue	
HSF-2β	HSF-2β expressed during early development	
HSF-3	HSF-3 also involved in stress-induced heat shock protein gene expression and has a higher threshold in response to heat shock than HSF-1	NA
HSF-4α	HSF-4α acts as a repressor of heat shock protein gene expression	NA
HSF-4β	HSF-4β is a transactivator of heat shock protein gene expression	

in regulating heat shock proteins in response to thermal stress, whereas HSF-2 and HSF-4 are involved in heat shock protein regulation in unstressed cells, and their levels are regulated in response to a wide variety of biological processes such as immune activation and cellular differentiation [8] (Figure 11.1). In general however, the stimuli that induce such alterations in heat shock protein gene expression under non-stress conditions are poorly characterised, and the mechanisms by which they act are also unclear. Heat shock proteins are not only

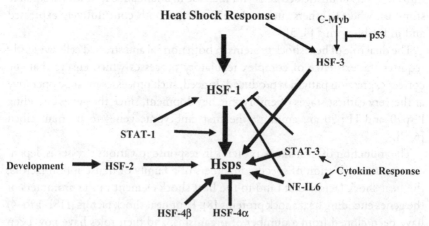

Figure 11.1. Heat shock factor pathways and their interaction or cooperation in modulating heat shock proteins.

regulated by heat shock factors, and this chapter will discuss heat shock factors and transcription factors that are able to interact or co-operate with HSF-1 to modulate the transcriptional regulation of heat shock proteins in response to non-stressful stimuli.

11.2. Transcriptional regulation of heat shock proteins by the HSF family

11.2.1. HSF-1

As mentioned earlier, HSF-1 has been identified as the heat shock factor that mediates stress-induced heat shock protein gene expression in response to environmental stressors. Such stresses induce HSF-1 oligomerisation and nuclear translocalisation followed by enhanced DNA binding on the heat shock protein DNA promoters. HSF-1 is negatively regulated by Hsp70 and Hsp90 which is suggestive of a negative-feedback loop for the regulation of Hsp70 and Hsp90 genes following a heat shock response [9, 10].

The phosphorylation of HSF-1 also modulates its activity, and constitutive phosphorylation is important for negatively regulating the activity of HSF-1 under normal growth conditions [11]. The kinases responsible for phosphorylating HSF-1 on several serine sites include glycogen synthase kinase 3β and c-jun N-terminal kinase [12, 13]. Although a positive role of HSF-1 phosphorylation in stress-induced activation of heat shock protein gene expression is also known to occur, the kinases involved and the phosphorylation sites on HSF-1 have not yet been characterised.

Cells from HSF-1 knock-out mice exhibit defects in heat shock protein induction following exposure to heat shock [14]. Moreover, cells lacking HSF-1 are susceptible to apoptotic cell death following exposure to heat stress [14]. Mice lacking HSF-1 also exhibit elevated levels of tumour necrosis factor-α, which results in an increased mortality after endotoxin and inflammatory challenge [14]. Interestingly, HSF-1 has also been shown to modulate other genes such as IL-1β and c-fos [15, 16], suggesting a role for HSF-1 in regulating other stress-responsive genes.

11.2.2. HSF-2

As mentioned earlier, heat shock protein gene expression is crucial not only for the survival of cells exposed to extracellular stress stimuli, but also for normal cellular physiological events such as embryonic development and cellular differentiation. HSF-2 has now been described as the factor involved in regulating heat shock proteins under non-stressful conditions. For example, Hsp70 expression is induced by haemin in K562 cells, and this causes them to differentiate; this process requires activation of HSF-2 [17]. HSF-2 exists as two isoforms, HSF-2α and HSF-2β, due to alternative splicing. The HSF-2α isoform is predominantly expressed in adult tissue, whereas the HSF-2β isoform is predominantly expressed in embryonic tissue [18]. HSF-2 DNA binding activity is high during early embryogenesis in tissues such as the heart, central nervous system and testis [18]. The importance of HSF-2 in development will become more apparent when the HSF-2 knock-out animals become available.

11.2.3. HSF-3 and HSF-4

HSF-3 was originally identified in avian cells and, like HSF-1, is also heat stress responsive [19]. However, the threshold temperatures required for the activation of HSF-3 and HSF-1 are different in that HSF-1 is activated by less severe heat shock than HSF-3 [19]. No reports have yet described HSF-3 in other organisms. Previously, HSF-3 was reported to bind to c-Myb, a transcription factor involved in cellular proliferation and required for the G1/S transition of the cell cycle, which also paralleled the expression of Hsp70 [20]. These studies suggest that HSF-3/cMyb interaction might be involved in cell-cycle-dependent expression of heat shock proteins. Furthermore, it has also been shown that HSF-3/c-Myb association is disrupted by direct binding of p53 tumour suppressor transcription factor, resulting in inhibition of Hsp70 expression [21].

In contrast to other heat shock factors, HSF-4 has been reported to function as a repressor of heat shock protein gene expression [22]. HSF-4 also exists as two isoforms, HSF-4α and HSF-4β [22, 23], and it was the HSF-4α isoform that was cloned and used in the original study reporting it to be a repressor. However, HSF-4β has subsequently been shown to activate heat shock protein gene expression, which suggests that the HSF-4 gene is able to generate both an activator and a repressor of heat shock genes [23].

11.3. The role of non-HSF transcription factors in modulating heat shock protein gene expression

The phenotype of mice lacking HSF-1 is normal in the absence of stress, and the expression of Hsp70 and Hsp90 in cells lacking HSF-1 is similar to that in wild-type cells, despite the fact that they exhibit a defect in the heat shock response following heat stress [14]. These studies suggest that other heat shock factors might compensate for the lack of HSF-1 and/or that other factors are also responsible for the expression of heat shock proteins under normal growth

Table 11.2. Functional role of STATs and their phenotype observed in the STAT knockout animal

	Induce	Knockout phenotype
STAT-1	Interferons, IL-6	Viable – defects in immune responses to microbes
STAT-2	Interferons	Viable, defects in INF responses
STAT-3	IL-6 family	Embryonic lethal
STAT-4	IL-12	Viable – defects in immune responses
STAT-5α	Numerous	Viable – defects in mammary gland development due to loss of responses to growth hormone
STAT-5β	Numerous	Viable – defects in responses to growth hormone and prolactin as well as defects in T cell responses
STAT-6	IL-4	Viable – IL-4 responses abolished resulting in defects in immune responses

conditions. Studies from our laboratory have unravelled a separate group of transcription factors that are activated by distinct cytokines and are able to modulate Hsp70 and Hsp90 gene expression. These factors include STAT-1, STAT-3 and NF-IL6, and their functional roles are described in the next section.

11.3.1. STATs

The signal transducers and activators of transcription (STATs) are a family of cytoplasmic transcription factors which mediate intracellular signalling initiated at cytokine cell-surface receptors and transmitted to the nucleus (Table 11.2). STATs are activated by phosphorylation on conserved tyrosine and serine residues on their C-terminal domains by the Janus kinases (JAKs) and mitogen-activated protein kinase families, respectively. These allow the STATs to dimerise and translocate to the nucleus and thereby regulate gene expression (for review see [24]). Interferon-γ is a potent activator of STAT-1, whilst the interleukin-6 (IL-6) family members including IL-6, leukaemia inhibitory factor and CT-1 primarily activate STAT-3 [24].

Studies from our laboratory have shown STAT-1 and STAT-3 to have opposing action on apoptotic cell death in various cell types [25]. We have reported that over-expression of STAT-1 is able to enhance apoptotic cell death in cardiac myocytes exposed to ischaemia-reperfusion, whereas over-expression of STAT-3 plus STAT-1 reduces the level of STAT-1–induced cell death following ischaemia-reperfusion by modulating the expression of pro- and anti-apoptotic genes [26]. Furthermore, these effects on apoptosis require serine-[727] (but not tyrosine-[701])

phosphorylation on the C-terminal transactivation domain of STAT-1 [27, 28]. We have subsequently shown that STAT-1 is able to modulate the activity of p53 and its effects on apoptosis [29]. These effects involve STAT-1/p53 protein–protein interaction with STAT-1 acting as a co-activator for p53 [29].

11.3.2. NF-IL6

The cytokine IL-6 is known to stimulate two distinct signalling pathways which results in the activation of two different classes of cellular transcription factors [30] (Figure 11.2). Thus, initial studies showed that a variety of IL-6–inducible genes contained binding sites for a transcription factor named NF-IL6 (nuclear factor IL-6), which showed high homology with the rat-liver nuclear factor C/EBP (CCAAT-enhancer-binding protein), and is therefore also known as C/EBPβ [31]. Subsequently, a second member of the C/EBP family, known as NF-ILβ or C/EBPδ, was identified and shown to form heterodimers with NF-IL6, resulting in a synergistic transcriptional effect [32]. After exposure of cells to IL-6, NF-IL6 is phosphorylated, resulting in its enhanced ability to stimulate transcription [32], whereas NF-IL6β is synthesised *de novo* [32]. As mentioned earlier, the second pathway that is stimulated by IL-6 is the JAK/STAT-3 signalling pathway.

It is generally accepted that the NF-IL6/NF-IL6β and STAT-3 signalling pathways allow IL-6 to activate two distinct sets of genes, each of which is responsive to one of these pathways. Thus, class 1 acute-phase proteins (such as α_1-acid glycoprotein, haptoglobin, C-reactive protein and serum amyloid) contain response elements for NF-IL6 and NF-IL6β, and these factors have been shown to be involved in the activation of these genes following IL-6 treatment [33]. In agreement with this idea, these genes are stimulated by exposure of cells to IL-1, which also stimulates NF-IL6/NF-IL6β activity without affecting STAT-3 [33]. In contrast, type 2 acute-phase genes such as fibrinogen, thiostatin and α_2-microglobulin, are not inducible by IL-1 and lack binding sites for NF-IL6/NF-IL6β. Instead, these genes contain binding sites allowing binding of STAT-3, which is responsible for activation of these genes in response to IL-6 [33].

11.4. Role of STAT-1, STAT-3 and NF-IL6 factors in modulating heat shock proteins

We have reported [34] that IL-6 can induce increased expression of Hsp90 in a variety of different cell types. The Hsp90β gene promoter is responsive to IL-6

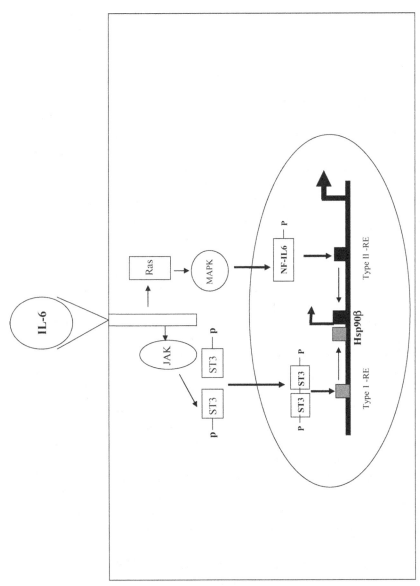

Figure 11.2. Distinct pathways activated by IL-6 that stimulate either type I or type II genes and heat shock proteins which are modulated by both pathways.

185

and can also be activated by NF-IL6 or NF-IL6β [34]. Moreover, a short region of the promoter containing an NF-IL6 binding site is essential for activation of the promoter by both IL-6 and NF-IL6 and can confer responsiveness both to IL-6 and to over-expression of NF-IL6 on a heterologous promoter. These findings suggest that Hsp90 is a member of the class of IL-6–responsive genes that are activated by NF-IL6/NF-IL6β.

Interestingly, this short region of the promoter also contains binding sites for STAT-3, and the Hsp90 promoter can also be activated by this factor. Moreover, over-expression of NF-IL6 and STAT-3 has a synergistic effect on the Hsp90 promoter and both these signalling pathways appear to be required for activation of the Hsp90 promoter by IL-6 [35]. However, despite their synergistic action in IL-6 signalling, these two pathways have opposite effects on the heat shock–mediated regulation of the Hsp90 promoter. Thus, STAT-3 reduces the stimulatory effect of heat shock, whereas NF-IL6 enhances it. When applied together, heat shock and IL-6 produce only weak activation of the Hsp90 promoter compared with either stimulus alone, indicating that the inhibitory effect of STAT-3 on heat shock factor predominates under these conditions [35]. In contrast, IL-1, which activates only the NF-IL6 pathway, synergises with heat shock to produce strong activation of Hsp90 [35]. These results therefore open up a new aspect of Hsp90 gene regulation which is additional to, and interacts with, the heat shock–activated pathway.

We have also examined whether STAT-1 is able to modulate heat shock protein expression. We have shown that IFN-γ treatment increases the levels of Hsp70 and Hsp90 and also enhances the activity of the Hsp70 and Hsp90β promoters, with these effects being dependent on activation of the STAT-1 transcription factor by IFN-γ [36]. These effects were not seen in a STAT-1–deficient cell line, indicating that IFN-γ modulates heat shock protein induction via a STAT-1–dependent pathway. The effect of IFN-γ/STAT-1 is mediated via a short region of the Hsp70/Hsp90 promoters, which also mediates the effects of NF-IL6 and STAT-3 and can bind STAT-1 [36].

This region also contains a binding site for the stress-activated transcription factor HSF-1. We have shown that STAT-1 and HSF-1 interact with one another via a protein–protein interaction to produce a strong activation of transcription [36]. This is in contrast to our previous finding that STAT-3 and HSF-1 antagonise each other, and we showed that STAT-3 and HSF-1 do not interact directly. To our knowledge, this was the first report of HSF-1 interacting directly via a protein–protein interaction with another transcription factor. Such protein–protein interactions and the binding of a number of different stress and cytokine-activated transcription factors to a short region of the Hsp90 and Hsp70 gene promoters are likely to play a very important role in heat shock protein gene activation by non-stressful stimuli and the integration of these responses with the stress response of these genes.

11.5. Linking STAT-1, STAT-3 and NF-IL6 elevation to pathological states

A number of disease states have been shown to exhibit elevated levels of heat shock proteins [37]. This includes patients with systemic lupus erythematosus (SLE), in whom elevated levels of Hsp90 are present. Interestingly, elevated levels of circulating IL-6 have also been reported in patients with SLE [38], and levels have been shown to be correlated with disease activity, being highest in patients with active disease. Moreover, spontaneous production of IgG by normal and SLE-derived B lymphocytes in culture can be enhanced by the addition of exogenous IL-6 and inhibited by antibody to IL-6 [39]. These findings therefore suggest that IL-6 might play a role in the pathogenesis of autoimmune diseases and the infusion of an antibody to IL-6 can relieve disease symptoms in lupus-prone NZB/NZW F1 mice [40].

In order to directly test the role of IL-6 in regulating Hsp90 expression *in vivo* we have used mice that have been artificially engineered to express elevated levels of IL-6 either by being made transgenic for extra copies of the IL-6 gene [41] or by inactivation of the gene encoding the transcription factor C/EBPβ, which also results in elevation of IL-6 levels in these mice [42]. In these experiments, elevated levels of Hsp90 were observed in both the IL-6 transgenic and the C/EBPβ knock-out mice [43]. Hence, the elevated IL-6 levels induced in these animals are indeed paralleled by increased levels of Hsp90 compared to normal control mice. In addition, elevated Hsp90 was associated with the specific production of autoantibodies to Hsp90 in both IL-6 transgenic and C/EBPβ knock-out animals. It is also of interest that inactivation of the IL-6 gene in the C/EBPβ knock-out mice results in the suppression of Castleman-like disease normally observed in these animals and a reduction in the production of autoantibodies.

Furthermore, elevated levels of Hsp90 in SLE correlates with levels of IL-6 and autoantibodies to Hsp90 [44]. These results support a model in which elevated levels of IL-6 in SLE patients induce elevated levels of Hsp90 protein, which in turn results in the production of autoantibodies to this protein. Additionally, IL-10 is also elevated in SLE, and IL-10 was demonstrated to enhance Hsp90 gene expression [45]. Therefore, these studies strongly suggest that IL-6 and IL-10 are likely to play a critical role in the regulation of Hsp90 levels and autoantibody production, both in autoimmune disease states and potentially in normal cells *in vivo*.

11.6. Conclusion

In this chapter, studies demonstrating the modulation of heat shock proteins by a group of transcription factors other than the traditional heat shock factor family under normal non-stressful conditions and also in several disease states have been reviewed. The finding that the responses to these factors occur around the HSF DNA binding site suggests that HSF-1, as well as other heat shock factors, are able to interact or cooperate with STATs or NF-IL6 family members. Further studies to identify novel protein interacting partners for heat shock factors will also provide insight into the regulation of heat shock proteins. Unravelling the mechanistic basis to this cooperation will undoubtedly enhance our understanding of the interdependent relationship between distinct heat shock factors and their interaction with other factors in these complex regulatory processes, which ensure that the correct heat shock protein expression pattern is produced under different physiological states.

REFERENCES

1. Ellis R J and van der Vies S M. Molecular chaperones. Ann Rev Biochem 1991, 60: 321–347.
2. Mathew A and Morimoto R I. Role of the heat-shock response in the life and death of proteins. Ann NY Acad Sci 1998, 851: 99–111.
3. Ritossa F A. A new puffing pattern induced by temperature shock and DNP in *Drosophila*. Experientia 1962, 18: 571–573.
4. Morimoto R I, Kline M P, Bimston D N and Cotto J J. The heat-shock response: regulation and function of heat-shock proteins and molecular chaperones. Essays Biochem 1997, 32: 17–29.
5. Jolly C and Morimoto R I. Role of the heat shock response and molecular chaperones in oncogenesis and cell death. J Natl Cancer Inst 2000, 92: 1564–1572.
6. Neidhardt F C, VanBogelen R, A, and Vaughn V. The genetics and regulation of heat-shock proteins. Ann Rev Genet 1984, 18: 295–329.
7. Walsh D, Li Z, Wu Y and Nagata K. Heat shock and the role of the HSPs during neural plate induction in early mammalian CNS and brain development. Cell Mol Life Sci 1997, 53: 198–211.
8. Pirkkala L, Nykanen P and Sistonen L. Roles of the heat shock transcription factors in regulation of the heat shock response and beyond. FASEB J 2001, 15: 1118–1131.
9. Shi Y, Mosser D D and Morimoto R I. Molecular chaperones as HSF1-specific transcriptional repressors. Genes Develop 1998, 12: 654–666.
10. Ali A, Bharadwaj S, O'Carrol R and Ovsenek N. HSP90 interacts with and regulates the activity of heat shock factor 1 in Xenopus oocytes. Mol Cell Biol 1998, 18: 4949–4960.
11. Xia W and Voellmy R. Hyperphosphorylation of heat shock transcription factor 1 is correlated with transcriptional competence and slow dissociation of active factor trimers. J Biol Chem 1997, 272: 4094–4102.

12. Xavier I J, Mercier P A, McLoughlin C M, Ali A, Woodgett J R and Ovsenek N. Glycogen synthase kinase 3β negatively regulates both DNA-binding and transcriptional activities of heat shock factor 1. J Biol Chem 2000, 275: 29147–29152.

13. Park J and Liu A Y. JNK phosphorylates the HSF1 transcriptional activation domain: role of JNK in the regulation of the heat shock response. J Cell Biochem 2001, 82: 326–338.

14. Xiao X, Zuo X, Davis A A, McMillan D R, Curry B B, Richardson J A and Benjamin I J. HSF1 is required for extra-embryonic development, postnatal growth and protection during inflammatory responses in mice. EMBO J 1999, 18: 5943–5952.

15. Xie Y, Chen C, Stevenson M A, Auron P E and Calderwood S K. Heat shock factor 1 represses transcription of the IL-1β gene through physical interaction with the nuclear factor of interleukin 6. J Biol Chem 2002, 277: 11802–11810.

16. Chen C, Xie Y, Stevenson M A, Auron P E and Calderwood S K. Heat shock factor 1 represses Ras-induced transcriptional activation of the c-fos gene. J Biol Chem 1997, 272: 26803–26806.

17. Sistonen L, Sarge K D, Phillips B, Abravaya K and Morimoto R I. Activation of heat shock factor 2 during hemin-induced differentiation of human erythroleukemia cells. Mol Cell Biol 1992, 12: 4104–4111.

18. Goodson M L, Park-Sarge O K and Sarge K D. Tissue-dependent expression of heat shock factor 2 isoforms with distinct transcriptional activities. Mol Cell Biol 1995, 15: 5288–5293.

19. Nakai A and Morimoto R I. Characterization of a novel chicken heat shock transcription factor, heat shock factor 3, suggests a new regulatory pathway. Mol Cell Biol 1993, 13: 1983–1997.

20. Kanei-Ishii C, Tanikawa J, Nakai A, Morimoto R I and Ishii S. Activation of heat shock transcription factor 3 by c-Myb in the absence of cellular stress. Science 1997, 277: 246–248.

21. Tanikawa J, Ichikawa-Iwata E, Kanei-Ishii C, Nakai A, Matsuzawa S, Reed J C and Ishii S. p53 suppresses the c-Myb-induced activation of heat shock transcription factor 3. J Biol Chem 2000, 275: 15578–15585.

22. Nakai A, Tanabe M, Kawazoe Y, Inazawa J, Morimoto R I and Nagata K. HSF4, a new member of the human heat shock factor family which lacks properties of a transcriptional activator. Mol Cell Biol 1997, 17: 469–481.

23. Tanabe M, Sasai N, Nagata K, Liu X D, Liu P C, Thiele D J and Nakai A. The mammalian HSF4 gene generates both an activator and a repressor of heat shock genes by alternative splicing. J Biol Chem 1999, 274: 27845–27856.

24. Ihle J N. The Stat family in cytokine signaling. Cur Opin Cell Biol 2001, 13: 211–217.

25. Battle T E and Frank D A. The role of STATs in apoptosis. Cur Mol Med 2002, 2: 381–392.

26. Stephanou A, Brar B K, Scarabelli T, Jonassen A K, Yellon D M, Marber M S, Knight R A and Latchman D S. Ischaemia-induced STAT-1 expression and activation plays a critical role in cardiac myocyte apoptosis. J Biol Chem 2000, 275: 10002–10008.

27. Stephanou A, Scarabelli T, Brar B K, Nakanishi Y, Matsumura M, Knight R A and Latchman D S. Induction of apoptosis and Fas/FasL expression by ischaemia/reperfusion in cardiac myocytes requires serine 727 of the STAT1 but not tyrosine 701. J Biol Chem 2001, 276: 28340–28347.

28. Stephanou A, Scarabelli T, Townsend P A, Bell R, Yellon D M, Knight R A and Latchman D S. The carboxyl-terminal activation domain of the STAT-1 transcription factor enhances ischaemia/reperfusion-induced apoptosis in cardiac myocytes. FASEB J 2002, 16: 1841–1843.

29. Townsend P, Scarabelli T M, Davidson S M, Knight R A, Latchman D S and Stephanou A. STAT-1 interacts with p53 to enhance DNA damage-induced apoptosis. J Biol Chem 2004, 279: 5811–5822.

30. Kishimoto T, Akira S, Narazaki M and Taga T. Interleukin-6 family of cytokines and gp130. Blood 1995, 86: 1243–1254.

31. Nakajima T, Kinoshita S, Sasagawa T, Sasaki K, Naruto M, Kishimoto T and Akira S. Phosphorylation at threonine-235 by ras-dependent mitogen-activated protein kinase cascade is essential for transcription factor NF-IL6. Proc Natl Acad Sci USA 1993, 90: 2207–2211.

32. Kinoshita S, Akira S and Kishimoto T. A member of the C/EBP family, NF-IL6 beta, forms a heterodimer and transcriptionally synergizes with NF-IL6. Proc Natl Acad Sci USA 1992, 89: 1473–1476.

33. Ganter U, Arcone R, Toniatti C, Morrone G and Ciliberto G. Dual control of C-reactive protein gene expression by interleukin-1 and interleukin-6. EMBO J 1989, 8: 3773–3779.

34. Stephanou A, Amin V, Isenberg D A, Akira S, Kishimoto T and Latchman D S. IL-6 activates heat shock protein 90 gene expression. Biochem J 1997, 321: 103–106.

35. Stephanou A, Isenberg D A, Akira S, Kishimoto T and Latchman D S. NF-IL6 and STAT-3 signalling pathways co-operate to mediate the activation of the Hsp90β gene by IL-6 but have opposite effects on its inducibility by heat shock. Biochem J 1998, 330: 189–195.

36. Stephanou A, Isenberg D A, Nakajima K and Latchman D S. Signal transducer and activator of transcription-1 and heat shock factor-1 interact and activate the transcription of the Hsp-70 and Hsp-90β gene promoters. J Biol Chem 1999, 274: 1723–1728.

37. Twomey B M, Dhillon V B, McCallum S, Isenberg D A and Latchman D S. Elevated levels of the 90kD heat shock protein in patients with systemic lupus erythematosus are dependent upon enhanced transcription of the Hsp90 gene. J Autoimmunity 1993, 6: 495–506.

38. De Benedetti F, Massa M, Robbion R, Ravelli A, Burgio G R and Martini A. Correlation of serum IL-6 levels with joint involvement and thrombocytosis in systemic juvenile rheumatoid arthritis. Arthritis Rheum 1991, 34: 1158–1163.

39. Linker-Israeli M, Deans R J, Wallace D J, Prehn J, Ozeri-Chen T and Kinenberg J R. Elevated levels of endogenous IL-6 in SLE. A putative role in pathogenesis. J Immunol 1991, 147: 117–123.

40. Finck B K, Chan B and Wofsy D. Interleukin 6 promotes murine lupus in NZB/NZW F1 mice. J Clin Invest 1994, 945: 585–591.

41. Suematsu S, Matsuda T, Aozasa K, Akira S, Nakano N, Ohno S, Miyazaki J-I, Yamamura K-I, Hirano T and Kishimoto T. IgG1 plasmacytosis in IL-6 transgenic mice. Proc Natl Acad Sci USA 1989, 86: 7547–7551.

42. Screpanti I, Romani L, Musiani P, Modesti A, Fattoro E, Lazzaro D, Sellitto C, Scarpa S, Bellavia D, Lattanzio G, Bistoni F, Frati L, Cortese R, Gulino A, Ciliberto G,

Costani F and Poli V. Lymphoproliferative disorder and imbalanced T-helper response in C/EBP-deficient mice. EMBO J 1995, 14: 1932–1941.

43. Stephanou A, Conroy S, Isenberg D A, Poli V, Ciliberto G and Latchman D S. Elevation of IL-6 in transgenic mice results in increased levels of the 90KD heat shock protein and production of the anti-Hsp90 antibodies. J Autoimmunity 1998, 11: 249–253.

44. Ripley B J, Isenberg D A and Latchman D. S. Elevated levels of the 90 kDa heat shock protein [hsp90] in SLE correlate with levels of IL-6 and autoantibodies to hsp90. J Autoimmunity 2001, 17: 341–346.

45. Ripley B J, Stephanou A, Isenberg D A and Latchman D S. Interleukin-10 activates heat-shock protein 90β gene expression. Immunology 1999, 97: 226–231.

Extracellular Biology of Molecular Chaperones: Physiological and Pathophysiological Signals

12

Heat Shock Protein Release and Naturally Occurring Exogenous Heat Shock Proteins

Johan Frostegård and A. Graham Pockley

12.1. Introduction

Although for many years the perception has been that mammalian heat shock proteins are intracellular molecules that are only released into the extracellular environment in pathological situations such as necrotic cell death, it is now known that these molecules can be released from a variety of viable (non-necrotic) cell types [1–4]. Moreover, we and a number of others have reported Hsp60 and/or Hsp70 to be present in the peripheral circulation of normal individuals [5–12]. These observations have profound implications for the perceived role of these proteins as pro-inflammatory intercellular 'danger' signalling molecules and have prompted a re-evaluation of the functional significance and role(s) of these ubiquitously expressed and highly conserved families of molecules. The reader should refer to Chapter 2, which discusses the intracellular dispositions of molecular chaperones and also touches on the release of heat shock proteins, and Chapter 3, in which novel pathways of protein release are described.

The mechanism(s) leading to the release of heat shock proteins are unknown, as is the source of circulating heat shock proteins in the peripheral circulation and their physiological and pathophysiological role(s). The inverse relationship between levels of circulating Hsp70 and the progression of carotid atherosclerosis [13], or the presence of coronary artery disease (CAD) [14], appears to be inconsistent with the concept that this molecule is a danger signal and an *in vitro* activator of innate and pro-inflammatory immunity [15]. Although a great deal of attention has focussed on the capacity of exogenous heat shock proteins to act as inflammatory activators of innate and adaptive immunity (discussed in detail in many of the chapters in this volume), exogenous heat shock proteins have also been shown to have a number of anti-inflammatory and non-immunological, cytoprotective effects on a variety of cells types (see, for example, Chapters 13, 14 and 16).

This chapter reviews the evolving evidence that heat shock proteins are present in, and can be released into, the extracellular compartment under physiological conditions and summarises the functional versatility of such exogenous proteins. Further insight into the functionality and significance of actively released and circulating heat shock proteins might reveal hitherto unknown physiological and pathophysiological roles for these ubiquitously expressed families of proteins.

12.2. Heat shock protein release – *in vitro* studies

12.2.1. Historical perspective

During the course of experiments between 1996 and 1997 in which the Pockley laboratory was investigating the influence of different physicochemical stressors on stress protein induction and expression by human peripheral blood mononuclear cells [16], it became apparent that the heat shock proteins Hsp60 and Hsp70 were present in the plasma of normal individuals. These findings were counterintuitive to the proposition that heat shock proteins were only present in the extracellular milieu in the event of pathological processes that involved cellular necrosis and were viewed with scepticism. However, the literature revealed that these and other heat shock proteins can be released from a variety of intact cells and that this release appears to be via active, rather than passive, processes.

One of the earliest papers documenting heat shock protein release came from Tytell and colleagues, who reported the transfer of glia-axon transfer proteins, which include Hsp70, Hsc70 and Hsp100, from adjacent glial cells into the squid giant axon [17]. This finding prompted the suggestion that the release of such proteins might be a mechanism by which glial cells, which are capable of generating effective stress protein-mediated resistance to physical and metabolic insults, can protect adjacent neuronal cells, which exhibit a deficient response to stress. The capacity of glial cells to export Hsp70 has subsequently been demonstrated in a human system using T98G human glioma cells and stress-sensitive, differentiated LA-N-5 human neuroblastoma cells [18].

One of the seminal studies in the area of heat shock protein release came from Hightower and Guidon (see Chapter 19) when they reported that heat shock proteins could be released from cultured rat embryo cells [1]. Heat treatment increased the number of proteins released from a small set, which included the constitutively expressed Hsc70, to include the inducible Hsp70 and Hsp110 molecules. Although uncertain, the proposition was that the release of heat shock proteins might have resulted from changes in pH and gas tension, a disruption of the diffusion layer at the cell surface or by mechanical stresses that were associated with the manipulations that are an inevitable consequence of prolonged *in vitro* cell culture techniques [1]. The release of heat shock proteins did not appear to be mediated via the common secretory pathway, because it was not blocked by the inhibitors colchicine and monensin [1]. Nor was heat shock protein release due to cell lysis, because Hsp70 was not readily released from cells exposed to low concentrations of non-ionic detergents [1]. Rather, a selective release mechanism was suggested, and this was supported by the observation that Hsp70 synthesised in the presence of the lysine amino acid analogue aminoethyl cysteine was not released from cells. This was probably due to an alteration in the structure and/or function of the molecule which prevents its correct interaction with the specific release mechanism [1]. A number of studies have since reported the release of heat shock proteins from a range of cell types. See Chapter 3 for a detailed discussion of protein secretion mechanisms.

12.2.2. Heat shock proteins are released from a number of different cell types

Heat shock induces a four-fold increase in the levels of Hsp60 in the medium from cultured human islet cells [2], and an Hsp60-like protein is released from insulin-secreting β-cells [19].

An Hsp60-like protein has also been detected in conditioned media derived from cultured rat cortical astrocytes and a human neuroblastoma cell line [3]. In the case of neuroblastoma cells, extracellular Hsp60-like immunoreactivity is increased three-fold in the presence of the neuropeptide vasoactive intestinal peptide (VIP), and this increase occurs concomitantly with a two- to three-fold reduction in intracellular levels. Levels of exogenous Hsp60 are also increased two-fold after temperature elevation, and the effects are additive when VIP and thermal stress are combined [3]. As with most studies, no lactate dehydrogenase activity, an exclusively intracellular enzyme, was observed in the extracellular compartment, thus excluding the presence of cellular necrosis/damage [3]. The ability of cells to secrete heat shock proteins appears to be dependent on the cell type, because the levels of extracellular Hsp60-like protein generated by a human keratinocyte-derived cell line are at least 10-fold less than those generated by human neuroblastoma cells [3]. Readers should refer to Chapter 15 for further discussion of the interactions among neuronal cells, VIP and Hsp60.

Heat shock proteins are also released from cultured vascular smooth muscle cells subjected to oxidative stress by treatment with the naphthoquinolinedione LY83583. Sequential chromatography and tandem mass spectrometry has identified several proteins that are specifically secreted in response to such a stress, one of which is Hsp90α [4]. The release of heat shock proteins appears to be selective, because Hsp90β is not secreted under such conditions [4].

Murine and human prostate cancer cell lines secrete Hsp70, and secretion can be increased by forcing the expression of the protein by transfecting cells with a vector coding for murine Hsp70 [20]. Data from a number of sources would suggest that the release of such proteins from tumour cells might have implications for the development of tumour immunity (see Chapter 18), and indeed the forced over-expression of Hsp70 delays tumour growth and extends the survival of mice administered such Hsp70 transformed cells [20]. Another study has shown that IFN-γ can induce the active release of Hsc70 from K562 erythroleukaemic cells and that this was mirrored by a concomitant reduction in the amount of Hsc70 present on the surface of these cells [21]. The impact that such a release might have on the development of immunity is unclear given the evidence that Hsc70 appears to be unable to mediate the induction of tumour-specific immunity [22].

As detailed in Chapter 14, the presence of cell-free BiP has been reported in the synovial fluid and serum of patients with rheumatoid arthritis, and also in the synovial fluid of patients with other joint diseases. Hsp70 levels are dramatically increased in the synovial fluid of patients with rheumatoid arthritis and, to a much lesser extent, in the synovial fluid of patients with osteoarthritis and gout [23]. Given that BiP has anti-inflammatory actions, the release of this protein in rheumatoid arthritis might be part of a natural anti-inflammatory mechanism involving molecular chaperones.

12.3. Mechanisms of heat shock protein release

Although cellular necrosis inevitably leads to the non-specific release of intra-cellular proteins, the mechanism(s) via which heat shock proteins are actively released, either constitutively or in response to various factors from viable (non-necrotic) cells, has yet to be fully elucidated. As described in Chapter 3, 'non-classical' secretion of proteins that lack the typical N-terminal signal peptide sequences has been observed for a number of proteins such as fibroblast growth factors 1 and 2, IL-1 as well as viral proteins [24] and the mechanisms involved in such a process are reviewed in that chapter. A number of mechanisms might be involved in the release of heat shock proteins.

12.3.1. Release of heat shock proteins via classical or non-classical secretory pathways?

The original report of heat shock protein release from rat embryo cells suggested that it was not influenced by the inhibitors of the common secretory pathway, colchicine and monensin [1]. Inhibiting the common secretory pathway using brefeldin A also appears to have no effect on the release of Hsp70 from neuroblastoma cells treated with VIP [3], nor does it influence the release of Hsp70 from prostate cancer cell lines [20].

One study showing that a pharmacological inhibitor of phospholipase C activity (U731222) induces the release of Hsp70 from the A431 human carcinoma cell line suggests that phospholipase C inhibition might be one such mechanism [25]. The release of Hsp70 induced by inhibiting phospholipase C activity might occur via vesicular transport, because in the same publication the authors refer to unpublished data that indicate that the inhibition of vesicular transport with brefeldin A prevents Hsp70 release [25].

Another observation has been that a large proportion of the Hsp70 released by A431 human carcinoma cells is ubiquitinylated [25]. In addition to signalling for proteosome-dependent degradation, ubiquitination has been shown to serve as a trigger for different transport events [26, 27].

12.3.2. Release of heat shock proteins via exosomes?

Exosomes are small membrane vesicles that form within late endocytic compartments called multi-vesicular bodies (MVBs) and are distinct to apoptotic vesicles in that they differ in their mode of production and protein composition [28]. Further details on exosomes can be found in Chapter 3. The fusion of MVBs with the plasma membrane leads to the release of exosomes into the extracellular space. Various haematopoietic and non-haematopoietic cell types secrete exosomes, including reticulocytes, B and T lymphocytes, mast cells, platelets, macrophages, alveolar lung cells, tumour cells, intestinal epithelial cells and professional antigen-presenting cells (APCs) such as dendritic cells (DCs), and their function in different physiological processes depends on their origin [29]. DC- and tumour-derived exosomes are enriched in Hsp70, Hsc70 and Hsp90 [30, 31] and exosomes released from reticulocytes contain Hsp70 [32]. It might be that the release of heat shock proteins from cells is achieved via such a route.

12.3.3. Release of heat shock proteins via lipid rafts?

Lipid rafts might also be involved in the localisation of Hsp70 to the cell surface and its secretion into the extracellular environment [33]. Lipid rafts are specialised membrane domains enriched in sphingolipids, cholesterol and proteins that have been primarily characterised in polarised epithelial cells. Many functions have been attributed to lipid rafts, including cholesterol transport, membrane sorting, endocytosis and signal transduction [34, 35], and they can be isolated as detergent-resistant microdomains (DRMs) [36]. Hsp70 and Hsp90 and other molecules that are implicated in lipopolysaccharide (LPS)-mediated cellular activation are present in DRMs following LPS stimulation [37].

In unstressed Caco-2 human colonic adenocarcinoma epithelial cells, heat shock proteins (especially Hsp70) are present in a major Triton X-100 soluble form and a minor detergent insoluble form, which is associated with DRMs. Levels of Hsc70 and chaperones that are typically resident in the endoplasmic reticulum are low or undetectable in DRMs [33]. The translocation of Hsp70 into DRMs can be enhanced by heat shock or by increasing intracellular Ca^{2+} levels [33]. Although the incorporation of Hsp70 into the DRMs and the release of Hsp70 from Caco-2 cannot be inhibited by blockade of the common secretory pathway using brefeldin A or monensin, Hsp70 release can be blocked by disrupting lipid rafts using methyl-β-cyclodextrin [33].

Hypotheses about the extracellular biology of molecular chaperones find lack of support from referees and granting bodies because of a perception that because there are no defined mechanisms for the secretion of these proteins they, therefore, cannot be secreted. This position has been clearly defined by John Ellis, one of the pioneers of chaperone biology, in Chapter 1. However, it must be recognised that we know surprisingly little about protein secretions. It is only within the past decade or so that a number of extremely important and novel pathways of protein secretion in bacteria have been identified. The secretion pathways by which key mediators like IL-1 are released from eukaryotic cells are still not fully identified. The development of the biology of extracellular molecular chaperones will depend on the elucidation of the pathways by which these proteins are released from cells, and the authors hope that readers of this volume will take up this challenge.

12.4. Circulating heat shock proteins in health and disease

12.4.1. Circulating heat shock proteins in normal individuals

Hsp60 and Hsp70 are present in the serum of clinically normal individuals, in some instances at levels that are likely to elicit biological effects (>1000 ng/mL; [5, 8, 10, 12]. Circulating Hsp60 levels are not associated with cardiovascular risk factors such as body mass index, blood pressure and smoking status [5]; however, higher levels of circulating Hsp60 have been noted in individuals exhibiting an unfavourable lipid profile, as indicated by a low HDL cholesterol and a high total/HDL cholesterol ratio [10]. Levels are also associated with levels of the inflammatory cytokine tumour necrosis factor (TNF)-α [10]. In another study, serum Hsp60, but not Hsp70, levels have been shown to be associated with VLDL and triglyceride levels [5].

Exercise has been shown to induce the release of Hsp70 into the peripheral circulation of normal individuals [38, 39]. In one study, the observed increase in Hsp70 levels preceded any increases in Hsp70 protein and gene expression in contracting muscle, thereby arguing against contracting muscle being the source of the circulating Hsp70 [38]. Subsequent studies have demonstrated that the release of Hsp70 from splanchnic tissues during exercise is responsible, in part at least, for the elevated systemic concentrations of this protein [39].

12.4.2. Circulating heat shock proteins and ageing

Increasing age is associated with a reduced capacity to maintain homeostasis in all physiological systems, and it might be that this results, in part at least, from a parallel and progressive decline in the ability to produce heat shock proteins. If this is so, an attenuated heat shock protein response could contribute to the increased susceptibility to environmental challenges and the more prevalent morbidity and mortality which is seen in aged individuals [40, 41].

In vitro studies have shown that Hsp70 expression in heat-stressed lung cells [42], hepatocytes and liver [43, 44], splenocytes [45], myocardium [46] and mononuclear cells is reduced with increasing age [40], as is the induction of Hsp70 expression in response to ischaemia [47] and mitogenic stimulation [48]. Hsp70 gene expression declines during normal aging in human retina [49], and heat shock–induced Hsp70 expression is decreased in senescent and late-passage cells, both of which suggest that the process of aging itself might be associated with reduced Hsp70 production [50–52].

In keeping with the reduced capacity of cells and organisms to generate stress responses with aging, Hsp70 levels in peripheral blood lymphocytes decline with age, as do serum levels of Hsp60 and Hsp70 [9, 53]. The biological and physiological relevance of declining levels of circulating heat shock proteins with increasing age are unclear; however, intuitively, one consequence might be a reduced resistance to stress and the accumulation of damage.

12.4.3. Circulating heat shock proteins in cardiovascular disease

Evidence suggests that the inflammatory component to atherosclerosis might, at least in part, involve immune reactivity to heat shock proteins [54, 55], and a number of investigators have measured circulating levels of heat shock proteins in a variety of cardiovascular disease states. Hsp60 and Hsp70 are present in the serum of clinically normal individuals [5, 8, 10], and we and others have shown that serum Hsp60 levels are associated with early atherosclerosis in such individuals [5, 8]. Hsp60 has also been detected in the circulation of patients with acute coronary syndromes and chronic stable angina [56], and Hsp70 levels are elevated in patients with peripheral and renal vascular disease [57].

12.4.3.1. Circulating heat shock proteins as attenuators of cardiovascular disease?

A study of 218 subjects with established hypertension has shown that increases in carotid intima-media (IM) thicknesses (a measure of atherosclerosis) at a four-year follow-up are significantly less prevalent (odds ratio 0.42; $p < 0.008$) in those individuals having high serum Hsp70 levels (75th percentile) at enrollment [13]. A similar, albeit non-significant, trend for Hsp60 levels (odds ratio 0.6; $p = 0.10$) has also been observed. The relationship between Hsp70 levels and changes in IM thickness is independent of age, smoking habits and blood lipids.

A cross-sectional study that measured serum Hsp70 levels in 421 individuals evaluated for CAD by coronary angiography found that serum Hsp70 levels are significantly higher in patients without evidence of CAD, which supports our findings [14]. Again, the association of high Hsp70 levels with lack of CAD was independent of any relationship with traditional risk factors [14]. These findings indicate that circulating Hsp70 levels predict the development of atherosclerosis, at least in subjects with established hypertension and, arguably more importantly, suggest that Hsp70 influences its progression. The mechanism(s) by which such effects are manifested are currently unclear. However, atherosclerosis is an inflammatory condition, and Hsp70 is known to be capable of attenuating inflammatory responses by inducing self-heat shock protein–specific Th2-type CD4$^+$ T cells producing the regulatory cytokines IL-4 and IL-10 [58–60], and mycobacterial Hsp70 induces the secretion of IL-10 from peripheral blood monocytes [61].

Although we have previously reported elevated levels of Hsp70 in the peripheral circulation of patients with peripheral and renal vascular disease [57], we have not observed any relationship between Hsp70 levels and IM thickness in subjects with established hypertension [11], nor have we observed any relationship between Hsp70 levels and IM thickness in subjects with borderline hypertension [5]. Elevated levels of Hsp70 in peripheral and renal vascular disease [57] might result from the inflammatory response that is associated with established atherosclerotic disease. This proposition is supported by the observation that, although higher than controls, Hsp70 levels in patients with localised renal vascular disease are significantly lower than those in patients with more disseminated peripheral vascular disease [57]. It appears to be difficult to draw parallels between the events leading to elevated Hsp70 levels in overt and clinically established symptomatic vascular disease with those involved in the more subtle changes associated with increases in IM thickness.

12.4.3.2. Circulating heat shock proteins as promoters of cardiovascular disease?

A subset of $CD4^+$ T cells which lacks expression of the CD28 co-stimulation antigen ($CD4^+CD28^{null}$) is expanded in the circulation of patients with unstable angina and can comprise up to 50% of this cell population [62]. That these cells exhibit characteristics of natural killer cells, can produce high levels of IFN-γ and are present in ruptured atherosclerotic plaques suggests that they might have a role in the events that lead to plaque destabilisation and acute coronary syndrome (ACS) [63]. Human Hsp60 (which is present in the peripheral circulation of these individuals [56]) induces $CD4^+CD28^{null}$ T cells from patients with ACS to express mRNA for IFN-γ and the cytolytic molecule perforin, whereas $CD4^+CD28^{null}$ cells obtained from normal individuals or patients with chronic stable angina do not respond to Hsp60 [56]. The influence of Hsp60-reactive $CD4^+CD28^{null}$ cells on the events leading to ACS remains to be more fully determined.

12.4.4. Circulating heat shock proteins in diabetes

12.4.4.1. Type 1 diabetes

Although they did not detect them in plasma from normal individuals, Finotti and colleagues have reported that Hsp70 and grp94 (gp96) are present at high concentrations in the plasma of patients with type 1 diabetes and that they are complexed with IgG and albumin [64, 65]. Vascular complications in patients with type 1 diabetes are reflected, often independently, by (i) glycaemic control, (ii) alterations in proteolytic enzyme action and inhibition and (iii) a higher than normal proteolytic activity of plasma [66]. Grp94 entirely accounts for the proteolytic activity of plasma from diabetic patients and the proteolytic form of circulating grp94 appears to lack the glycosylation exhibited by its ER-derived counterpart [64, 65]. Alpha$_1$-anti-trypsin (α_1AT) is the most important circulating inhibitor of serine protease activity and is complexed with grp94 in the plasma of patients with type 1 diabetes [64, 65].

12.4.4.2. Type 2 diabetes

The presence of Hsp70 in the plasma of patients with type 2 diabetes has been reported by Williams and colleagues [67]. Patients with type 2 diabetes are subject to oxidative stress as a consequence of their hyperglycaemic state, and this might contribute to the vascular complications that are experienced by these individuals [68, 69]. Oxidative stress also results from the elevated homocysteine levels that are often found in patients with type 2 diabetes, high levels of which are a significant risk factor for diabetes [70]. Hsp70 levels in patients that are not taking insulin are higher than those that are present in patients taking insulin, and the reduction of serum homocysteine levels by the administration of the antioxidant folic acid, which reduces oxidative stress *in vivo*, significantly lowers circulating Hsp70 levels [67]. Hsp70 might therefore be a suitable marker of the severity of this clinical condition and be useful for monitoring patients with type 2 diabetes.

12.4.5. Circulating heat shock proteins and stress

During the course of a study investigating levels of circulating Hsp60 in the plasma of 229 healthy British civil servants taking part in the Whitehall II study, a prospective study aimed at identifying risk factors for coronary heart disease [71], Henderson and colleagues identified a significant association between elevated levels of Hsp60, low socioeconomic status and social isolation in males and females, as well as psychological distress in women [10]. Some insight into the mechanism by which this occurs has been provided by experimental animal studies.

Psychological stress induced by exposing male rats to a cat without physical contact increases serum levels of Hsp70, concomitant with an induction of intracellular expression of Hsp70 in the hypothalamus and dorsal vagal complex [72]. This effect appears to be mediated by adrenal hormones, as the induction on intracellular expression and circulating levels of Hsp70 elicited by cat exposure does not occur, or is attenuated in adrenalectomised animals [72].

12.4.6. Circulating heat shock proteins and infection

Elevated serum levels of Hsp70 have been found in patients with acute infections, and Hsp70 levels correlate with levels of the inflammatory markers IL-6 and TNF-α, as well as with levels of the anti-inflammatory cytokine IL-10 [12].

12.4.7. Circulating heat shock proteins after surgery and trauma

Surgical procedures increase circulating levels of heat shock proteins. Plasma concentrations of Hsp70 and IL-6 markedly increase in patients undergoing liver resection and are significantly associated with post-operative infection [73]. Hsp70 is also associated with hepatic ischaemic time and with the degree of post-operative organ dysfunction [73]. Although the observed relationship between Hsp70 and organ dysfunction would suggest that, rather than being cytoprotective, circulating Hsp70 is involved in the development of organ dysfunction, another study has shown there to be no relationship between Hsp70 levels and organ dysfunction, nor between Hsp70 levels and the severity of the post-injury inflammatory response following severe trauma [74]. Indeed, in the latter study, high levels of Hsp70 appear to be associated with an improved survival under such circumstances [74].

Hsp70 is released into the circulation following coronary artery bypass grafting (CABG) [75]. The observation that levels of Hsp70 peak immediately after surgery and before those of IL-6 (which peak at 5 hours) prompted the suggestion that Hsp70 release might contribute to the inflammatory response which is reflected by elevated IL-6 levels [75]. The release of Hsp70 following CABG appears to be related to the use of a heart-lung machine, because on-pump procedures result in plasma levels of Hsp70 that are approximately four times greater than those present after off-pump procedures [76]. Nevertheless, Hsp70 levels after off-pump procedures are significantly higher than pre-operative levels [76]. Interestingly, on-pump procedures also induce high levels of IL-10, and the highest levels of IL-10 are present in those individuals with the highest levels of Hsp70 [76]. Although this would appear to be counterintuitive given the perceived role of Hsp70 as an inflammatory agent, it is consistent with the observations that Hsp70 can induce regulatory Th2-type CD4$^+$ T cells producing the cytokines IL-4 and IL-10 ([58–60] and see Section 12.7) and that mycobacterial Hsp70 can induce the production of IL-10 from monocytes [61].

12.5. Sequence versus functional conservation in the heat shock protein families

The ability of heat shock proteins to influence the activities of the innate and adaptive immune systems independently of chaperoned peptides has been demonstrated using both microbial- and endogenously derived (self) heat shock proteins (micHsp and enHsp, respectively). One of the dogmas of heat shock protein biology is that the high degree of sequence homology between equivalent heat shock protein family members derived from prokaryotes and eukaryotes (~50%) is reflected in a high degree of functional conservation. However, the rigidity of this concept is questioned by a number of studies because, despite their high degree of phylogenetic conservation, the biological activities of highly homologous heat shock proteins can differ considerably (see Section 6.4 in Chapter 6). Immune responses to micHsp and enHsp are tightly controlled, differentially controlled, and quantitatively and qualitatively different. A small number of bacteria, one of which is *Mycobacterium tuberculosis*, contain multiple genes encoding Hsp60 (chaperonin 60, cpn) genes, and despite having greater than 73% amino acid similarity, mycobacterial Cpn60.1 is between 10- and 100-fold more active in inducing cytokine secretion than Cpn60.2

(otherwise known as Hsp65) [77]. In addition, whereas Cpn60.3 from *Rhizobium leguminosarum* induces the production of a range of cytokines from human monocytes, Cpn60.1, which has a 74% amino acid sequence homology with Cpn60.3, exhibits no cytokine-inducing activity [78]. Thirdly, Hsp60 from the oral bacterium *Actinobacillus actinomycetemcomitans* and *Escherichia coli* are potent stimulators of bone resorption [79, 80], whereas equivalent molecules from mycobacteria are not [79, 81].

Despite having similar potency for stimulating TNF-α production in mouse macrophages, phylogenetically separate Hsp60 species interact with murine macrophages via different recognition systems [82]. The same might be true for members of other heat shock protein families, and it has been shown that human Hsp70 binds to murine macrophages via the CD40 molecule, but at a binding site which is distinct to that used by the bacterial Hsp70 homologue DnaK [83].

T cells appear to be capable of distinguishing enHsp60 and micHsp60, because the phenotype of T cells responding to eukaryotic and prokaryotic Hsp60 and their cytokine secretion profile differ. Whereas human Hsp60 activates CD45RA$^+$RO$^-$ (naïve) human peripheral blood T cells, bacterial-specific peptides activate CD45RA$^-$RO$^+$ (memory) T cells, and bacterial Hsp60 (which contains both conserved (human) and non-conserved (bacterial) sequences) activates CD45RA$^+$RO$^-$ and CD45RA$^-$RO$^+$ T-cells [84].

The phenotype of the immune response to enHsp and micHsp60 also differs because T cells isolated from the synovial fluid of patients with rheumatoid arthritis respond to enHsp60 by predominantly producing regulatory Th2-type cytokine responses, whereas the response to micHsp60 produces higher levels of IFN-γ, which is consistent with a pro-inflammatory Th1-type response [85]. In addition, T cell lines generated from the synovial fluid of patients with rheumatoid arthritis in response to enHsp60 suppress the production of the pro-inflammatory cytokine TNF-α by peripheral blood mononuclear cells, whereas cells generated using mycobacterial Hsp65 have no such regulatory effect [85].

12.6. Functional consequences of exogenous heat shock proteins: immunological

Much attention has focussed on the capacity of heat shock proteins to interact with and influence the activities of innate and adaptive immune cells via a number of different receptors, including those of the Toll-like receptor family, and these activities are reviewed in great detail elsewhere in this volume (see Chapters 5 to 11).

Heat shock protein expression and immune reactivity towards heat shock proteins have been implicated in autoimmune diseases such as arthritis [86–88], multiple sclerosis [89–91], diabetes [92–94] and cardiovascular disease [55], and the administration of mammalian Hsp70 has been shown to prevent the induction of tolerance and promote the development of autoimmune disease *in vivo* [95].

The general perception is that exogenous heat shock proteins act as inflammatory mediators. However, from an evolutionary perspective it would not seem reasonable that mammalian responses to bacterial heat shock proteins (which presumably evolved as a defence) should also occur against ubiquitously expressed mammalian heat shock proteins, especially given that these proteins are present in the extracellular compartment under non-pathological conditions. A number of observations question the proposition that immune reactivity to self-heat shock proteins necessarily has a direct pro-inflammatory role in inflammatory disease.

The induction of T cell reactivity to enHsp60 and enHsp70 down-regulates disease in a number of experimental arthritis models, by a mechanism that appears to involve the induction of Th2-type $CD4^+$ T cells producing the regulatory cytokines IL-4 and IL-10 [58–60, 96–100]. The clinical relevance of these findings has been confirmed by studies that have reported an inverse association between the severity of disease and the production of regulatory cytokines such as IL-4 and IL-10 by T cells stimulated with Hsp60 in patients with rheumatoid arthritis [101–103]. The anti-inflammatory capacity of heat shock proteins also appears to be effective at the level of the APC, because mycobacterial Hsp70 induces the production of IL-10 by synoviocytes from patients with arthritis, and from monocytes from both patients and healthy controls [61]. IL-10 production by synoviocytes is accompanied by a decrease in the production of TNF-α [61].

The anti-inflammatory capacity of enHsp60 reactivity appears to dominate, because the administration of whole mycobacterial Hsp65, which contains the epitope that induces T cell activation and can induce arthritis in rats when administered alone, does not induce the disease. It therefore appears that the concomitant presence of conserved (self) epitopes can dominantly down-regulate the arthritogenic capacity of the non-conserved (non-self) epitopes [98]. The capacity of heat shock proteins to regulate inflammatory disease and the potential mechanisms by which this is achieved are reviewed in Chapter 16.

Much less studied has been the regulatory role of heat shock proteins in transplant rejection [104]; however, immunising recipient mice with enHsp60, or Hsp60 peptides that have the capacity to shift Hsp60 reactivity from a pro-inflammatory Th1 phenotype towards a regulatory Th2 phenotype, can delay murine skin allograft rejection [105]. In the clinical situation, it also appears that the development of immune responses to enHsp60 can regulate the allograft rejection response, in that in the late post-transplantation period (longer than one year) IL-10 production in response to enHsp60 peptides is increased [106]. At this time, the recognition of peptides from the intermediate and C-terminal regions of the protein appears to dominate [106].

A number of other heat shock proteins have been shown to exhibit anti-inflammatory activity. Human Hsp10 is now known to be early pregnancy factor [107, 108] and has been shown to be capable of inhibiting inflammation in the animal model of multiple sclerosis, experimental allergic encephalomyelitis [109, 110] (see Chapter 5). Extracellular Hsp27 stimulates IL-10 secretion by monocytes [111], and the anti-inflammatory properties of this protein are reviewed in Chapter 13. The protein BiP (grp78), a member of the Hsp70 family of molecules, is an autoantigen in rheumatoid arthritis [112], and its anti-inflammatory effects are reviewed in Chapter 14.

Taken together, these findings suggest that, rather than being pro-inflammatory, reactivity to endogenous (self) heat shock proteins is part of a normal immunoregulatory response that has the potential to dominantly control pro-inflammatory responses and inflammatory disease. The report that the treatment of human monocytes with enHsp60 suppresses their production of TNF-α following re-stimulation with enHsp60 or treatment with LPS, yet enhances their production of IL-1β, and that it down-regulates the expression of HLA-DR, CD86 and Toll-like receptor 4 [113], highlights the complexity of heat shock protein–mediated immunoregulation and the difficulties that will be encountered in attempting to understand the balance between the ability of these proteins to control inflammatory and regulatory responses.

12.7. Functional consequences of exogenous heat shock proteins: non-immunological

Although much attention has focussed on the capacity of exogenous heat shock proteins to interact with and influence the activities of innate and adaptive immune cells, a number of studies have also considered non-immunological consequences of heat shock protein interactions with a range of cell types. For example, exogenous Hsp/Hsc70 can change the differentiation patterns of the U937 promonocytic cell line [114] and has been shown to have a number of cytoprotective and other activities. Exogenous Hsp27, which is a member of the smaller heat shock protein family, inhibits *in vitro* culture-induced neutrophil apoptosis [115].

12.7.1. Cytoprotection

The cytoprotective effects of intracellular heat shock proteins have been appreciated for some considerable time [116]; however, extracellular heat shock proteins also appear to have cytoprotective effects (see Chapter 9). Some of the earliest evidence that heat shock proteins might have a therapeutic potential arose from the observations that exogenous members of the Hsp70 family protect spinal sensory neurons from axotomy-induced death and cultured aortic cells from heat stress [117, 118]. Subsequent work using a neonatal mouse model, in which spinal sensory (dorsal root ganglion) and motor neurons are induced to die by transection of their peripheral projections, demonstrated that exogenous Hsc70 can prevent axotomy-induced death of spinal sensory neurons [117]. The protective effect was selective, because treatment had no effect on the survival of motor neurons [117].

The capacity of glial cells to export Hsp70, and of Hsp70 to protect stressed neural cells, has also been demonstrated in a human system using T98G human glioma cells and stress-sensitive, differentiated LA-N-5 human neuroblastoma cells [18]. It might therefore be the case that heat shock protein release is an altruistic response on the part of one cell which is aimed at the protection of its more vulnerable neighbours [1].

Evidence that exogenous Hsp70 has cytoprotective effects on vascular-derived cells arose from the observations that exogenous Hsp70 protects heat-stressed cynomolgus macaque aortic cells [118] and serum-deprived rabbit arterial smooth muscle cells [119], the latter by a mechanism which involved cell association, but not internalisation. The mechanism by which such protection is induced is unknown, and the cell surface receptors involved have not been identified. However, some insight has been provided by studies that have shown exogenous Hsp70 to increase intracellular Hsp70 levels, which in turn delays the decline in viability of stressed cells [120], and that the accumulation of Hsp70 protects a range of cell types from apoptotic cell death induced by a number of apoptotic stimuli [121–125]. Hsp70 can inhibit apoptosis downstream of cytochrome c release, but upstream of caspase-3 cleavage, and the carboxyl-terminal region containing the peptide-binding domain is sufficient to inhibit caspase-3 activation [126].

Exogenous Hsp/Hsc70 renders neuroblastoma cells more resistant to staurosporine-induced apoptosis [18] and U937 pre-monocytes more resistant to cell death and apoptosis induced by TNF-α [114]. In human carcinoma cells, the release of Hsp70 induced by an inhibitor of phospholipase C activity leads to a concomitant reduction in intracellular levels of the protein, which in turn renders cells more sensitive to the apoptogenic effects of hydrogen peroxide

[25]. However, other studies in which Hsp70 expression has been enhanced by transfection have reported that elevated release of Hsp70 does not result in reduced intracellular levels [20].

12.7.2. Other activities

It must be emphasised that the biology of extracellular molecular chaperones is still in its infancy and the authors suggest that many and varied effects of these proteins will be reported. As examples from the current literature, *M. tuberculosis* Cpn10 is a potent inducer of bone resorption and the major osteolytic component of this organism [81] (see Chapter 5). As highlighted earlier, Hsp60 from the oral bacteria *A. actinomycetemcomitans* and *E. coli* are potent stimulators of bone resorption [79, 80], although equivalent molecules from mycobacteria are not [79, 81]. Human Hsp70, bacterial Hsp60 and their mycobacterial homologues induce ion-conducting pores across planar lipid bilayers at low or neutral pH [127]. A final example of the evolutionary plasticity of Hsp60 is the report that the endosymbiotic bacterium, *Enterobacter aerogenes*, present in the saliva of the antlion (a hunting insect), is the source of the neurotoxin produced by this insect. This insect toxin is none other than our old friend Hsp60 [128].

12.8. Conclusions

Heat shock proteins are extremely versatile and potent molecules, the importance of which to biological processes is highlighted by the high degree to which their structure and function are phylogenetically conserved. Our knowledge of the physiological role of heat shock proteins is currently limited; however, a better understanding of their function and thereby the acquisition of the capacity to harness their power might lead to their use as therapeutic agents and revolutionise clinical practice in a number of areas.

The observations that heat shock proteins can be released, and that they can directly or indirectly elicit potent immunoregulatory activities, requires that a new perspective on the roles of heat shock proteins and anti–heat shock protein reactivity in autoimmunity, transplantation, vascular disease and other conditions must be considered. It is the qualitative nature of the response to, or induced by, heat shock proteins rather than its presence *per se* that is important, and future experimental and clinical studies attempting to associate heat shock proteins in disease pathogenesis need to be structured and designed to address these issues. It is also important to definitively define the specificity of any responses, so that its outcome can be attributed to self- or non-self-reactivity. By doing this, the contribution of infective agents to pathogenic processes such as autoimmunity and vascular disease can be truly evaluated.

Acknowledgements

Work in the authors' laboratories has been funded by Boehringer-Ingelheim, Sweden, the Swedish Heart Lung Foundation, the King Gustav V 80th Birthday Fund, the Swedish Society of Medicine, the Swedish Rheumatism Association, the Söderberg Foundation, the Swedish Science Fund (JF), and the National Heart, Lung and Blood Institute, the Association for International Cancer Research and the British Heart Foundation (AGP).

REFERENCES

1. Hightower L E and Guidon P T. Selective release from cultured mammalian cells of heat-shock (stress) proteins that resemble glia-axon transfer proteins. J Cell Physiol 1989, 138: 257–266.
2. Child D F, Williams C P, Jones R P, Hudson P R, Jones M and Smith C J. Heat shock protein studies in type 1 and type 2 diabetes and human islet cell culture. Diabetic Med 1995, 12: 595–599.
3. Bassan M, Zamostiano R, Giladi E, Davidson A, Wollman Y, Pitman J, Hauser J, Brenneman D E and Gozes I. The identification of secreted heat shock 60-like protein from rat glial cells and a human neuroblastoma cell line. Neurosci Letters 1998, 250: 37–40.
4. Liao D-F, Jin Z-G, Baas A S, Daum G, Gygi S P, Aebersold R and Berk B C. Purification and identification of secreted oxidative stress-induced factors from vascular smooth muscle cells. J Biol Chem 2000, 275: 189–196.
5. Pockley A G, Wu R, Lemne C, Kiessling R, de Faire U and Frostegård J. Circulating heat shock protein 60 is associated with early cardiovascular disease. Hypertension 2000, 36: 303–307.
6. Pockley A G, Bulmer J, Hanks B M and Wright B H. Identification of human heat shock protein 60 (Hsp60) and anti-Hsp60 antibodies in the peripheral circulation of normal individuals. Cell Stress Chaperon 1999, 4: 29–35.
7. Pockley A G, Shepherd J and Corton J. Detection of heat shock protein 70 (Hsp70) and anti-Hsp70 antibodies in the serum of normal individuals. Immunol Invest 1998, 27: 367–377.
8. Xu Q, Schett G, Perschinka H, Mayr M, Egger G, Oberhollenzer F, Willeit J, Kiechl S and Wick G. Serum soluble heat shock protein 60 is elevated in subjects with atherosclerosis in a general population. Circulation 2000, 102: 14–20.
9. Rea I M, McNerlan S and Pockley A G. Serum heat shock protein and anti-heat shock protein antibody levels in aging. Exp Gerontol 2001, 36: 341–352.
10. Lewthwaite J, Owen N, Coates A, Henderson B and Steptoe A. Circulating human heat shock protein 60 in the plasma of British civil servants. Circulation 2002, 106: 196–201.

11. Pockley A G, de Faire U, Kiessling R, Lemne C, Thulin T and Frostegård J. Circulating heat shock protein and heat shock protein antibody levels in established hypertension. J Hypertension 2002, 20: 1815–1820.

12. Njemini R, Lambert M, Demanet C and Mets T. Elevated serum heat-shock protein 70 levels in patients with acute infection: use of an optimized enzyme-linked immunosorbent assay. Scand J Immunol 2003, 58: 664–669.

13. Pockley A G, Georgiades A, Thulin T, de Faire U and Frostegård J. Serum heat shock protein 70 levels predict the development of atherosclerosis in subjects with established hypertension. Hypertension 2003, 42: 235–238.

14. Zhu J, Quyyumi A A, Wu H, Csako G, Rott D, Zalles-Ganley A, Ogunmakinwa J, Halcox J and Epstein S E. Increased serum levels of heat shock protein 70 are associated with low risk of coronary artery disease. Arterioscler Thromb Vasc Biol 2003, 23: 1055–1059.

15. Todryk S M, Gough M J and Pockley A G. Facets of heat shock protein 70 show immunotherapeutic potential. Immunology 2003, 110: 1–9.

16. Bulmer J, Bolton A E and Pockley A G. Effect of combined heat, ozonation and ultraviolet light (VasoCare™) on heat shock protein expression by peripheral blood leukocyte populations. J Biol Reg Homeostatic Agents 1997, 11: 104–110.

17. Tytell M, Greenberg S G and Lasek R J. Heat shock-like protein is transferred from glia to axon. Brain Res 1986, 363: 161–164.

18. Guzhova I, Kislyakova K, Moskaliova O, Fridlanskaya I, Tytell M, Cheetham M and Margulis B. In vitro studies show that Hsp70 can be released by glia and that exogenous Hsp70 can enhance neuronal stress tolerance. Brain Res 2001, 914: 66–73.

19. Brudzynski K and Martinez V. Synaptophysin-containing microvesicles transport heat-shock protein Hsp60 in insulin-secreting β cells. Cytotechnology 1993, 11: 23–33.

20. Wang M H, Grossmann M E and Young C Y. Forced expression of heat-shock protein 70 increases the secretion of Hsp70 and provides protection against tumour growth. Brit J Cancer 2004, 90: 926–931.

21. Barreto A, Gonzalez J M, Kabingu E, Asea A and Fiorentino S. Stress-induced release of HSC70 from human tumors. Cell Immunol 2003, 222: 97–104.

22. Ménoret A, Patry Y, Burg C and Le Pendu J. Co-segregation of tumor immunogenicity with expression of inducible but not constitutive Hsp70 in rat colon carcinomas. J Immunol 1995, 155: 740–747.

23. Martin C A, Carsons S E, Kowalewski R, Bernstein D, Valentino M and Santiago-Schwarz F. Aberrant extracellular and dendritic cell (DC) surface expression of heat shock protein (Hsp)70 in the rheumatoid joint: possible mechanisms of Hsp/DC-mediated cross-priming. J Immunol 2003, 171: 5736–5742.

24. Cleves A E. Protein transports: the nonclassical ins and outs. Curr Biol 1997, 7: R318–R320.

25. Evdonin A L, Guzhova I V, Margulis B A and Medvedeva N D. Phospholipase C inhibitor, U73122, stimulates release of hsp-70 stress protein from A431 human carcinoma cells. Cancer Cell Int 2004, 4: 2.

26. Strous G J and Gent J. Dimerization, ubiquitylation and endocytosis go together in growth hormone receptor function. FEBS Letters 2002, 529: 102–109.

27. Katzmann D J, Odorizzi G and Emr S D. Receptor downregulation and multivesicular-body sorting. Nat Rev Mol Cell Biol 2002, 3: 893–905.

28. Théry C, Boussac M, Véron P, Ricciardi-Castagnoli P, Raposo G, Garin G and Amigorena S. Proteomic analysis of dendritic cell-derived exosomes: a secreted subcellular compartment distinct from apoptotic vesicles. J Immunol 2001, 166: 7309–7318.

29. Denzer K, Kleijmeer M J, Heijnen H F, Stoorvogel W and Geuze H J. Exosome: from internal vesicle of the multivesicular body to intercellular signalling device. J Cell Sci 2000, 113 Pt 19: 3365–3374.

30. Théry C, Zitvogel L and Amigorena S. Exosomes: composition, biogenesis and function. Nat Rev Immunol 2002, 2: 569–579.

31. Chaput N, Taïeb J, Schartz N E, Andre F, Angevin E and Zitvogel L. Exosome-based immunotherapy. Cancer Immunol Immunother 2004, 53: 234–239.

32. Mathew A, Bell A and Johnstone R M. Hsp-70 is closely associated with the transferrin receptor in exosomes from maturing reticulocytes. Biochem J 1995, 308: 823–830.

33. Broquet A H, Thomas G, Masliah J, Trugnan G and Bachelet M. Expression of the molecular chaperone Hsp70 in detergent-resistant microdomains correlates with its membrane delivery and release. J Biol Chem 2003, 278: 21601–21606.

34. Pralle A, Keller P, Florin E L, Simons K and Horber J K. Sphingolipid-cholesterol rafts diffuse as small entities in the plasma membrane of mammalian cells. J Cell Biol 2000, 148: 997–1008.

35. Vereb G, Matko J, Vamosi G, Ibrahim S M, Magyar E, Varga S, Szollosi J, Jenei A, Gaspar R J, Waldmann T A and Damjanovich S. Cholesterol-dependent clustering of IL-2Rα and its colocalization with HLA and CD48 on T lymphoma cells suggest their functional association with lipid rafts. Proc Natl Acad Sci USA 2000, 97: 6013–3018.

36. Horejsi V, Cebecauer M, Cerny J, Brdicka T, Angelisova P and Drbal K. Signal transduction in leucocytes via GPI-anchored proteins: an experimental artefact or an aspect of immunoreceptor function? Immunol Lett 1998, 63: 63–73.

37. Triantafilou M, Miyake K, Golenbock D T and Triantafilou K. Mediators of innate immune recognition of bacteria concentrate in lipid rafts and facilitate lipopolysaccharide-induced cell activation. J Cell Sci 2002, 115: 2603–2611.

38. Walsh R C, Koukoulas I, Garnham A, Moseley P L, Hargreaves M and Febbraio M A. Exercise increases serum Hsp72 in humans. Cell Stress Chaperon 2001, 6: 386–393.

39. Febbraio M A, Ott P, Nielsen H B, Steensberg A, Keller C, Krustrup P, Secher N H and Pedersen B K. Exercise induces hepatosplanchnic release of heat shock protein 72 in humans. J Physiol 2002, 544: 957–962.

40. Richardson A and Holbrook N J. Aging and the cellular response to stress: reduction in the heat shock response. In Holbrook, N. J., Martin, G. R. and Lockshin, R. A. (Eds.) Cellular Aging and Cell Death. Wiley-Liss, New York 1996, pp 67–79.

41. Shelton D N, Chang E, Whittier P S, Choi D and Funk W D. Microarray analysis of replicative senescence. Curr Biol 1999, 9: 939–945.

42. Fargnoli J, Kunisada T, Fornace A J J, Schneider E L and Holbrook N J. Decreased expression of heat shock protein 70 mRNA and protein after heat treatment in cells of aged rats. Proc Natl Acad Sci USA 1990, 87: 846–850.

43. Heydari A R, Conrad C C and Richardson A. Expression of heat shock genes in hepatocytes is affected by age and food restriction in rats. J Nutrition 1995, 125: 410–418.

44. Hall D, Xu L, Drake V J, Oberley L W, Oberley T D, Museley P L and Kregel K C. Aging reduces adaptive capacity and stress protein expression in the liver after heat stress. J Appl Physiol 2000, 2: 749–759.
45. Pahlavani M A, Denny M, Moore S A, Weindruch R and Richardson A. The expression of heat shock protein 70 decreases with age in lymphocytes from rats and rhesus monkeys. Expl Cell Res 1995, 218: 310–318.
46. Gray C C, Amrani M, Smolenski R T, Taylor G I and Yacoub M H. Age dependence of heat stress mediated cardioprotection. Ann Thoracic Surg 2000, 2: 621–626.
47. Nitta Y, Abe K, Aoki M, Ohno I and Isoyama S. Diminished heat shock protein 70 mRNA induction in aged rats after ischemia. Am J Physiol 1994, 267: H1795–1803.
48. Faassen A E, O'Leary J J, Rodysill K J, Bergh N and Hallgren H M. Diminished heat-shock protein synthesis following mitogen stimulation of lymphocytes from aged donors. Exp Cell Res 1989, 183: 326–334.
49. Bernstein S L, Liu A M, Hansen B C and Somiari R I. Heat shock cognate-70 gene expression declines during normal aging of the primate retina. Invest Ophthalmol Vis Sci 2000, 10: 2857–2862.
50. Liu A Y, Lin Z, Choi H, Sorhage F and Li B. Attenuated induction of heat shock gene expression in aging diploid fibroblasts. J Biol Chem 1989, 264: 12037–12045.
51. Luce M C and Cristofalo V J. Reduction in heat shock gene expression correlates with increased thermosensitivity in senescent human fibroblasts. Exp Cell Res 1992, 202: 9–16.
52. Effros R B, Zhu X and Walford R L. Stress response of senescent T lymphocytes: reduced hsp70 is independent of the proliferative block. J Gerontol 1994, 49: B65–70.
53. Jin X, Wang R, Xiao C, Cheng L, Wang F, Yang L, Feng T, Chen M, Chen S, Fu X, Deng J, Wang R, Tang F, Wei Q, Tanguay R M and Wu T. Serum and lymphocyte levels of heat shock protein 70 in aging: a study in the normal Chinese population. Cell Stress Chaperon 2004, 9: 69–75.
54. Wick G, Kleindienst R, Schett G, Amberger A and Xu Q. Role of heat shock protein 65/60 in the pathogenesis of atherosclerosis. Int Arch Allergy Immunol 1995, 107: 130–131.
55. Pockley A G. Heat shock proteins, inflammation and cardiovascular disease. Circulation 2002, 105: 1012–1017.
56. Zal B, Kaski J C, Arno G, Akiyu J P, Xu Q, Cole D, Whelan M, Russell N, Madrigal J A, Dodi I A and Baboonian C. Heat-shock protein 60-reactive CD4+CD28null T cells in patients with acute coronary syndromes. Circulation 2004, 109: 1230–1235.
57. Wright B H, Corton J, El-Nahas A M, Wood R F M and Pockley A G. Elevated levels of circulating heat shock protein 70 (Hsp70) in peripheral and renal vascular disease. Heart Vessels 2000, 15: 18–22.
58. Kingston A E, Hicks C A, Colston M J and Billingham M E J. A 71-kD heat shock protein (hsp) from *Mycobacterium tuberculosis* has modulatory effects on experimental rat arthritis. Clin Exp Immunol 1996, 103: 77–82.
59. Tanaka S, Kimura Y, Mitani A, Yamamoto G, Nishimura H, Spallek R, Singh M, Noguchi T and Yoshikai Y. Activation of T cells recognizing an epitope of heat-shock protein 70 can protect against rat adjuvant arthritis. J Immunol 1999, 163: 5560–5565.
60. Wendling U, Paul L, van der Zee R, Prakken B, Singh M and van Eden W. A conserved mycobacterial heat shock protein (hsp) 70 sequence prevents adjuvant arthritis upon

nasal administration and induces IL-10-producing T cells that cross-react with the mammalian self-hsp70 homologue. J Immunol 2000, 164: 2711–2717.

61. Detanico T, Rodrigues L, Sabritto A C, Keisermann M, Bauer M E, Zwickey H and Bonorino C. Mycobacterial heat shock protein 70 induced interleukin-10 production: immunomodulation of synovial cell cytokine profile and dendritic cell maturation. Clin Exp Immunol 2004, 135: 336–342.

62. Liuzzo G, Kopecky S L, Frye R L, O' Fallon W M, Maseri A, Goronzy J J and Weyand C M. Perturbation of the T-cell repertoire in patients with unstable angina. Circulation 1999, 100: 2135–2139.

63. Nakajima T, Schulte S, Warrington K J, Kopecky S L, Frye R L, Goronzy J J and Weyand C M. T-cell-mediated lysis of endothelial cells in acute coronary syndromes. Circulation 2002, 105: 570–575.

64. Pagetta A, Folda A, Brunati A M and Finotti P. Identification and purification from the plasma of type 1 diabetic subjects of a proteolytically active Grp94. Evidence that Grp94 is entirely responsible for plasma proteolytic activity. Diabetologia 2003, 46: 996–1006.

65. Finotti P and Pagetta A. A heat shock protein 70 fusion protein with α_1-antitrypsin in plasma of type 1 diabetic subjects. Biochem Biophys Res Comm 2004, 315: 297–305.

66. Finotti P, Carraro P and Calderan A. Purification of proteinase-like and Na^+/K^+-ATPase stimulating substance from plasma of insulin-dependent diabetics and its identification as α_1-antitrypsin. Biochim Biophys Acta 1992, 1139: 122–132.

67. Hunter-Lavin C, Hudson P R, Mukherjee S, Davies G K, Williams C P, Harvey J N, Child D F and Williams J H H. Folate supplementation reduces serum Hsp70 levels in patients with type 2 diabetes. Cell Stress Chaperone, 2004, 9: 344–349.

68. Sampson M J, Gopaul N, Davies I R, Hughes D A and Carrier M J. Plasma F2 isoprostanes: direct evidence of increased free radical damage during acute hyperglycemia in type 2 diabetes. Diabetes Care 2002, 25: 537–541.

69. Spanheimer G R. Reducing cardiovascular risk in diabetes. Which factors to modify first? Postgrad Med 2001, 109: 33–36.

70. Stehouwer C D, Gall M A, Hougaard P, Jakobs C and Parving H H. Plasma homocysteine concentration predicts mortality in non-insulin-dependent diabetic patients with and without microalbuminuria. Kidney Int 1999, 55: 308–314.

71. Marmot M G, Smith G D, Stansfeld S, Patel C, North F, Head J, White I, Brunner E and Feeney A. Health inequalities among British civil servants: the Whitehall II study. Lancet 1991, 337: 1387–1393.

72. Fleshner M, Campisi J, Amiri L and Diamond D M. Cat exposure induces both intra- and extracellular Hsp72: the role of adrenal hormones. Psychoneuroendocrinology 2004, 29: 1142–1152.

73. Kimura F, Itoh H, Ambiru S, Shimizu H, Togawa A, Yoshidome H, Ohtsuka M, Shimamura F, Kato A, Nukui Y and Miyazaki M. Circulating heat-shock protein 70 is associated with postoperative infection and organ dysfunction after liver resection. Am J Surg 2004, 187: 777–784.

74. Pittet J F, Lee H, Morabito D, Howard M B, Welch W J and Mackersie R C. Serum levels of Hsp 72 measured early after trauma correlate with survival. J Trauma 2002, 52: 611–617.

75. Dybdahl B, Wahba A, Lien E, Flo T H, Waage A, Qureshi N, Sellevold O F, Espevik T and Sundan A. Inflammatory response after open heart surgery: release of

heat-shock protein 70 and signaling through Toll-like receptor-4. Circulation 2002, 105: 685–690.

76. Dybdahl B, Wahba A, Haaverstad R, Kirkeby-Garstad I, Kierulf P, Espevik T and Sundan A. On-pump versus off-pump coronary artery bypass grafting: more heat-shock protein 70 is released after on-pump surgery. Eur J Cardiothorac Surg 2004, 25: 985–992.

77. Lewthwaite J C, Coates A R M, Tormay P, Singh M, Mascagni P, Poole S, Roberts M, Sharp L and Henderson B. *Mycobacterium tuberculosis* chaperonin 60.1 is a more potent cytokine stimulator than chaperonin 60.2 (Hsp 65) and contains a CD14-binding domain. Infect Immun 2001, 69: 7349–7355.

78. Lewthwaite J, George R, Lund P A, Poole S, Tormay P, Sharp L, Coates A R M and Henderson B. *Rhizobium leguminosarum* chaperonin 60.3, but not chaperonin 60.1, induces cytokine production by human monocytes: activity is dependent on interaction with cell surface CD14. Cell Stress Chaperon 2002, 7: 130–136.

79. Kirby A C, Meghji S, Nair S P, White P, Reddi K, Nishihara T, Nakashima K, Willis A C, Sim R, Wilson M and Henderson B. The potent bone-resorbing mediator of *Actinobacillus actinomycetemcomitans* is homologous to the molecular chaperone GroEL. J Clin Invest 1995, 96: 1185–1194.

80. Reddi K, Meghji S, Nair S P, Arnett T R, Miller A D, Preuss M, Wilson M, Henderson B and Hill P. The *Escherichia coli* chaperonin 60 (groEL) is a potent stimulator of osteoclast formation. J Bone Miner Res 1998, 13: 1260–1266.

81. Meghji S, White P, Nair S P, Reddi K, Heron K, Henderson B, Zaliani A, Fossati G, Mascagni P, Hunt J F, Roberts M M and Coates A R. *Mycobacterium tuberculosis* chaperonin 10 stimulates bone resorption: a potential contributory factor in Pott's disease. J Exp Med 1997, 186: 1241–1246.

82. Habich C, Kempe K, van der Zee R, Burkart V and Kolb H. Different heat shock protein 60 species share pro-inflammatory activity but not binding sites on macrophages. FEBS Letters 2003, 533: 105–109.

83. Becker T, Hartl F U and Wieland F. CD40, an extracellular receptor for binding and uptake of Hsp70-peptide complexes. J Cell Biol 2002, 158: 1277–1285.

84. Ramage J M, Young J L, Goodall J C and Hill Gaston J S. T cell responses to heat shock protein 60: differential responses by CD4+ T cell subsets according to their expression of CD45 isotypes. J Immunol 1999, 162: 704–710.

85. van Roon J A G, van Eden W, van Roy J L A M, Lafeber F J P G and Bijlsma J W J. Stimulation of suppressive T cell responses by human but not bacterial 60-kD heat shock protein in synovial fluid of patients with rheumatoid arthritis. J Clin Invest 1997, 100: 459–463.

86. Res P C, Schaar C G, Breedveld F C, van Eden W, van Embden J D S, Cohen I R and De Vries R R P. Synovial fluid T cell reactivity against 65 kDa heat shock protein of mycobacteria in early chronic arthritis. Lancet 1988, ii: 478–480.

87. Gaston J S H, Life P F, Jenner P J, Colston M J and Bacon P A. Recognition of a mycobacteria-specific epitope in the 65kD heat shock protein by synovial fluid derived T cell clones. J Exp Med 1990, 171: 831–841.

88. de Graeff-Meeder E R, van der Zee R, Rijkers G T, Schuurman H J, Kuis W, Bijlsma J W J, Zegers B J M and van Eden W. Recognition of human 60 kD heat shock protein by mononuclear cells from patients with juvenile chronic arthritis. Lancet 1991, 337: 1368–1372.

89. Wucherpfennig K, Newcombe J, Li H, Keddy C and Cuzner M L. $\gamma\delta$ T cell receptor repertoire in acute multiple sclerosis lesions. Proc Natl Acad Sci USA 1992, 89: 4588–4592.
90. Georgopoulos C and McFarland H. Heat shock proteins in multiple sclerosis and other autoimmune diseases. Immunology Today 1993, 14: 373–375.
91. Stinissen P, Vandevyver C, Medaer R, Vandegaar L, Nies J, Tuyls L, Hafler D A, Raus J and Zhang J. Increased frequency of $\gamma\delta$ T cells in cerebrospinal fluid and peripheral blood of patients with multiple sclerosis: reactivity, cytotoxicity, and T cell receptor V gene rearrangements. J Immunol 1995, 154: 4883–4894.
92. Elias D, Markovits D, Reshef T, van der Zee R and Cohen I R. Induction and therapy of autoimmune diabetes in the non-obese diabetic mouse by a 65-kDa heat shock protein. Proc Natl Acad Sci USA 1990, 87: 1576–1580.
93. Child D, Smith C and Williams C. Heat shock protein and the double insult theory for the development of insulin-dependent diabetes. J Royal Soc Med (Eng) 1993, 86: 217–219.
94. Tun R Y M, Smith M D, Lo S S M, Rook G A W, Lydyard P and Leslie R D G. Antibodies to heat shock protein 65 kD in type 1 diabetes mellitus. Diabetic Medicine 1994, 11: 66–70.
95. Millar D G, Garza K M, Odermatt B, Elford A R, Ono N, Li Z and Ohashi P S. Hsp70 promotes antigen-presenting cell function and converts T-cell tolerance to autoimmunity *in vivo*. Nat Med 2003, 9: 1469–1476.
96. van den Broek M F, Hogervorst E J M, van Bruggen M C J, van Eden W, van der Zee R and van den Berg W. Protection against streptococcal cell wall induced arthritis by pretreatment with the 65kD heat shock protein. J Exp Med 1989, 170: 449–466.
97. Thompson S J, Rook G A W, Brealey R J, van der Zee R and Elson C J. Autoimmune reactions to heat shock proteins in pristane induced arthritis. Eur J Immunol 1990, 20: 2479–2484.
98. Anderton S M, van der Zee R, Prakken B, Noordzij A and van Eden W. Activation of T cells recognizing self 60-kD heat shock protein can protect against experimental arthritis. J Exp Med 1995, 181: 943–952.
99. Anderton S M and van Eden W. T lymphocyte recognition of Hsp60 in experimental arthritis. In van Eden, W. and Young, D. (Eds.) *Stress Proteins in Medicine*. Marcel Dekker, New York 1996, pp 73–91.
100. Paul A G A, van Kooten P J S, van Eden W and van der Zee R. Highly autoproliferative T cells specific for 60-kDa heat shock protein produce IL-4/IL-10 and IFN-γ and are protective in adjuvant arthritis. J Immunol 2000, 165: 7270–7277.
101. de Graeff-Meeder E R, van Eden W, Rijkers G T, Prakken B J, Kuis W, Voorhorst Ogink M M, van der Zee R, Schuurman H J, Helders P J and Zegers B J. Juvenile chronic arthritis: T cell reactivity to human HSP60 in patients with a favorable course of arthritis. J Clin Invest 1995, 95: 934–940.
102. van Roon J, van Eden W, Gmelig-Meylig E, Lafeber F and Bijlsma J. Reactivity of T cells from patients with rheumatoid arthritis towards human and mycobacterial hsp60. FASEB 1996, 10: A1312.
103. Macht L M, Elson C J, Kirwan J R, Gaston J S H, Lamont A G, Thompson J M and Thompson S J. Relationship between disease severity and responses by blood mononuclear cells from patients with rheumatoid arthritis to human heat-shock protein 60. Immunology 2000, 99: 208–214.

104. Pockley A G. Heat shock proteins, heat shock protein reactivity and allograft rejection. Transplantation 2001, 71: 1503–1507.
105. Birk O S, Gur S L, Elias D, Margalit R, Mor F, Carmi P, Bockova J, Altmann D M and Cohen I R. The 60-kDa heat shock protein modulates allograft rejection. Proc Nat Acad Sci USA 1999, 96: 5159–5163.
106. Caldas C, Spadafora-Ferreira M, Fonseca J A, Luna E, Iwai L K, Kalil J and Coelho V. T-cell response to self HSP60 peptides in renal transplant recipients: a regulatory role? Transplant Proc 2004, 36: 833–835.
107. Cavanagh A C. Identification of early pregnancy factor as chaperonin 10: implications for understanding its role. Rev Reprod 1996, 1: 28–32.
108. Rolfe B, Cavanagh A, Forde C, Bastin F, Chen C and Morton H. Modified rosette inhibition test with mouse lymphocytes for detection of early pregnancy factor in human pregnancy serum. J Immunol Methods 1984, 70: 1–11.
109. Zhang B, Walsh M D, Nguyen K B, Hillyard N C, Cavanagh A C, McCombe P A and Morton H. Early pregnancy factor treatment suppresses the inflammatory response and adhesion molecule expression in the spinal cord of SJL/J mice with experimental autoimmune encephalomyelitis and the delayed-type hypersensitivity reaction to trinitrochlorobenzene in normal BALB/c mice. J Neurol Sci 2003, 212: 37–46.
110. Athanasas-Platsis S, Zhang B, Hillyard N C, Cavanagh A C, Csurhes P A, Morton H and McCombe P A. Early pregnancy factor suppresses the infiltration of lymphocytes and macrophages in the spinal cord of rats during experimental autoimmune encephalomyelitis but has no effect on apoptosis. J Neurol Sci 2003, 214: 27–36.
111. De A K, Kodys K M, Yeh B S and Miller-Graziano C. Exaggerated human monocyte IL-10 concomitant to minimal TNF-α induction by heat-shock protein 27 (Hsp27) suggests Hsp27 is primarily an anti-inflammatory stimulus. J Immunol 2000, 165: 3951–3958.
112. Corrigall V M, Bodman-Smith M D, Fife M S, Canas B, Myers L K, Wooley P, Soh C, Staines N A, Pappin D J, Berlo S E, van Eden W, van der Zee R, Lanchbury J S and Panayi G S. The human endoplasmic reticulum molecular chaperone BiP is an autoantigen for rheumatoid arthritis and prevents the induction of experimental arthritis. J Immunol 2001, 166: 1492–1498.
113. Kilmartin B and Reen D J. HSP60 induces self-tolerance to repeated HSP60 stimulation and cross-tolerance to other pro-inflammatory stimuli. Eur J Immunol 2004, 34: 2041–2051.
114. Guzhova I V, Arnholdt A C, Darieva Z A, Kinev A V, Lasunskaia E B, Nilsson K, Bozhkov V M, Voronin A P and Margulis B A. Effects of exogenous stress protein 70 on the functional properties of human promonocytes through binding to cell surface and internalization. Cell Stress Chaperon 1998, 3: 67–77.
115. Sheth K, De A, Nolan B, Friel J, Duffy A, Ricciardi R, Miller-Graziano C and Bankey P. Heat shock protein 27 inhibits apoptosis in human neutrophils. J Surg Res 2001, 99: 129–133.
116. Hightower L E. Heat shock, stress proteins, chaperones and proteotoxicity. Cell 1991, 66: 191–197.
117. Houenou L J, Li L, Lei M, Kent C R and Tytell M. Exogenous heat shock cognate protein Hsc70 prevents axonomy-induced death of spinal sensory neurons. Cell Stress Chaperon 1996, 1: 161–166.

118. Johnson A D, Berberian P A and Bond M G. Effect of heat shock proteins on survival of isolated aortic cells from normal and atherosclerotic cynomolgus macaques. Atherosclerosis 1990, 84: 111–119.

119. Johnson A D and Tytell M. Exogenous Hsp70 becomes cell associated, but not internalised by stressed arterial smooth muscle cells. In Vitro Cell Dev Biol 1993, 29A: 807–812.

120. Berberian P, Johnson A and Bond M. Exogenous 70kD heat shock protein increases survival of normal and atheromatous arterial cells. FASEB J 1990, 4: A1031.

121. Jäättelä M, Wissing D, Bauer P A and Li G C. Major heat shock protein Hsp70 protects tumor cells from tumor necrosis factor cytotoxicity. EMBO J 1992, 11: 3507–3512.

122. Simon M M, Reikerstorfer A, Schwarz A, Krone C, Luger T A, Jäättelä M and Schwarz T. Heat shock protein 70 overexpression affects the response to ultraviolet light in murine fibroblasts. Evidence for increased cell viability and suppression of cytokine release. J Clin Invest 1995, 95: 926–933.

123. Samali A and Cotter T G. Heat shock proteins increase resistance to apoptosis. Exp Cell Res 1996, 223: 163–170.

124. Lasunskaia E B, Fridlianskaia I I, Guzhova I V, Bozhkov V M and Margulis B A. Accumulation of major stress protein 70kDa protects myeloid and lymphoid cells from death by apoptosis. Apoptosis 1997, 2: 156–163.

125. Mosser D D, Caron A W, Bourget L, Denis-Larose C and Massie B. Role of the human heat shock protein Hsp70 in protection against stress-induced apoptosis. Mol Cell Biol 1997, 17: 5317–5327.

126. Li C-Y, Lee J-S, Ko Y-G, Kim K-I and Seo J-S. Heat shock protein 70 inhibits apoptosis downstream of cytochrome c release and upstream of caspase-3 activation. J Biol Chem 2000, 275: 25665–25671.

127. Alder G M, Austen B M, Bashford C L, Mehlert A and Pasternak C A. Heat shock proteins induce pores in membranes. Biosci Rep 1990, 10: 509–518.

128. Yoshida N, Oeda K, Watanabe E, Mikami T, Fukita Y, Nishimura K, Komai K and Matsuda K. Protein function. Chaperonin turned insect toxin. Nature 2001, 411: 44.

13

Hsp27 as an Anti-inflammatory Protein

Krzysztof Laudanski, Asit K. De and Carol L. Miller-Graziano

13.1. Introduction

As discussed in other chapters in this volume, heat shock proteins are traditionally viewed as protein chaperones rather than immunomodulators [1-3]. However, recent data suggest that heat shock proteins might also be ancestral danger signals which activate adaptive and innate immune responses [3]. The majority of studies examining the immunomodulatory activities of heat shock proteins have focused on the large heat shock proteins, Hsp60, Hsp70 and gp96, and these proteins have been shown to stimulate the innate immune system via binding to a variety of cellular receptors, particularly on monocytes, and to play an important role in health and disease [1-6].

This chapter focuses on the small heat shock protein, Hsp27, which, although shown to have some role in resistance to chemotherapeutic drugs, cytokine-induced cytotoxicity and to have been described as a prognostic marker in serum of breast cancer patients, has not been well characterised as an immunomodulator [7-13]. Hsp27 has been reported as present in increased amounts in the serum of patients with several human diseases, as well as being necessary for activation of the signal transduction pathway leading to monocyte production of the anti-inflammatory and immunoinhibitory cytokine IL-10 [1-3, 14]. These data led to our interest in investigating the possible immunomodulatory activity of Hsp27 on different human monocyte functions, which are pivotal in both the development of inflammatory responses as well as the triggering of lymphocyte-specific immunity.

In contrast to the immunomodulatory effects of Hsp60 or Hsp70, the influence of which on monocytes is primarily pro-inflammatory, we found Hsp27 to have potent monocyte anti-inflammatory and immune inhibitory effects [15]. Since Hsp27 has been identified as circulating in the serum of breast cancer patients as well as burn victims, it has the potential to act as an exogenous mediator

[2–4, 7]. Exogenous Hsp27 could function as an anti-inflammatory and/or immunoinhibitory monocyte mediator through a number of mechanisms, which are not mutually exclusive. Hsp27 binding to leukocytes could activate other cytokines/mediators which then influence innate and specific immunity, in a similar manner to large heat shock proteins which induce monocyte production of pro-inflammatory cytokines such as tumour necrosis factor (TNF)-α [2, 3, 16, 17]. Hsp27 binding could also alter monocyte surface receptor expression, thereby changing their activation potential or differentiation capacity. Our data support the concept that Hsp27 has all of these activities. In the next chapter (Chapter 14) a similar set of actions are ascribed to the large chaperone, BiP.

Cytokine production by monocytes is key to their inflammatory and lymphocyte and antigen-presenting cell (APC) immune-activating activities, as well as their differentiation to either the most potent APC, the dendritic cell (DC), or to the end-stage inflammatory macrophage. Consequently, Hsp27 modulation of cytokine production by monocytes could dramatically impact their inflammatory and immune functions.

13.2. Hsp27 as an inducer of anti-inflammatory/immunomodulatory cytokines

We first assessed Hsp27 for its stimulation of monocyte mediators, because large heat shock proteins induce highly elevated pro-inflammatory cytokine production by monocytes [2, 3]. All assays were performed in polymyxin B–containing media to prevent any lipopolysaccharide (LPS) present in the recombinant Hsp27 from binding to lipopolysaccharide binding protein (LBP) and/or to the LPS receptor complex of CD14 and Toll-like receptor 4 (TLR4), thereby stimulating LPS rather than Hsp27-mediated responses [2]. Hsp27 induced exaggerated production of IL-10 by monocytes, but equivalent low levels of TNF-α as compared to other non-LPS bacterial stimuli such as *Staphylococcus aureus* enterotoxin B (SEB) or muramyl dipeptide (MDP), both of which are also unaffected by polymyxin B (Figure 13.1). Hsp27 induction of IL-10 is dose-dependent and inhibited by addition of a neutralising antibody to Hsp27, thereby further confirming that Hsp27 induces monocyte cytokine production (Figure 13.1). We have also shown that the activation of monocyte p38 mitogen-activated protein (MAP) kinase signalling pathway is prolonged when compared to LPS-induced activation [9]. This prolongation is crucial for the stimulation of large quantities of IL-10 by its alteration of the IL-10 induction kinetics [15, 18].

Hsp27-induced production of IL-10 by monocytes can inhibit the production of other pro-inflammatory cytokines such as IL-1β, TNF-α, IL-6 and

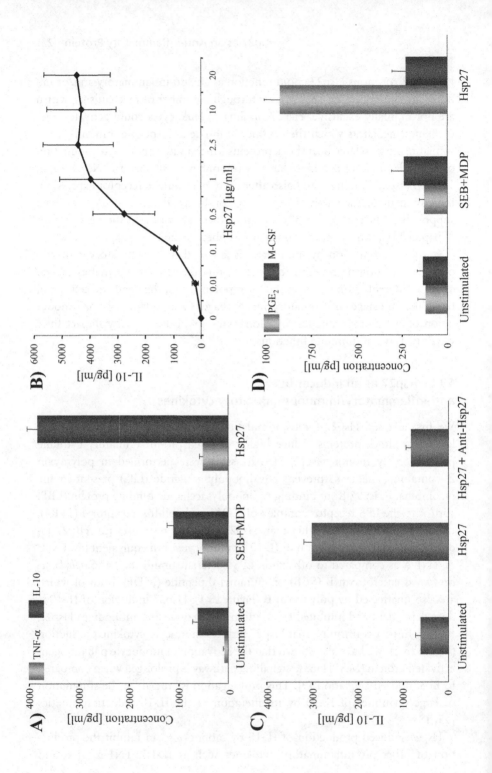

IL-12 and can decrease the expression of co-stimulatory molecules on DCs, diminish immunoglobulin production by B cells and prevent development of APCs [18]. These effects are all crucially important activities for the development and persistence of inflammatory responses. Therefore, Hsp27-mediated induction of IL-10 can result in a profound and widespread inhibition of inflammation and immunity. Hsp27 has been shown to induce high levels of monocyte prostaglandin E_2 (PGE_2), which is another mediator having well-described anti-inflammatory properties [19] (Figure 13.1). PGE_2 suppresses antigen- and mitogen-induced proliferation of T and B cells, as well as antibody production by B cells [19]. PGE_2 also decreases oxidative and phagocytic responses of monocytes, and their cytokine production. These activities suggest that the negative modulation of immune response after bacterial stimulation is one of its primary roles in host defence [19].

Finally, Hsp27 induces monocyte production of M-CSF. M-CSF has both pro- and anti-inflammatory actions on monocytes. M-CSF boosts phagocytosis, superoxide production, cytotoxicity and secondary cytokine-secreting macrophages, and suppresses immune response in pregnancy, facilitating HIV infection and depleting APC precursors from the peripheral blood pool of monocytes [20, 21]. M-CSF drives the emergence of terminally differentiated, activated, tissue macrophages from the monocyte population, thereby decreasing the DC precursor population by depleting available monocyte precursors. This can result in immunoparalysis and contribute to the monocyte-related pathology which has been described in trauma and in other patient groups [20–22].

13.3. Hsp27 can modulate the expression of monocyte receptors

Both monocyte activation and differentiation are dependent on the level and combination of surface receptors expressed [23–26]. Among the crucial receptors influencing monocyte activation and differentiation are the pattern recognition receptors such as the TLRs [23, 27].

Many authors have demonstrated the binding of large heat shock proteins to TLR4 [5, 28], and the interested reader should refer to Chapters 7, 8 and

Figure 13.1. Anti-inflammatory properties of Hsp27 in comparison to bacterial stimuli (SEB – staphylococcus enterotoxin B; MDP – muramyl dipeptide): (A) Hsp27 induced higher level of IL-10 in comparison to SEB + MDP, with only a minimal induction of TNF. (B) Induction of IL-10 is dose-dependent, reaching maximum levels at 5 µg/ml. (C) Neutralising antibody to Hsp27 significantly reduces Hsp27-mediated IL-10 induction. (D) Hsp27 is also a potent inducer of PGE and M-CSF in human monocytes.

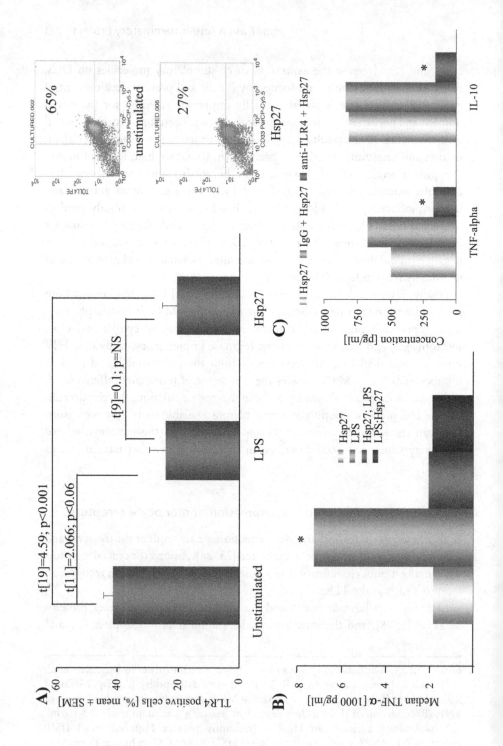

10 for more details. TLR4 is the receptor for not only LPS but also for other circulating mediators [24]. The variable detection and transduction of different ligand-specific signals by monocytes depends on the formation of co-receptor clusters around TLR4 [23–25, 29]. The specificity of this TLR4 receptor combination determines how the surface signal from LPS, Hsp60 or other TLR4 ligands is transduced and results in differential activation of kinases followed by signal-specific cytokine release [24, 29]. Based on the ability of large heat shock proteins to signal through TLR4 or other innate immune receptors, we hypothesise that Hsp27 might also interact with a TLR4 receptor cluster, thereby resulting in the production of its unique cytokine pattern, which is distinctive for this small chaperone. Binding of LPS to TLR4 on monocytes down-regulates this receptor and contributes to the development of LPS tolerance [24, 29], in which a subsequent dose of LPS elicits diminished cytokine production due to receptor changes, activation of intracellular inhibitory pathways, and the release of inhibitory cytokines (IL-10, TGF-β). LPS tolerance plays an important role in sepsis, autoimmune disease and allergy [1, 24–26]. Although it is a beneficial mechanism, under certain circumstances it can impair the ability to respond to pathogenic challenge [26, 30].

We have shown that exogenous Hsp27 is at least as potent as LPS in decreasing surface expression of the TLR4 receptor (Figure 13.2). In contrast, surface expression of TLR1 remains unchanged after monocyte stimulation with exogenous Hsp27, and TLR2 receptor expression is slightly up-regulated. The decrease of TLR4 levels on monocytes after Hsp27 stimulation *in vitro* might reflect classical down-regulation of surface expression, since intracellular staining of Hsp27-treated monocytes reveals no significant differences in total intracellular TLR4 concentrations. The precise mechanism by which Hsp27 influences TLR4 expression is undefined, although there are several possibilities.

Hsp27 actions on TLR4 could be mediated by direct binding to TLR4 followed by internalisation of the TLR4–Hsp27 complex into the cytoplasmic compartment. Although several authors have demonstrated that TLR4 is the binding receptor for large heat shock proteins, there is no conclusive proof to show that the same is true for Hsp27 [2, 3, 5, 24, 31]. Alternatively, Hsp27 could

---

Figure 13.2. Hsp27 may interact with TLR4 expressed on monocytes. Hsp27 induces a profound down-regulation of TLR4 receptors similar to that produced by LPS stimulation. Pre-treatment with Hsp27 modulates the magnitude of monocyte LPS TNF responses. Hsp27 added 2 hours prior to LPS stimulation reduces TNF production to 27% of LPS alone, whereas LPS pre-treatment did not affect Hsp27-induced TNF levels. Blocking anti-TLR4 antibody reduced Hsp27-induced pro- and anti-inflammatory cytokine production (* p < 0.05).

bind to other monocyte receptors such as RAGE, CD91, CD36 and chemokine receptors, and this might then indirectly stimulate down-regulation of TLR4 expression on the surface of monocytes.

Since TLR4 surface expression is significantly down-regulated after stimulation of monocytes with exogenous Hsp27, the monocyte response on subsequent exposure to LPS might also be diminished in a manner similar to LPS-induced tolerance, should TLR4 be a major Hsp27 receptor on monocytes [3, 26, 30]. When human monocytes are treated with Hsp27 followed by LPS (2-hour interval), or pre-treated with LPS followed by Hsp27 *in vitro*, the TNF-α response pattern is the same as that of Hsp27 alone, thereby demonstrating a reduced LPS response, which is similar to that observed for LPS tolerance [26] (Figure 13.2). Hsp27 pre-treatment or post-treatment reduces LPS-induced monocyte TNF-α levels, whereas Hsp27-induced TNF-α levels remain constant. One explanation for this finding is that Hsp27 induced early increases (8 hours after stimulation) of IL-10, which down-regulate LPS-induced TNF-α levels. LPS-induced IL-10 normally peaks 18 hours after stimulation, which is considerably later than TNF-α induction, thereby allowing an enhanced TNF-α induction [18]. However, treating monocytes with a neutralising antibody to TLR4 diminishes their Hsp27-induced IL-10 and TNF-α production, which indicates some type of Hsp27–TLR4 interactions (Figure 13.2).

The failure of monocyte LPS pre-treatment to reduce Hsp27-induced TNF-α implies that additional or other elements of the LPS binding receptor cluster might be affected. Antibody to TLR4 could inhibit Hsp27 by steric hindrance of the LPS receptor complex, rather than by directly interfering with Hsp27–TLR4 binding. The LPS receptor cluster involves TLR4, CD14, MD-2, CD11c, CD18 and CD36 arranged in a specific spatial pattern [24, 29]. Stimulation with other TLR4 ligands besides LPS results not only in different spatial composition of co-receptors for LPS receptor cluster, but also different composition of co-stimulatory receptors [22]. Whatever the mechanism, the ability of Hsp27 stimulation to decrease monocyte TLR4 levels will attenuate monocyte stimulation by other TLR4-dependent mediators and further influence the anti-inflammatory/inhibitory effects of Hsp27.

13.4. Hsp27 stimulation alters monocyte differentiation

Monocytes are pluripotent cells that can differentiate into a range of myeloid cell populations including immature DCs (iDCs) and macrophages, depending on the local microenvironmental signals [22]. IL-4 plus granulocyte macrophage-colony stimulating factor (GM-CSF), or similar cytokine combinations, drive monocytes to differentiate into iDCs, whereas stimulation by M-CSF, GM-CSF

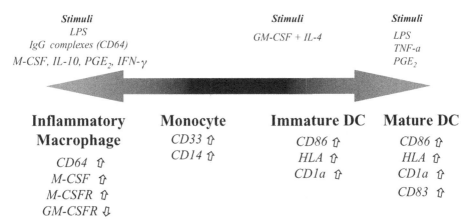

Figure 13.3. Mediator effects on differentiation of monocytes into terminally differentiated macrophage or DCs.

alone, or LPS induces monocyte-to-macrophage differentiation (Figure 13.3) [6, 12, 27]. The emergence of mature DCs (mDCs) is a two-phase process which requires distinct signals that induce differentiation to iDCs and their subsequent maturation. IL-4 and GM-CSF stimulate the differentiation of the iDCs, and these undergo terminal specialisation following exposure to LPS, TNF-α, PGE$_2$, or other TLR ligands. A relative lack of IL-4 or GM-CSF, or the presence of LPS, PGE$_2$, IL-10 or TNF-α at the initial differentiation stage of iDC development, results in the induction of apoptosis or a failure of the differentiation of monocytes into iDCs [22]. The absence of maturation stimuli for iDC development into fully competent mDCs also results in an aborted differentiation and iDC apoptosis. Surprisingly, mediators that inhibit the initial differentiation of monocytes into iDCs often subsequently induce the maturation of iDCs, and the timing of stimulation must therefore be well orchestrated. Monocyte-derived DCs play a pivotal role in T lymphocyte activation [22]. DCs are the only APC which is capable of activating naïve T cells, and they are crucial for the transition of activated T lymphocytes to memory T cells. DC defects have been described as major contributors to immunoaberrations in the regulation of T cells and the attenuation of specific immunity [21, 22, 32].

We have demonstrated that Hsp27 is a potent inducer of M-CSF, PGE$_2$ and IL-10. All of these mediators inhibit monocyte differentiation into DCs while promoting the differentiation of monocytes into macrophages. Hsp27 might therefore exert significant immunosuppressive effects on monocytes by inhibiting their differentiation into DCs. In a series of experiments, we have shown that the stimulation of monocytes with Hsp27 inhibits their differentiation into

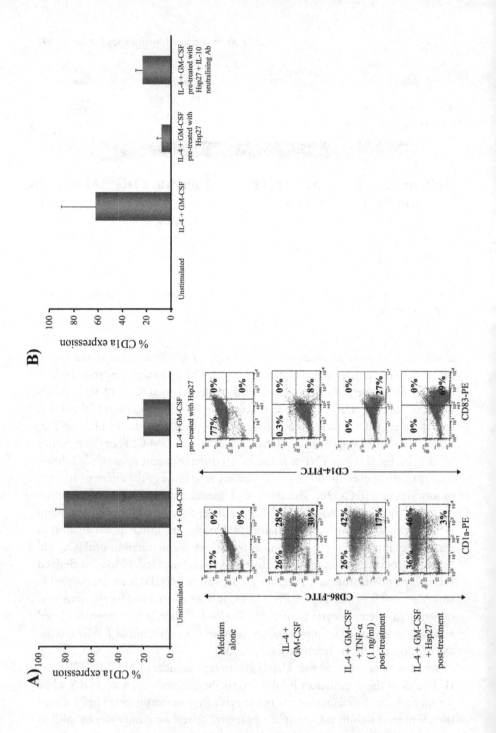

iDCs (CD1a$^+$ cells) after IL-4/GM-CSF co-stimulation, but promotes iDC maturation into mDCs (CD83$^+$ cells) (Figure 13.4). Hsp27-induced monocyte production of M-CSF, PGE$_2$ and IL-10 could be responsible for the inhibition of iDC differentiation after IL-4 plus GM-CSF stimulation [21, 22, 32]. IL-10 and M-CSF promote the differentiation of monocytes to macrophages and are secreted by monocytes after exposure to Hsp27 (Figures 13.2, 13.4). The action of M-CSF is especially critical since this cytokine induces autocrine secretion of both itself (positive feedback loop) and IL-10 [20, 21]. In healthy volunteers, neutralising antibody to IL-10 partially reverses the inhibitory effect of Hsp27 on iDC differentiation (Figure 13.4). This suggests that an increase in IL-10 levels is at least partially responsible for the diminished differentiation of monocytes to iDC, which is mediated by Hsp27. However, the inhibitory effects of Hsp27 on monocyte differentiation to iDCs naturally involve other inhibitory effects, perhaps elevated PGE$_2$ and M-CSF levels.

In contrast to its inhibitory effect on the differentiation of monocytes into iDCs, Hsp27 is a potent inducer of iDC maturation (Figure 13.4). Neither M-CSF nor IL-10 are particularly potent maturation factors for iDC, and PGE$_2$ is only a co-stimulator of maturation [19, 22, 32]. LPS stimulation is suggested to act as a DC maturation signal via TLR4 induction of TNF-α, IL-1β and, to a lesser degree, TGF-β. Since Hsp27 is a relatively poor inducer of TNF-α compared to LPS, the mechanism by which it induces DC maturation awaits further dissection. Nevertheless, these data illustrate that Hsp27 has both immunosuppressive and immunostimulatory capacities *in vitro*, and these will be dependent on when and where it interacts with differentiating monocytes.

13.5. Hsp27 in disease

We have shown exogenous Hsp27 to have potent inhibitory effects on monocyte inflammatory cytokine production and immunostimulatory activities *in vitro*. However, it is unclear whether Hsp27 has similar activities *in vivo*. Hsp27 is predominantly an intracellular molecule [3] and, to date, the secretion of Hsp27 under physiological conditions has not been demonstrated. However, it is released from cells undergoing necrosis [3, 7, 14, 33, 34]. This implies that Hsp27 released into tissue or serum after cell injury or death could then interact with leukocytes.

Figure 13.4. Hsp27 treatment affects differentiation and maturation of DCs: (A) Exogenous Hsp27 significantly inhibits the differentiation of iDCs from monocytes but augments the maturation of mature DCs from iDCs. (B) Addition of antibodies to monocytes during iDC differentiation partially reduces the inhibitory effects of Hsp27.

Elevated serum levels of Hsp27 have been demonstrated in breast cancer patients as well as in sufferers of other malignancies [7, 11]. Apoptosis and necrosis are increased in malignant cells and this possibly contributes to the release of Hsp27 and elevated circulating levels. Circulating Hsp27 also appears to trigger an immune response, as anti-Hsp27 antibodies are also detected in the serum of some cancer patients [9, 13]. Interestingly, increased serum titres of anti-Hsp27 antibodies correlate with an improved survival, which suggests that exogenous circulatory Hsp27 might itself be exerting activities that suppress anti-cancer immune functions. The capacity of Hsp27 to induce the immunosuppressive mediators IL-10, M-CSF and PGE_2, and to inhibit the differentiation of monocytes into DCs, might impair anti-tumour immunity. Hsp27 has also been implicated as a regulator of oestrogen receptor expression in breast cancer patients [35]. Further characterisation of the role of Hsp27 in cancer host defence is clearly required.

Since elevated circulatory Hsp27 levels are apparent in subjects with extensive tissue damage, we hypothesise that large quantities of Hsp27 can be released into the bloodstream in trauma. In a small pilot study, we have shown that Hsp27 can be detected in the serum of trauma patients at levels approximately three times higher than those found in healthy volunteers (24 ± 5.38 vs. 7.32 ± 5.1 ng/ml, respectively; $p < 0.05$). This finding suggests a new investigative area in which the capacity of circulating Hsp27 to modulate inflammatory events and immune function in the severely injured patient should be defined. Trauma patients experience monocyte paralysis, which is typified by LPS tolerance and a significant defect in the differentiation of monocytes to iDCs [21, 26]. Down-regulation of TLR4 as well as other monocyte receptors is also characteristic of these patients. Our laboratory is investigating the contribution of elevated exogenous Hsp27 in the development and maintenance of these monocyte aberrations in trauma patients.

13.6. Conclusions

Hsp27 is unusual among heat shock proteins in that it has the potential to suppress aspects of the immune system by multiple mechanisms. Induction of monocyte production of predominantly inhibitory cytokines (IL-10, M-CSF, PGE_2) with a relatively small release of pro-inflammatory molecules (TNF-α) clearly distinguishes Hsp27 from Hsp60, Hsp70 and other members of the large shock protein family. Hsp27 might also interfere with the response to bacterial endotoxins by down-regulating TLRs. However, the clinical importance of these activities remains to be established. Hsp27 is a potent inhibitor of monocyte differentiation into iDCs, and this activity could profoundly impair the induction of specific adaptive immunity. Further studies are necessary to determine the precise mechanisms by which exogenous Hsp27 elicits immunoinhibitory and anti-inflammatory activities and their impact. Nevertheless, like many members of the large heat shock protein family, Hsp27 is emerging as a major modulator of host defence as well as a molecular chaperone and substrate in the p38 MAP kinase signalling pathway.

Acknowledgements

This work was supported by grant GM036214 from the National Institutes of Health, U.S.A.

REFERENCES

1. Wick G, Knoflach M and Xu Q. Autoimmune and inflammatory mechanisms in atherosclerosis. Ann Rev Immunol 2004, 22: 361–403.
2. Tsan M F and Gao B. Cytokine function of heat shock proteins. Am J Physiol Cell Physiol 2004, 286: C739–C744.
3. Feder M E and Hofmann G E. Heat-shock proteins, molecular chaperones, and the stress response: evolutionary and ecological physiology. Ann Rev Physiol 1999, 61: 243–282.
4. Jäättelä M and Wissing D. Emerging role of heat shock proteins in biology and medicine. Ann Med 1992, 24: 249–258.
5. Ohashi K, Burkart V, Flohé S and Kolb H. Heat shock protein 60 is a putative endogenous ligand of the Toll-like receptor-4 complex. J Immunol 2000, 164: 558–561.
6. Zanin-Zhorov A, Nussbaum G, Franitza S, Cohen I R and Lider O. T cells respond to heat shock protein 60 via TLR2: activation of adhesion and inhibition of chemokine receptors. FASEB J 2003, 17: 1567–1569.
7. Fanelli M A, Cuello Carrion F D, Dekker J, Schoemaker J and Ciocca D R. Serological detection of heat shock protein hsp27 in normal and breast cancer patients. Canc Epidemiol Biomarkers Prev 1998, 7: 791–795.
8. Wissing D and Jäättelä M. HSP27 and HSP70 increase the survival of WEHI-S cells exposed to hyperthermia. Int J Hyperthermia 1996, 12: 125–138.
9. Korneeva I, Bongiovanni A M, Girotra M, Caputo T A and Witkin S S. IgA antibodies to the 27-kDa heat-shock protein in the genital tracts of women with gynecologic cancers. Int J Cancer 2000, 87: 824–828.
10. Garrido C, Ottavi P, Fromentin A, Hammann A, Arrigo A P, Chauffert B and Mehlen P. Hsp27 as a mediator of confluence-dependent resistance to cell death induced by anticancer drugs. Cancer Res 1997, 57: 2661–2667.
11. Korneeva I, Caputo T A and Witkin S S. Cell-free 27 kDa heat shock protein (hsp27) and hsp27-cytochrome c complexes in the cervix of women with ovarian or endometrial cancer. Int J Cancer 2002, 102: 483–486.

12. Jäättelä M and Wissing D. Heat-shock proteins protect cells from monocyte cytotoxicity: possible mechanism of self-protection. J Exp Med 1993, 177: 231–236.

13. Conroy S E, Sasieni P D, Amin V, Wang D Y, Smith P, Fentiman I S and Latchman D S. Antibodies to heat-shock protein 27 are associated with improved survival in patients with breast cancer. Br J Cancer 1998, 77: 1875–1879.

14. Carter Y, Liu G, Stephens W B, Carter G, Yang J and Mendez C. Heat shock protein (HSP72) and p38 MAPK involvement in sublethal hemorrhage (SLH)-induced tolerance. J Surg Res 2003, 111: 70–77.

15. De A K, Kodys K M, Yeh B S and Miller-Graziano C. Exaggerated human monocyte IL-10 concomitant to minimal TNF-α induction by heat-shock protein 27 (Hsp27) suggests Hsp27 is primarily an anti-inflammatory stimulus. J Immunol 2000, 165: 3951–3958.

16. Asea A, Rehli M, Kabingu E, Boch J A, Baré O, Auron P E, Stevenson M A and Calderwood S K. Novel signal transduction pathway utilized by extracellular HSP70. Role of Toll-like receptor (TLR) 2 and TLR4. J Biol Chem 2002, 277: 15028–15034.

17. Srivastava P. Interaction of heat shock proteins with peptides and antigen presenting cells: chaperoning of the innate and adaptive immune responses. Ann Rev Immunol 2002, 20: 395–425.

18. Moore K W, de Waal Malefyt R, Coffman R L and O'Garra A. Interleukin-10 and the interleukin-10 receptor. Ann Rev Immunol 2001, 19: 683–765.

19. Hwang D. Fatty acids and immune responses – a new perspective in searching for clues to mechanism. Ann Rev Nutr 2000, 20: 431–456.

20. Hashimoto S, Yamada M, Motoyoshi K and Akagawa K S. Enhancement of macrophage colony-stimulating factor-induced growth and differentiation of human monocytes by interleukin-10. Blood 1997, 89: 315–321.

21. De A K, Laudanski K and Miller-Graziano C L. Failure of monocytes of trauma patients to convert to immature dendritic cells is related to preferential macrophage-colony-stimulating factor-driven macrophage differentiation. J Immunol 2003, 170: 6355–6362.

22. Banchereau J, Briere F, Caux C, Davoust J, Lebecque S, Liu Y J, Pulendran B and Palucka K. Immunobiology of dendritic cells. Ann Rev Immunol 2000, 18: 767–811.

23. Dobrovolskaia M A and Vogel S N. Toll receptors, CD14, and macrophage activation and deactivation by LPS. Microbes Infect 2002, 4: 903–914.

24. Triantafilou M and Triantafilou K. Lipopolysaccharide recognition: CD14, TLRs and the LPS-activation cluster. Trends Immunol 2002, 23: 301–304.

25. van Amersfoort E S, van Berkel T J and Kuiper J. Receptors, mediators, and mechanisms involved in bacterial sepsis and septic shock. Clin Microbiol Rev 2003, 16: 379–414.

26. West M A and Heagy W. Endotoxin tolerance: a review. Crit Care Med 2002, 30(1 Supp): S64–S73.

27. Kirschning C J and Schumann R R. TLR2: cellular sensor for microbial and endogenous molecular patterns. Curr Top Microbiol Immunol 2002, 270: 121–144.

28. Habich C, Baumgart K, Kolb H and Burkart V. The receptor for heat shock protein 60 on macrophages is saturable, specific, and distinct from receptors for other heat shock proteins. J Immunol 2002, 168: 569–576.

29. Heine H, El-Samalouti V T, Notzel C, Pfeiffer A, Lentschat A, Kusumoto S, Schmitz G, Hamann L and Ulmer A J. CD55/decay accelerating factor is part of the lipopolysaccharide-induced receptor complex. Eur J Immunol 2003, 33: 1399–1408.

30. Volk H D, Reinke P and Docke W D. Clinical aspects: from systemic inflammation to 'immunoparalysis'. Chem Immunol 2000, 74: 162–177.

31. Vabulas R M, Ahmad-Nejad P, da Costa C, Miethke T, Kirschning C J, Hacker H and Wagner H. Endocytosed HSP60s use toll-like receptor 2 (TLR2) and TLR4 to activate the toll/interleukin-1 receptor signaling pathway in innate immune cells. J Biol Chem 2001, 276: 31332–31339.

32. Guermonprez P, Valladeau J, Zitvogel L, Théry C and Amigorena S. Antigen presentation and T cell stimulation by dendritic cells. Ann Rev Immunol 2002, 20: 621–667.

33. Stevens T R, Winrow V R, Blake D R and Rampton D S. Circulating antibodies to heat-shock protein 60 in Crohn's disease and ulcerative colitis. Clin Exp Immunol 1992, 90: 271–274.

34. Gao Y L, Raine C S and Brosnan C F. Humoral response to hsp 65 in multiple sclerosis and other neurological conditions. Neurology 1994, 44: 941–946.

35. O'Neill P A, Shaaban A M, West C R, Dodson A, Jarvis C, Moore P, Davies M P, Sibson D R and Foster C S. Increased risk of malignant progression in benign proliferating breast lesions defined by expression of heat shock protein 27. Br J Cancer 2004, 90: 182–188.

14

BiP, a Negative Regulator Involved in Rheumatoid Arthritis

Valerie M. Corrigall and Gabriel S. Panayi

14.1. Introduction

The heat shock protein (Hsp) 70 family is a collection of evolutionarily conserved, ubiquitous proteins that are either constitutively expressed and/or stress induced and which are nominally defined by their molecular weight (Hsp70, Hsc73, BiP (binding immunoglobulin protein, or glucose regulated protein (grp) 78)). Historically, these proteins have been perceived to function as intracellular molecular chaperones that ensure the correct folding of nascent proteins and are involved in the translocation of proteins and assist in protein degradation through the proteasome [1]. At times of physical or chemical stress, such chaperones are upregulated by the unfolded protein response and provide protection against the accumulation and aggregation of denatured proteins.

In contrast to this long-standing perception, there is now increasing interest in an intercellular signalling role for these proteins and, as a consequence, they have been termed 'chaperokines' in light of their cytokine-like qualities [2, 3]. The interaction between heat shock proteins and specific cell surface receptors that signal the release of inflammatory mediators has revealed a link between the innate and adaptive immune response. A wide range of extracellular receptors for human Hsp70 has been identified. These include CD14 [2, 4, 5], Toll-like receptor (TLR) 4, TLR2 [4, 6], CD91 [7, 8] and CD40 [9, 10] on monocytes, and scavenger receptors such as LOX-1 on dendritic cells (DCs) [11]. The role of these receptors is detailed in Chapters 7 and 10. Hsp70 and Hsp60 stimulation of monocytes via these receptors induces predominantly pro-inflammatory cytokine/chemokine production, including TNF-α, IL-1β and IL-12 [6, 12].

The present assessment of heat shock proteins is that they act as 'danger signals' alerting the adaptive immune system to raise a Th1 immune response [13]. However, immunological homeostasis needs to be maintained. For this, other proteins, including heat shock proteins, must induce a counter-regulatory

Th2 response or the production of cytokines and other factors with anti-inflammatory properties. Our preliminary studies have shown that BiP does not bind to any of the previously identified receptors for Hsp70 despite the high degree of homology between the two molecules. There is now accumulating evidence that BiP, like Hsp27 (reviewed in Chapter 13) [14], might serve such an immunomodulatory role partially through the secretion of IL-10.

This chapter proposes that BiP is an immunoregulator of the innate and adaptive immune systems which may prevent inappropriate damaging responses to antigenic challenge. These immune functions of BiP have only recently been identified by us, and information is limited. The evidence reviewed in the following indicates that BiP is a natural stimulator of anti-inflammatory cytokines from mononuclear cells. BiP may also prevent DC maturation and stimulate regulatory T cells, all of which may collectively play a key role in regulating the immune system and thus maintaining homeostasis.

14.2. Glucose regulated protein 78 or binding immunoglobulin protein (BiP)

Glucose regulated protein 78 (grp78) or binding immunoglobulin protein (BiP) is classified as a member of the Hsp70 family [1, 15, 16]. As evident from its name, BiP was first identified as the chaperone protein that was involved in the folding of the H and L chains of the immunoglobulin molecule to generate the complete immunoglobulin molecule prior to its exit from the endoplasmic reticulum (ER) [1, 16, 17]. It is now known that BiP is involved in the correct folding of all nascent proteins [1, 17]. BiP binds to the sequences of seven amino acids with exposed hydrophobic residues that are only seen in denatured proteins to prevent intracellular damage [1, 17]. It is from this function, perhaps, that the misconception arose that BiP binds to all proteins. In fact, BiP binds only to the hydrophobic regions of the polypeptide chains that are obscured as the protein is folded [1, 17].

BiP is regulated at two levels: firstly, to maintain a basal constitutive level which is sufficient for intracellular protein folding functions, and secondly, at an induced level which is upregulated when the cell is stressed. Constitutive BiP is tethered in the lumen of the ER by the 3' carboxyl terminal amino acid sequence, KDEL, which binds to the ERD2 receptor [18]. Upregulation occurs when the cell is stressed, particularly under conditions of reduced glucose or oxygen levels, calcium flux or increased concentrations of reactive oxygen species [19]. These conditions are often associated with inflammation, especially with the late stages of acute inflammation or chronic inflammation, and are also prevalent within the joints of patients with rheumatoid arthritis (RA) [20–22].

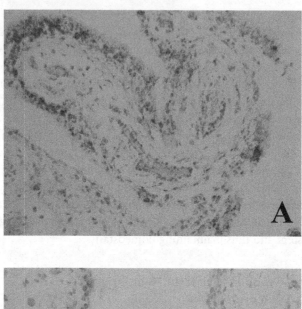

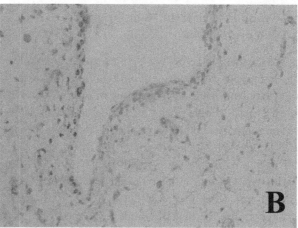

Figure 14.1. BiP is over-expressed in RA synovial membrane. Synovial membrane from patients with (A) rheumatoid arthritis or (B) osteoarthritis was stained with 1/100 dilution of anti-BiP antibody with vector red alkaline phosphatase substrate kit (Dako). Magnification 20×.

Indeed, we and Blass and colleagues [23] have shown that BiP is over-expressed in the synovial membrane of patients with rheumatoid arthritis when compared with membranes from osteoarthritis patients (Figure 14.1).

Upregulation of BiP by stress *in vitro* causes its translocation from the ER and its expression on the cell surface. Delpino and colleagues [24] have reported cell surface BiP expression in response to thapsigargin treatment, which is an inhibitor of ER calcium ATPase and thus a powerful stress inducer. In addition, Gagnon and colleagues [25] have shown that, during endocytosis, ER proteins may become exposed on the cell surface as the pinocytotic cup becomes inverted at the point of internalisation (see Chapter 3). There is *in vivo* evidence for cell surface expression of BiP and heat shock proteins by tumour cells [26–28], and the secretion of heat shock proteins at levels that can be detected by enzyme immunoassays in sera has also been reported [29]. These findings reinforce the view that these molecules have important functions as intercellular messengers outwith their established intracellular functions.

In vitro studies into the extracellular functions of the Hsp70 family have used either recombinant human/bacterial heat shock proteins [6, 10, 12, 30] produced in a bacterial system or mammalian heat shock proteins isolated from cell lysates [23, 31–33]. Protein glycosylation and tertiary structure-dependent differences in their functional ability have not been directly compared. Furthermore, the role of contaminating endotoxin in the heat shock protein preparations used has not always been adequately addressed (a point debated in many chapters in this volume). Much work on heat shock proteins has been performed in animal models [30, 32] or has used mouse and/or human cell lines [6, 10, 12, 30, 33]. Although this approach has considerably simplified the analysis and reduced the variability of the results obtained, extrapolation of these findings into an entirely human system, especially *in vivo*, must be made with caution. At present, data from experiments assessing the interactions of human heat shock proteins with primary human cells are limited and much of the work cited in this chapter involves human peripheral blood mononuclear cells (PBMCs) in conjunction with a recombinant human BiP (rhuBiP) from an *Escherichia coli* source [34]. This same preparation of rhuBiP has been used throughout the animal studies.

14.3. Studies in animals

14.3.1. Induction of arthritis

On identification of BiP as an autoantigen in RA, we determined whether BiP was arthritogenic. A wide range of concentrations of BiP in complete Freund's adjuvant was administered to several strains of rats and mice. However, all failed to induce arthritis [34]. There are two possible explanations for this. The first is that the correct genetic strain of mouse or rat had not been chosen, because it is known that the induction of arthritis exhibits a high genetic dependency. The second possibility is that BiP is unable to stimulate the induction of arthritogenic Th1 T cells.

Human *in vitro* studies provided the clue that the second possibility was the most likely. We showed that, although BiP stimulated the proliferation of T cells from the RA joint, the proliferation was low and was not accompanied by the secretion of IFN-γ [34]. It is therefore possible that BiP is not a 'pathogenic' autoantigen but rather plays some other, possibly regulatory, role in RA synovitis. This hypothesis was supported by the finding that intravenous BiP prevented collagen-induced arthritis (CIA) in the DBA/1 or HLA-DR1$^{+/+}$ transgenic mouse [34].

14.3.2. Prevention of arthritis

Work with CIA and adjuvant arthritis (AA) animal models confirmed that BiP may have anti-inflammatory properties. A single intravenous injection of BiP totally protects mice from the onset of CIA and reduces the incidence and severity of AA in rats when compared with animals similarly treated with a control recombinant protein (β-galactosidase) or the PBS vehicle [34].

14.3.3. Therapy of collagen-induced arthritis

Although protection from disease is of scientific interest, it is of limited clinical value. Therapeutic studies were therefore initiated in which a single dose of BiP was administered via three different routes, namely intravenous (IV), subcutaneous (SC) and intranasal (IN). In these studies, BiP was administered to DBA/1 or HLA-DR$^{+/+}$ transgenic mice in a single dose at the time of onset of CIA, as judged by paw swelling. IV- and SC-administered BiP caused significant therapeutic activity with the minimal effective IV dose being 1 μg/mouse, and that for SC dose being 50 μg/mouse [35]. IN administration was without therapeutic effect. The reason for the failure is presently unknown.

14.3.4. Animal studies – *in vitro* experiments

14.3.4.1. Cellular studies
In vitro experiments on spleen and lymph node single-cell preparations from CIA mice were carried out in parallel with the disease studies. Incubation of cells from the DBA/1 mice (not previously injected with BiP) with rhuBiP stimulated a strong proliferative response when compared to that induced by a control protein (bovine serum albumin). Although interferon (IFN)-γ was present in the supernatants, the level of IL-4 was significantly increased. It should be noted that it is difficult to induce IL-4 secretion from T cells of this strain of mouse. Analysis of the splenocyte and lymph node cell responses to either

rhuBiP or collagen type II (CII) showed that those animals that had been treated with BiP produced high levels of the Th2 cytokines IL-10, IL-4 and IL-5, spontaneously and when re-stimulated with BiP or CII [35, 36]. In contrast, the response to CII was characterised by a significantly greater secretion of IFN-γ.

Ongoing work in our laboratory is aimed at determining whether BiP-treated cells can transfer protection and evaluating whether this is dependent on the production of anti-inflammatory cytokines or by cell-to-cell contact.

14.3.4.2. Antibody studies

The analysis of serum from the mice used in the CIA prevention study not only demonstrated that anti-CII antibody levels were reduced in BiP-treated mice but also that these were predominantly of the IgG$_1$ isotype, an isotype that is typically associated with Th2-driven immune responses [34].

14.4. Human studies

If BiP has intercellular signalling functions, then two requirements must be satisfied. Firstly, cell-free BiP must be found *in vivo*; secondly, there must be the presence of a cell surface receptor capable of binding to BiP and transducing intracellular signals.

14.4.1. Cell-free BiP in synovial fluid

Cell-free BiP, in either synovial fluid (SF) or serum, could originate from cells by several different routes. In the SF, it is known that cell death occurs and this would certainly lead to the release of BiP. In addition, BiP can be expressed on the cell surface [26] and has been detected on the cell surface of single-cell suspensions from RA synovial membranes (unpublished results). Proteolytic enzymes, which are plentiful in the SF, may cleave BiP from the cell surface. Alternatively, upregulation of BiP expression under stress may induce alternative splicing of the BiP gene which, by omitting the retaining KDEL sequence, allows extracellular release. BiP has been detected in the SF of 74% of patients with RA and 38% of individuals with other inflammatory joint diseases. Interestingly, no KDEL was detected on the BiP, despite the fact that other KDEL-containing proteins were detected. Thus, it appears that the mechanism leading to the presence of BiP in the SF is distinct from that used by other ER proteins. The possibilities for such differences include cleavage, and alternative splicing of BiP was discussed earlier.

14.4.2. BiP receptor-like molecule (BiPRL)

We have shown that a variety of cells express a BiP-binding, receptor-like molecule (BiPRL) on their surface such that >95% monocytes, <29% B cells, <10% T cells and 85% fibroblast-like synoviocytes bind fluorescein isothiocyanate-conjugated BiP (BiP.FITC). Specificity of binding is suggested by the difference in percentage of BiPRL$^+$ cells of the different cell types. No competitive binding for BiPRL has been seen with α_2-macroglobulin or antibodies to TLR2, TLR4, CD40, CD91 or CD14 (manuscript submitted). The latter have already been identified as receptors for human Hsp70, grp94 and Hsp60 (see preceding discussion). Following binding to the BiPRL on monocytes, BiP.FITC is internalised and co-localises with the vesicle membrane protein LAMP-1 (CD107a) (manuscript submitted). This internalisation is similar to the process that has been seen with Hsp70 [33] and grp94 [31].

The localisation of BiP within the endosome suggests that BiP could be degraded and processed and peptides derived therefrom presented to T cells by major histocompatibility complex (MHC) class II molecules. The pathway used for the degradation of BiP, either exogenous or endogenous, has not yet been elucidated; however, it is known that there is leakage in this process between peptides that are eventually presented either by MHC class I molecules to CD8$^+$ T cells or by MHC class II molecules to CD4$^+$ T cells. This important observation would explain the development of both CD4$^+$ and CD8$^+$ clones from BiP-stimulated human (PBMCs [37]; (see following text). Chapter 6 also describes the role of internalisation of chaperones in myeloid cell modulation.

14.4.3. T cells

The stimulation of T cells *in vitro* is used as a measure of T cell responsiveness to nominal antigen. However, the length of time taken for the response to become apparent must be carefully considered when reviewing such results. Proliferation that is maximal after seven days is most likely to be due to the induction of a primary response during the culture. In contrast, proliferation that occurs in less than seven days may be considered to be due to the induction of a secondary response by previously primed memory T cells. These considerations are important when reviewing work with BiP. Thus Blass and colleagues [23] have demonstrated borderline *primary* proliferative responses by peripheral blood CD45RA$^+$ T cells to BiP that were HLA-DR restricted. However, the low stimulation index of these cells renders the interpretation of the results difficult. We have shown that RA SF T cells proliferate to exogenous rhuBiP with the kinetics of a *secondary* response. In contrast to Blass et al.'s study we found little, or no, proliferative response by normal PBMCs and a limited proliferative response in PBMCs, even those from RA patients whose SF T cells showed a response [34]. The response to BiP appeared to be specifically in RA SF T cells, because those from other inflammatory diseases, including ankylosing spondylitis and psoriatic arthritis, showed no proliferative response to BiP. One peculiarity of this response was the lack of either IFN-γ or IL-2 secretion; the predominant cytokines produced were IL-10 and IL-4 [34].

There are at least four possible interpretations for these findings: firstly, the depressed proliferation was due to the IL-10 produced; secondly, BiP was activating Th2 T cells; thirdly, BiP stimulated the development of regulatory T cells; or fourthly, BiP drives alternative activation of monocytes to produce IL-10 and anti-inflammatory mediators and promote Th2 cell differentiation. With respect to the first possibility, there are many reports in the literature that describe depressed T cell proliferation caused by excess IL-10 [38]. We have shown that there is a correlation between PBMC proliferation to tuberculin purified protein derivative (PPD) and the ratio of IL-2 to IL-10 produced. This was particularly striking in patients with RA, the PBMC cultures from whom exhibited a reduced IL-2 production [39]. Because the cultures stimulated by BiP produced no IL-2 but high IL-10 levels, it is not surprising that proliferation was low. However, the addition of neutralising anti–IL-10 antibody to these cultures did not totally reverse the poor proliferation, thereby indicating that BiP had anti-proliferative properties that are independent of IL-10 [40]. Interestingly, in cultures in which IL-10 was produced by alternatively activated monocytes, the reduction of PBMC proliferation remained unchanged following the addition of neutralising anti–IL-10 antibody [41].

With respect to the second possibility – that BiP was generating a Th2 response – this is at present being investigated. Evidence from the animal work and the fact that CD4$^+$ and CD8$^+$ BiP specific T cell clones are secreting Th2 cytokines [37] would support this contention. As to the third possibility, we are currently investigating the regulatory capacity of BiP activated T cells. The fourth option, the stimulation of alternatively activated monocytes, is discussed next.

14.4.4. CD4 and CD8 human T cell clones

Additional evidence for the priming of T cells by BiP peptides is provided by the isolation of CD4$^+$ and CD8$^+$ BiP-specific T cell clones from normal PBMCs [37]. The generation of these clones suggests that BiP is processed via both endogenous and exogenous pathways, as has been found for Hsp70 [42]. The CD8$^+$ clones produce IL-10, IL-4 and IL-5, strongly suggesting that they are Th2 cells [37].

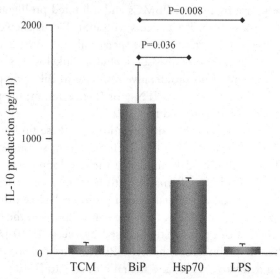

Figure 14.2. IL-10 production stimulated by BiP and Hsp70. IL-10 production by peripheral blood mononuclear cells following 24 hours of stimulation with BiP, Hsp70 (both at 20 μg/ml) or LPS (20 ng/ml). Polymyxin B (10 μg/ml) was added to all cultures.

14.4.5. Monocytes

An important function of BiP is its ability to stimulate human monocytes. As already noted, >95% of monocytes express BiPRL. In PBMC cultures, BiP-stimulated monocytes secrete large amounts of IL-10, significantly greater than that induced by an equivalent amount of Hsp70 (Figure 14.2). In addition, preliminary cytokine gene array analysis of BiP-stimulated, purified peripheral blood monocytes has indicated that BiP up- and/or downregulates many cytokine/chemokine/cytokine receptor genes. The pattern of genes activated indicates that BiP may be alternatively activating the monocytes [43] to produce an anti-inflammatory macrophage producing IL-10. This is similar to the process described with Hsp27 in Chapter 13. The natural inhibitors of IL-1β and TNF-α, IL-1 receptor antagonist and soluble TNF receptor II, respectively, are both upregulated by BiP stimulation (Table 14.1), as is macrophage migration inhibition factor and alternative monocyte activated chemokine (AMAC). In addition, IL-8, granulocyte macrophage-colony stimulating factor (GM-CSF) and epithelial neutrophic activating peptide (ENA)-78, all of which have the potential to increase the inflammatory influx of cells into the synovium, are downregulated following BiP stimulation. In addition, BiP induced a prolonged downregulation of CD86 and HLA-DR expression, in contrast to the intense but transient

Table 14.1. Cytokine gene array data analysed by densitometry

Upregulation			Downregulation		
Cytokine mRNA	BiP	PMA + IONO	Cytokine mRNA	BiP	PMA + IONO
TIMP	65.5 ± 19	30.7	GROα	29.8 ±18.6	57
IL-6	17 ± 18.2	Not detected	GROβ	35.4 ± 26	59
MIF	20.2 ± 10.6	3.4	GROγ	35.7 ± 25	64
TNF RII	4.1 ± 2.2	Not detected	CCL1	Not detected	8.8
			IL-8	57 ± 20.4	85.8
			GMCSF	3.9 ± 3.2	17.2
			Osteopontin	Not detected	14.7
			Urokinase R	50.8 ± 11	80.4
			LIGHT	6.6 ± 1.1	19.2
	BiP	Unstimulated		BiP	Unstimulated
IL-1Ra	26.7 ± 19.7	12.2	ENA-78	4.2 ± 16.2	55
Chem23	15.3 ± 15.9	7.2	GROγ	35.6 ± 25	47
IL6	17 ± 18.2	6.7	IL-8	57.8 ± 20.4	80
AMAC	18.3 ± 5.3	not detected	LDGF	5.4 ± 0.7	52
TNF RII	4.1 ± 2.2	not detected			

Note: Peripheral blood monocytes were isolated and stimulated with either BiP (20 μg/ml) or phorbol myristic acid (PMA; 20 ng/ml) + inomycin (IONO; 500 ng/ml) or left unstimulated for 24 hours. The cytokine gene array autoradiographs were analysed by densitometry and normalised to give a percentage expression of maximum (100%). The results shown are from two subjects for BiP-stimulated monocytes and one subject each for unstimulated and maximally stimulated cells (PMA + IONO). The results are shown as either upregulation or downregulation of BiP-stimulated cytokine mRNA, in both samples, when compared with the control cells.

downregulation observed in the presence of rhuIL-10 [40]. Thus, BiP may have direct anti-inflammatory effects outwith its ability to secrete anti-inflammatory mediators such as IL-10.

14.4.6. Dendritic cells

Maturation of monocytes into DCs is of paramount importance to the development of an efficient adaptive immune system. *In vitro* studies, which use GM-CSF plus IL-4 to drive the differentiation of immature DCs (iDCs), have been performed in the presence and absence of BiP. In concurrence with the hypothesised immunoregulatory functions of BiP, BiP inhibited the differentiation of monocytes to iDCs [44]. Failure of this development was accompanied by a depressed ability to induce the proliferation of allogeneic T cells. This corresponded with the production of high levels of IL-10 and could be reversed either by neutralising IL-10 or by blocking the IL-10 receptor by neutralising monoclonal antibodies. In contrast to BiP suppression of antigen stimulation of PBMCs, the reduced allogeneic proliferative response was reversed by the neutralisation of IL-10 [44]. Interestingly, this property of BiP is in complete contrast to Hsp60. Flohé and colleagues [45] have shown that a mouse macrophage cell line cultured with Hsp60 upregulates CD86 and CD40 and enhances iDC maturation. The antigen-presenting quality of the Hsp60-stimulated cells was increased [45].

14.5. Summary of BiP functions

In this chapter we have attempted to provide an overview of our studies with BiP. In summary, we have found that BiP, which is found cell-free in synovial fluid and can therefore exert intercellular activity, has immunoregulatory functions mediated via a receptor-like molecule that is different from those described for other members of the human Hsp70 family, which include

- anti-inflammatory properties, in distinction from other members of the Hsp70 family;
- the stimulation of PBMCs to produce IL-10 and T cell clones to produce IL-10, IL-4 and IL-5;
- the suppression of differentiation of iDCs from monocytes; and
- the prevention and treatment of CIA in DBA/1 and HLA-DR1$^{+/+}$ mice.

14.6. Conclusion

In conclusion, all the data gathered from the animal *in vivo* and *in vitro* studies and the human *in vitro* studies indicate that BiP is an immunomodulatory protein. This is in contrast to the data obtained with Hsp70, a molecule with which BiP has regions of high homology. Competitive binding studies show that BiP does not bind to any of the confirmed Hsp70 receptors. This supports the fact that BiP is also different functionally. We hypothesise, therefore, that in normal circumstances cell surface expressed and/or secreted BiP will regulate the development and cytokine profile of immune cells maintaining homeostasis and prevent an inappropriate inflammatory immune reaction (Figure 14.3). However, in a chronic inflammatory focus, such as the RA synovium, the overwhelming presence of pro-inflammatory agents (cytokines, chemokines, reactive oxygen species) will prevent upregulated BiP from being an effective means of reducing inflammation. In these circumstances, addition of exogenous BiP might prove to be an effective immunotherapy.

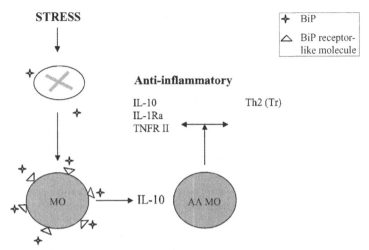

Figure 14.3. Stress, such as hypoxia, induces BiP release from the stressed cell. Cell-free BiP, after binding to a proposed BiP receptor-like molecule stimulates monocytes (MOs) to secrete IL-10. This deviates monocyte differentiation to alternatively activated monocytes (aaMOs). AaMOs induce an anti-inflammatory response via the secretion of anti-inflammatory molecules (such as IL-10, IL-1Ra, TNFR II) and regulatory T cells, thus controlling the inflammatory response.

REFERENCES

1. Gething M J. Role and regulation of the ER chaperone BiP. Semin Cell Dev Biol 1999, 10: 465–472.
2. Asea A, Kraeft S-K, Kurt-Jones E A, Stevenson M A, Chen L B, Finberg R W, Koo G C and Calderwood S K. Hsp70 stimulates cytokine production through a CD14-dependent pathway, demonstrating its dual role as a chaperone and cytokine. Nat Med 2000, 6: 435–442.
3. Asea A, Kabingu E, Stevenson M A and Calderwood S K. HSP70 peptide-bearing and peptide-negative preparations act as chaperokines. Cell Stress Chaperon 2000, 5: 425–431.
4. Kol A, Lichtman A H, Finberg R W, Libby P and Kurt-Jones E A. Heat shock protein (HSP) 60 activates the innate immune response: CD14 is an essential receptor for HSP60 activation of mononuclear cells. J Immunol 2000, 164: 13–17.
5. Lewthwaite J C, Coates A R M, Tormay P, Singh M, Mascagni P, Poole S, Roberts M, Sharp L and Henderson B. *Mycobacterium tuberculosis* chaperonin 60.1 is a more potent cytokine stimulator than chaperonin 60.2 (hsp 65) and contains a CD14-binding domain. Infect Immun 2001, 69: 7349–7355.
6. Asea A, Rehli M, Kabingu E, Boch J A, Baré O, Auron P E, Stevenson M A and Calderwood S K. Novel signal transduction pathway utilized by extracellular HSP70. Role of Toll-like receptor (TLR) 2 and TLR4. J Biol Chem 2002, 277: 15028–15034.

7. Basu S, Binder R J, Ramalingam T and Srivastava P K. CD91 is a common receptor for heat shock proteins gp96, hsp90, hsp70 and calreticulin. Immunity 2001, 14: 303–313.
8. Binder R J, Han D K and Srivastava P K. CD91: a receptor for heat shock protein gp96. Nat Immunol 2000, 1: 151–155.
9. Becker T, Hartl F U and Wieland F. CD40, an extracellular receptor for binding and uptake of Hsp70-peptide complexes. J Cell Biol 2002, 158: 1277–1285.
10. Wang Y, Kelly C G, Karttunen T, Whittall T, Lehner P J, Duncan L, MacAry P, Younson J S, Singh M, Oehlmann W, Cheng G, Bergmeier L and Lehner T. CD40 is a cellular receptor mediating mycobacterial heat shock protein 70 stimulation of CC-chemokines. Immunity 2001, 15: 971–983.
11. Delneste Y, Magistrelli G, Gauchat J, Haeuw J, Aubry J, Nakamura K, Kawakami-Honda N, Goetsch L, Sawamura T, Bonnefoy J and Jeannin P. Involvement of LOX-1 in dendritic cell-mediated antigen cross-presentation. Immunity 2002, 17: 353–362.
12. Habich C, Baumgart K, Kolb H and Burkart V. The receptor for heat shock protein 60 on macrophages is saturable, specific, and distinct from receptors for other heat shock proteins. J Immunol 2002, 168: 569–576.
13. Matzinger P. The Danger Model: A renewed sense of self. Science 2002, 296: 301–305.
14. De A K, Kodys K M, Yeh B S and Miller-Graziano C. Exaggerated human monocyte IL-10 concomitant to minimal TNF-α induction by heat-shock protein 27 (Hsp27) suggests Hsp27 is primarily an anti-inflammatory stimulus. J Immunol 2000, 165: 3951–3958.
15. Haas I G. BiP (GRP78), an essential hsp70 resident protein in the endoplasmic reticulum. Experientia 1994, 50: 1012–1020.
16. Kozutsumi Y, Normington K, Press E, Slaughter C, Sambrook J and Gething M J. Identification of immunoglobulin heavy chain binding protein as glucose-regulated protein 78 on the basis of amino acid sequence, immunological cross-reactivity, and functional activity. J Cell Sci Suppl 1989, 11: 115–137.
17. Knarr G, Gething M J, Modrow S and Buchner J. BiP binding sequences in antibodies. J Biol Chem 1995, 270: 27589–27594.
18. Janson I M, Toomik R, O'Farrell F and Ek P. KDEL motif interacts with a specific sequence in mammalian erd2 receptor. Biochem Biophys Res Commun 1998, 247: 447–451.
19. Kozutsumi Y, Segal M, Normington K, Gething M J and Sambrook J. The presence of malfolded proteins in the endoplasmic reticulum signals the induction of glucose-regulated proteins. Nature 1988, 332: 462–464.
20. Mapp P I, Grootveld M, C, and Blake D R. Hypoxia, oxidative stress and rheumatoid arthritis. Br Med Bull 1995, 51: 419–436.
21. Tak P P, Zvaifler N J, Green D R and Firestein G S. Rheumatoid arthritis and p53: how oxidative stress might alter the course of inflammatory diseases. Immunol Today 2000, 21: 78–82.
22. Maurice M M, Nakamura H, van der Voort E A, van Vliet A I, Staal F J, Tak P P, Breedveld F C and Verweij C L. Evidence for the role of an altered redox state in hyporesponsiveness of synovial T cells in rheumatoid arthritis. J Immunol 1997, 158: 1458–1465.

23. Blass S, Union A, Raymackers J, Schumann F, Ungethum U, Muller-Steinbach S, De Keyser F, Engel J M and Burmester G R. The stress protein BiP is overexpressed and is a major B and T cell target in rheumatoid arthritis. Arthritis Rheum 2001, 44: 761–771.

24. Delpino A and Castelli M. The 78 kDa glucose-regulated protein (GRP78/BIP) is expressed on the cell membrane, is released into cell culture medium and is also present in human peripheral circulation. Biosci Reports 2002, 22: 407–420.

25. Gagnon E, Duclos S, Rondeau C, Chevet E, Cameron P H, Steele-Mortimer O, Paiement J, Bergeron J J and Desjardins M. Endoplasmic reticulum-mediated phago-cytosis is a mechanism of entry into macrophages. Cell Biol Int 2002, 110: 119–131.

26. Shin B K, Wang H, Yim A M, Le Naour F, Brichory F, Jang J H, Zhao R, Puravs E, Tra J, Michael C W, Misek D E and Hanash S M. Global profiling of the cell surface proteome of cancer cells uncovers an abundance of proteins with chaperone function. J Biol Chem 2003, 278: 7607–7616.

27. Barreto A, Gonzalez J M, Kabingu E, Asea A and Fiorentino S. Stress-induced release of HSC70 from human tumors. Cell Immunol 2003, 222: 97–104.

28. Dai J, Liu B, Caudill M, Zheng H, Qiao Y, Podack E R and Li Z. Cell surface expression of heat shock protein gp96 enhances cross-presentation of cellular anti-gens and the generation of tumor-specific T cell memory. Cancer Immun 2003, 3: 1–5.

29. Lewthwaite J, Owen N, Coates A, Henderson B and Steptoe A. Circulating human heat shock protein 60 in the plasma of British civil servants. Circulation 2002, 106: 196–201.

30. Habich C, Kempe K, van der Zee R, Burkart V and Kolb H. Different heat shock pro-tein 60 species share pro-inflammatory activity but not binding sites on macrophages. FEBS Lett 2003, 533: 105–109.

31. Wassenberg J J, Dezfulian C and Nicchitta C V. Receptor mediated and fluid phase pathways for internalization of the ER Hsp90 chaperone grp94 in murine macrophages. J Cell Sci 1999, 112: 2167–2175.

32. Binder R J, Harris M L, Ménoret A and Srivastava P K. Saturation, competition, and specificity in interaction of heat shock proteins (hsp) gp96, hsp90, and hsp70 with CD11b+ cells. J Immunol 2000, 165: 2582–2587.

33. Arnold-Schild D, Hanau D, Spehner D, Schmid C, Rammensee H-G, de la Salle H and Schild H. Receptor-mediated endocytosis of heat shock proteins by professional antigen-presenting cells. J Immunol 1999, 162: 3757–3760.

34. Corrigall V M, Bodman-Smith M D, Fife M S, Canas B, Myers L K, Wooley P, Soh C, Staines N A, Pappin D J, Berlo S E, van Eden W, van der Zee R, Lanchbury J S and Panayi G S. The human endoplasmic reticulum molecular chaperone BiP is an autoantigen for rheumatoid arthritis and prevents the induction of experimental arthritis. J Immunol 2001, 166: 1492–1498.

35. Brownlie R, Sattar Z, Corrigall V M, Bodman-Smith M D, Panayi G S and Thompson S. Immunotherapy of collagen induced arthritis with BiP. Rheumatology (Oxford) 2003, 42 suppl: 13.

36. Sattar Z, Brownlie R, Corrigall V M, Bodman-Smith M D, Staines N A, Panayi G S et al. CD4+ T cells specific for the stress protein BiP modulate the development of collagen induced arthritis. Rheumatology (Oxford) 2003, 42 suppl: 124.

37. Bodman-Smith M D, Corrigall V M, Kemeny D M and Panayi G S. BiP, a putative autoantigen in rheumatoid arthritis, stimulates IL-10-producing CD8$^+$ T cells from normal individuals. Rheumatology (Oxford) 2003, 42: 637–644.
38. Krakauer T. Differential inhibitory effects of interleukin-10, interleukin-4, and dexamethasone on staphylococcal enterotoxin-induced cytokine production and T cell activation. J Leuk Biol 1995, 57: 450–454.
39. Corrigall V M, Garyfallos A and Panayi G S. The relative proportions of secreted interleukin-2 and interleukin-10 determine the magnitude of rheumatoid arthritis T-cell proliferation to the recall antigen tuberculin purified protein derivative. Rheumatology (Oxford) 1999, 38: 1203–1207.
40. Corrigall V M, Bodman-Smith M D, Brunst M, Cornell H and Panayi G S. Inhibition of antigen-presenting cell function and stimulation of human peripheral blood mononuclear cells to express an antiinflammatory cytokine profile by the stress protein BiP: relevance to the treatment of inflammatory arthritis. Arthritis Rheum 2004, 50: 1164–1171.
41. Schebesch C, Kodelja V, Muller C, Hakij N, Bisson S, Orfanos C E and Goerdt S. Alternatively activated macrophages actively inhibit proliferation of peripheral blood lymphocytes and CD4$^+$ T cells in vitro. Immunology 1997, 92: 478–486.
42. Castellino F, Boucher P E, Eichelberg K, Mayhew M, Rothman J E, Houghton A N and Germain R N. Receptor-mediated uptake of antigen/heat shock protein complexes results in major histocompatibility complex class I antigen presentation via two distinct pathways. J Exp Med 2000, 191: 1957–1964.
43. Gordon S. Alternative activation of macrophages. Nat Rev Immunol 2003, 3: 23–35.
44. Vittecoq O, Corrigall V M, Bodman-Smith M D and Panayi G S. The molecular chaperone BiP (GRP78) inhibits the differentiation of normal human monocytes into immature dendritic cells. Rheumatology (Oxford) 2003, 42 suppl: 43.
45. Flohé S B, Bruggemann J, Lendemans S, Nikulina M, Meierhoff G, Flohé S and Kolb H. Human heat shock protein 60 induces maturation of dendritic cells versus a Th1-promoting phenotype. J Immunol 2003, 170: 2340–2348.

Extracellular Biology of Molecular Chaperones: Molecular Chaperones as Therapeutics

15

Neuroendocrine Aspects of the Molecular Chaperones ADNF and ADNP

Illana Gozes, Inna Vulih, Irit Spivak-Pohis and Sharon Furman

15.1. Introduction

Vasoactive intestinal peptide (VIP), which was originally discovered in the intestine as a 28–amino acid peptide and shown to induce vasodilation, was later found to be a major brain peptide with neuroprotective activities *in vivo* [1–5]. To exert neuroprotective activity in the brain, VIP requires glial cells that secrete protective proteins such as activity-dependent neurotrophic factor (ADNF [6]). ADNF, isolated by sequential chromatographic methods, was named activity-dependent neurotrophic factor because it protects neurons from death associated with the blockade of electrical activity.

ADNF is a 14-kDa protein, and structure-activity studies have identified femtomolar-active neuroprotective peptides, ADNF-14 (VLGGGSALLRSIPA) [6] and ADNF-9 (SALLRSIPA) [7]. ADNF-9 exhibits protective activity in Alzheimer's disease–related systems (β-amyloid toxicity [7], presenilin 1 mutation [8], apolipoprotein E deficiencies [9] – genes that have been associated with the onset and progression of Alzheimer's disease (AD)). Other studies have indicated protection against oxidative stress via the maintenance of mitochondrial function and a reduction in the accumulation of intracellular reactive oxygen species [10]. In the target neurons, ADNF-9 regulates transcriptional activation associated with neuroprotection (nuclear factor-κB [11]), promotes axonal elongation through transcriptionally regulated cAMP-dependent mechanisms [12] and increases chaperonin 60 (Cpn60/Hsp60) expression, thereby providing cellular protection against the β-amyloid peptide [13].

Longer peptides that include the ADNF-9 sequence (e.g., ADNF-14) activate protein kinase C and mitogen-associated protein kinase kinase and protect developing mouse brain against excitotoxicity [14]. In neocortical synaptosomes, ADNF-9 enhances basal glucose and glutamate transport and attenuates oxidative impairment of glucose and glutamate transport induced by the β-amyloid peptide and Fe^{2+} [15]. In hippocampal neurons, ADNF-9 stimulates synapse formation as demonstrated by glutamate responses of excitatory neurons and morphological development [16]. In this hippocampal culture system, ADNF-9 induces the secretion of neurotrophin 3 (NT-3). Because both NT-3 and ADNF-9 regulate the NMDA receptor subunits 2A (NR2A) and NR2B, these results suggest *in vivo* effects of ADNF-9 on learning and behaviour in the adult nervous system. Indeed, in a rat model of cholinodeficiency, intranasal ADNF-9 enhances performance in a water maze, which is indicative of spatial learning and memory [17].

Antibody studies suggest that ADNF-like molecules mediate VIP neuroprotective and neurotrophic activities [12, 16, 18]. Preparation of ADNF-9–like analogues have resulted in the discovery of neuroprotective activity in an all D-amino acids ADNF-9 (D-ADNF-9, D-Ser-D-Ala-D-Leu-D-Leu-D-Arg-D-Ser-D-Ile-D-Pro-D-Ala) which suggests a non-chiral mode of action [17, 19]. Studies on ADNF-9 originated in our laboratory and Dr. D.E. Brenneman's laboratory [6, 7, 20, 21]. ADNF neuroprotective activity, at very low concentrations in models relevant to AD in particular and neurodegeneration in general, were independently corroborated in laboratories all over the world, for example, by Mattson and colleagues [8, 11, 15, 21], Gressens and colleagues [14], Hashimoto and colleagues [22] and Ramirez and colleagues [23].

Activity-dependent neuroprotective protein (ADNP) is another glial mediator of VIP-associated neuroprotection [9]. Antibodies that recognise ADNF-9 also recognise ADNP, and this has allowed the isolation of ADNP cDNA by expression cloning. An active eight–amino acid peptide (NAP, NAPVSIPQ) derived from ADNP, which shares structural and functional similarities with ADNF-9 in cell culture, has thus been identified [9]. ADNP was implicated in the maintenance of cell survival via a modulation of p53 expression [24]. A 100-fold more potent VIP analogue which provides neuroprotection is stearyl-Nle17-VIP (SNV [2–5]) and recent studies have now identified ADNP as a molecule which may mediate protection offered by SNV against ischaemic cell death [25].

As for NAP, *in vitro* experiments have shown that NAP protects neurons against numerous toxins and cellular stresses [9, 18, 26–30] including the AD neurotoxin (the β-amyloid peptide), excitotoxicity, the toxic envelope protein of the human immunodeficiency virus [9], electrical blockade [9], oxidative stress [18], dopamine toxicity [26], decreased glutathione [26], glucose deprivation [27] and tumour necrosis factor–associated toxicity [30].

NAP also has neuroprotective activity in a variety of animal models including the learning-deficient apolipoprotein E knockout mice (a model related to atherosclerosis [4, 9]), mouse paradigms of traumatic head injury (a risk factor for AD which exhibits some similar stroke-like secondary outcomes [30, 31]) and fetal alcohol syndrome (a model of oxidative damage [32]). In two rat paradigms, a model of cholinotoxicity and normal middle-aged animals treated daily by intranasal NAP administration, significant improvements in short-term spatial memory have been observed [17, 29]. NAP has a short structure, is active at exceptionally low concentrations (femtomolar), is water soluble, is bioavailable, is easily delivered via intranasal inhalation and is unusually stable. No NAP toxicity has been observed to date [33].

NAP has been studied in several independent laboratories. Busciglio and colleagues [34] have shown that the *in vitro* degeneration of Down syndrome neurons is prevented by ADNF-derived peptides; Shohami and colleagues have shown protection in head trauma [30], Brenneman and colleagues have shown protection in fetal alcohol syndrome [32], Leker and colleagues have shown protection in a model of mid cerebral artery occlusion [35], Offen and colleagues have shown protection against glutathione depletion [26] and Smith-Swintosky and colleagues have shown that NAP promotes neurite outgrowth in rat hippocampal and cortical cultures [36]. All of these findings corroborate the original description of NAP's neuroprotective properties [9]. The efficacy of NAP administration (µg to mg/kg, depending on the indication) has been demonstrated in animals using a variety of administration routes including intranasal [17], intraperitoneal [32], intravenous [35] and subcutaneous [31], and intact NAP has been detected in the brain 30 minutes and even 1 hour after administration [17, 32, 35].

Significant steps have been made recently towards understanding the mode of action of NAP. These include the initial identification of specific binding molecules and cells [37, 38], the identification of potential signal transduction pathways such as cGMP production [39], an interference with inflammatory mechanisms [30, 31] and a protective effect against apoptotic processes [35]. Protection against oxidative stress [18, 26, 32] and glucose deprivation [27] also suggests interference with fundamental processes.

The primary interest of the current chapter is the relationship between ADNF-9 and NAP with the chaperone family of proteins. Results have shown that ADNF-9, while having a structure similar to a short sequence in Hsp60, is directly associated with Hsp60 metabolism in the cell. These results are further discussed with regards to NAP which is highly homologous to ADNF-9 and provides an extracellular chaperone function. Because of the complexity of the systems used a brief description of the experimental methods used has been provided.

15.2. NAP and Chaperonin 60: the practical base

15.2.1. Cell culture and anti-sense oligodeoxynucleotide treatment

Cerebral cortical astrocytes were derived from newborn rat brains by trypsinisation of the cortices and growing the cells in conventional culture medium [9]. To determine the role of Hsp60, the anti-sense oligodeoxynucleotide (5'-TGT GGG TAG TCG AAG CAT-3') which has a sequence that is complementary to the 6–24 position on the rat liver Hsp60 cDNA and has previously been used to inhibit the neosynthesis of endogenous Hsp60 [40], was used. Anti-sense oligodeoxynucleotides (5 μM) were added to the culture medium (in the presence of serum) for 3 days with repeated additions at 24-hour intervals. Water was added to the control group. Culture medium was replaced only once, to prevent augmentation of Hsp60 expression due to the stress caused by medium replacement. Experiments were terminated 24 hours after the last addition. Cells were washed with phosphate buffered saline (PBS) and further incubated at room temperature for 3 hours with PBS containing 0.1 nM VIP. After the incubation, the conditioned medium was collected and the cells were harvested and subjected to protein extraction.

The intracellular protein content was analysed by gel electrophoresis and Western blotting using specific antibodies for Hsp60 (SPA-804, StressGen Biotechnologies Corp., Victoria, Canada) and ADNF [9]. The detected signals were analysed by densitometric scan and compared to the values of the untreated controls. Actin content was measured using actin-specific antibodies (anti-rabbit, Sigma), diluted 1:500, as a reference point of total protein content in each sample.

15.2.2. Hsp60 over-expression

Late passage C6 glioma cells (50–60), which exhibit an astrocytic phenotype, were used in this study. Cells (1×10^5 cells/ml) were seeded on tissue culture flasks [41] and were co-transfected either with the mouse Hsp60 expression construct [41] and neomycin expression vector using LipofectAMINE Plus (Life Technologies Inc.) at a ratio of 1:20 or with the neomycin expression vector by itself. Experiments were routinely carried out on a clone of the transfected cells, and all the results were confirmed on a number of individual clones expressing mouse Hsp60 and neomycin resistance and control clones expressing neomycin resistance only. Intracellular protein content of Hsp60 over-expressing clones was analysed by gel electrophoresis and Western blotting [13] [42] using polyclonal rabbit anti-Hsp60 antibody (StressGen, diluted 1:1000); polyclonal rabbit anti-ADNF antibody, diluted 1:250 [20] which was affinity purified against ADNF-9 [9]; and polyclonal rabbit anti-actin antibody (Sigma), diluted 1:500.

15.2.3. Immunoprecipitation

Intracellular protein mixtures and conditioned medium of Hsp60–over-expressing C-6 glioma cell clones were immunoprecipitated. Cells were washed in PBS (Biological Industries) and incubated for 3 hours in PBS. The conditioned medium was harvested, dialysed and lyophilised. The product was dissolved in anti-protease–containing buffer and was immunoprecipitated using the ADNF-9–specific antibodies [9]. Ten micrograms of affinity-purified antibody were added to each milligram of protein and allowed to conjugate in 4 °C for 1 hour. The conjugates were then precipitated by incubating at 4 °C for 2 hours with Protein A/G Plus-Agarose Beads (Santa Cruz Biotechnology). The washed precipitate was boiled, and the supernatant was collected and analysed by gel electrophoresis and Western blotting as described previously [14, 20].

15.3. The relationship of intracellular Hsp60 to ADNF

15.3.1. Hsp60 anti-sense oligodeoxynucleotides reduce ADNF-like expression

Although VIP treatment appeared to increase intracellular Hsp60 in astrocytes, albeit insignificantly, an apparent decrease in intracellular ADNF-like immunoreactivity was observed. As expected, anti-sense oligodeoxynucleotide treatment (Hsp60-specific) reduced Hsp60 expression, and this was paralleled by an even more pronounced reduction in ADNF-like immunoreactivity. Actin levels remained constant or were increased (Figure 15.1). Hence, the results suggest that Hsp60 anti-sense oligodeoxynucleotides reduce ADNF-like expression.

15.3.2. Over-expression of Hsp60 increases intracellular ADNF expression: Western blot analysis

Co-transfection generated 18 Hsp60 over-expressing clones, all of which exhibited an enhanced immunoreactivity in the 14-kDa ADNF-like band. Seven additional clones showed only enhanced ADNF-9–like 14-kDa immunoreactivity. Some of these clones were analysed further. In clone 6, a seven-fold increase in Hsp60 immunoreactivity was observed (60 kDa), compared to a

A

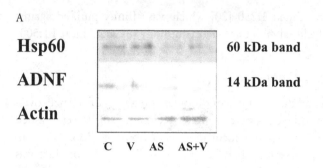

Hsp60 — 60 kDa band

ADNF — 14 kDa band

Actin

C   V   AS   AS+V

B

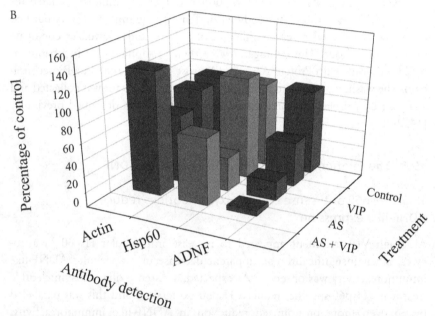

Figure 15.1. (A) Hsp60 anti-sense oligodeoxynucleotides reduce ADNF-like expression: Western blot. Antibodies used were anti-Hsp60 (Hsp60), ADNF-9 and actin. C = control, V = VIP treated, AS = anti-sense oligodeoxynuclotide (Hsp60), AS + V = anti-sense oligodeoxynuclotide + VIP. (B) Changes were quantified by densitometry.

control neo1-clone (which lacks the Hsp60 vector). Clone 13 exhibited a 30-fold greater immunoreactivity with the ADNF-14–specific antibody (14 kDa [9, 20]; Figure 15.2). In clone 14 there was a 30-fold greater immunoreactivity with anti-Hsp60 antibodies and a 258-fold greater reactivity with ADNF-14 antibodies. All of these analyses were performed on the same Western blot. Similar relationships between Hsp60 and ADNF were observed in all the clones analysed.

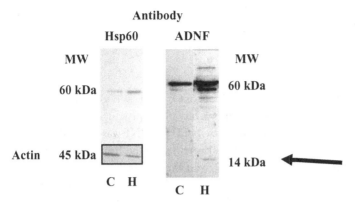

Figure 15.2. Over-expression of Hsp60 increases intracellular ADNF expression: Western blot analysis. MW = molecular weight, C = control, H = transfected with Hsp60 cDNA. Arrow indicates 14-kDa ADNF-like immunoreactivity.

15.3.3. Over-expression of Hsp60 increases intracellular ADNF expression: immunoprecipitation

A specific 14-kDa band was detected by Western blotting following the immunoprecipitation experiments only in the Hsp60-transfected clone lanes. No clear band of this size could be detected in the control lanes (Figure 15.3). The antibody used for both immunoprecipitation and immunodetection by Western blotting was the antibody recognising ADNF. The immunoprecipitation results

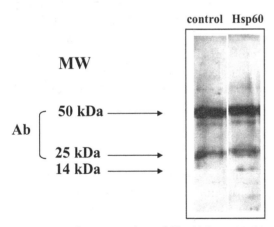

Figure 15.3. Over-expression of Hsp60 increases intracellular ADNF expression: immunoprecipitation experiments. A Western blot is shown and abbreviations are as in Figures 15.1 and 15.2.

of proteins in the extracellular milieu showed several ADNF-like bands at the lower molecular weight range (data not shown).

15.4. Discussion

The aforementioned results suggest a direct association between the expression of Hsp60 and of ADNF. ADNF exhibits a very close structural homology with Hsp60 which implies a chaperone-like activity [21]. ADNF was originally discovered as a 14-kDa potent neuroprotective protein secreted from glial cells in the presence of VIP [6]. Hsp60 is also secreted in the presence of VIP [43]. Hsp60 antibodies induce neuronal cell death which is inhibited by ADNF [6]. Furthermore, VIP-induced secretion of ADNF is associated with enhanced synapse formation [16]. Other studies have shown that over-expression of Hsp60/10 protects against ischaemia/reperfusion injury [44]. ADNF is a secreted protein. Hsp60 might also be secreted and the possibility that peptides derived from these proteins provide extracellular chaperonin activity is intriguing. The possibility of functional–structural interactions between Hsp60 and ADNF is also intriguing.

All the studies just described refer to Hsp60 and ADNF intracellular relationships and emanate from our previous findings of extracellular ADNF and Hsp60 sequences [6, 16, 43]. Because ADNF acts at femtomolar concentrations, high secretion levels are not required and hence the detection of ADNF in the extracellular milieu requires high quantities of conditioned medium. Our preliminary results (not shown) suggest that extracellular immunoreactive ADNF-like proteins in the lower molecular weight range are related to Hsp60. Interestingly, ADNF-9 (SALLRSIPA) is structurally homologous to NAP (NAPVSIPQ), an eight–amino acid peptide derived from ADNP [9]. ADNP is essential for brain formation [45] and NAP, which provides a broad range of neuroprotection [37, 46, 47], has been shown to inhibit the aggregation of the β-amyloid peptide, an aggregation that is associated with extracellular toxicity in AD [48]. Thus, NAP and related peptides may act as extracellular chaperones.

Acknowledgements

We thank Drs. Douglas E. Brenneman, Haya Brody and Ohad Birk for their help. Specifically, we thank Dr. Douglas E. Brenneman for his continuous support, Dr. Birk for the Hsp60 construct and Dr. Brody for her help with the glioblastoma Hsp60 over-expression colonies. These studies were supported by the United States–Israel Binational Science Foundation, the Israel Science Foundation, the Lily and Avraham Gildor Chair for the Investigation of Growth Factors, and the Institute for the Study of Aging and Allon Therapeutics, Inc.

NOTES ADDED IN PROOF: Several relevant new findings have been published or are in press since the original preparation of this chapter.

Divinski I, Mittelman L and Gozes I. A femtomolar acting octapeptide interacts with tubulin and protects astrocytes against zinc intoxication. J Biol Chem 2004, 279: 28531–28538

Gozes I and Divinski I. The femtomolar-acting NAP interacts with microtubules: Novel aspects of astrocyte protection. J Alzheimers Dis 2004, 6: S37–S41.

Furman S, Steingart R A, Mandel S, Hauser J M, Brenneman D E and Gozes I. Subcellular localization and secretion of activity-dependent neuroprotective protein in astrocytes. Neuron Glia Biology 2005, in press.

Brenneman D E, Spong C Y, Hauser J M, Abebe D, Pinhasov A, Golian T and Gozes I. Protective peptides that are orally active and mechanistically nonchiral. J Pharmacol Exp Ther 2004, 309: 1190–1197.

Wilkemeyer M F, Chen S Y, Menkari C E, Sulik KK and Charness M E. Ethanol antagonist peptides: structural specificity without stereospecificity. J Pharmacol Exp Ther 2004, 309: 1183–1189.

Zhou F C, Sari Y, Powrozek T A and Spong C Y. A neuroprotective peptide antagonizes fetal alcohol exposure-compromised brain growth. J Mol Neurosci 2004, 24: 189–199.

Chiba T, Hashimoto Y, Tajima H, Yamada M, Kato R, Niikura T, Terashita K, Schulman H, Aiso S, Kita Y, Matsuoka M and Nishimoto I. Neuroprotective effect of activity-dependent neurotrophic factor against toxicity from familial amyotrophic lateral sclerosis-linked mutant SOD1 in vitro and in vivo. J Neurosci Res 2004, 78: 542–552.

REFERENCES

1. Gozes I and Brenneman D E. VIP: molecular biology and neurobiological function. Mol Neurobiol 1989, 3: 201–236.
2. Gozes I, Fridkin M, Hill J M and Brenneman D E. Pharmaceutical VIP: prospects and problems. Cur Med Chem 1999, 6: 1019–1034.
3. Gozes I, Bardea A, Reshef A, Zamostiano R, Zhukovsky S, Rubinraut S, Fridkin M and Brenneman D E. Neuroprotective strategy for Alzheimer disease: intranasal administration of a fatty neuropeptide. Proc Natl Acad Sci USA 1996, 93: 427–432.
4. Gozes I, Bachar M, Bardea A, Davidson A, Rubinraut S, Fridkin M and Giladi E. Protection against developmental retardation in apolipoprotein E-deficient mice by a fatty neuropeptide: implications for early treatment of Alzheimer's disease. J Neurobiol 1997, 33: 329–342.
5. Gozes I, Perl O, Giladi E, Davidson A, Ashur-Fabian O, Rubinraut S and Fridkin M. Mapping the active site in vasoactive intestinal peptide to a core of four amino acids: neuroprotective drug design. Proc Natl Acad Sci USA 1999, 96: 4143–4148.
6. Brenneman D E and Gozes I. A femtomolar-acting neuroprotective peptide. J Clin Invest 1996, 97: 2299–2307.
7. Brenneman D E, Hauser J, Neale E, Rubinraut S, Fridkin M, Davidson A and Gozes I. Activity-dependent neurotrophic factor: structure-activity relationships of femtomolar-acting peptides. J Pharmacol Exp Therap 1998, 285: 619–627.

8. Guo Q, Sebastian L, Sopher B, Miller M W, Glazner G W, Ware C B, Martin G M and Mattson M. Neurotrophic factors [activity-dependent neurotrophic factor (ADNF) and basic fibroblast growth factor (bFGF)] interrupt excitotoxic neurodegenerative cascades promoted by a PS1 mutation. Proc Natl Acad Sci USA 1999, 96: 4125–4130.

9. Bassan M, Zamostiano R, Davidson A, Pinhasov A, Giladi E, Perl O, Bassan H, Blat C, Gibney G, Glazner G, Brenneman D E and Gozes I. Complete sequence of a novel protein containing a femtomolar-activity-dependent neuroprotective peptide. J Neurochem 1999, 72: 1283–1293.

10. Glazner G W, Boland A, Dresse A E, Brenneman D E, Gozes I and Mattson M P. Activity-dependent neurotrophic factor peptide (ADNF9) protects neurons against oxidative stress-induced death. J Neurochem 1999, 73: 2341–2347.

11. Glazner G W, Camandola S and Mattson M P. Nuclear factor-kappaB mediates the cell survival-promoting action of activity-dependent neurotrophic factor peptide-9. J Neurochem 2000, 75: 101–108.

12. White D M, Walker S, Brenneman D E and Gozes I. CREB contributes to the increased neurite outgrowth of sensory neurons induced by vasoactive intestinal polypeptide and activity-dependent neurotrophic factor. Brain Res 2000, 868: 31–38.

13. Zamostiano R, Pinhasov A, Bassan M, Perl O, Steingart R A, Atlas R, Brenneman D E and Gozes I. A femtomolar-acting neuroprotective peptide induces increased levels of heat shock protein 60 in rat cortical neurons: a potential neuroprotective mechanism. Neurosci Lett 1999, 264: 9–12.

14. Gressens P, Marret S, Bodenant C, Schwendimann L and Evrard P. Activity-dependent neurotrophic factor-14 requires protein kinase C and mitogen-associated protein kinase kinase activation to protect the developing mouse brain against excitotoxicity. J Mol Neurosci 1999, 13: 199–210.

15. Guo Z H and Mattson M P. Neurotrophic factors protect cortical synaptic terminals against amyloid and oxidative stress-induced impairment of glucose transport, glutamate transport and mitochondrial function. Cereb Cortex 2000, 10: 50–57.

16. Blondel O, Collin C, McCarran W J, Zhu S, Zamostiano R, Gozes I, Brenneman D E and McKay R D. A glia-derived signal regulating neuronal differentiation. J Neurosci 2000, 20: 8012–8020.

17. Gozes I, Giladi E, Pinhasov A, Golian T, Romano J and Brenneman D E. Activity-dependent neurotrophic factor: comparison of intranasal and oral administration of femtomolar-acting L and D peptides to improve memory. Soc Neurosci Abstract 2000: 223.

18. Steingart R A, Solomon B, Brenneman D E, Fridkin M and Gozes I. VIP and peptides related to activity-dependent neurotrophic factor protect PC12 cells against oxidative stress. J Mol Neurosci 2000, 15: 137–145.

19. Brenneman D E, Hauser J and Gozes I. Synergistic and non-chiral characteristics in dissociated cerebral cortical test cultures. Soc Neurosci Abstract 2000 223–224.

20. Gozes I, Davidson A, Gozes Y, Mascolo R, Barth R, Warren D, Hauser J and Brenneman D E. Antiserum to activity-dependent neurotrophic factor produces neuronal cell death in CNS cultures: immunological and biological specificity. Brain Res Dev Brain Res 1997, 99: 167–175.

21. Gozes I and Brenneman D E. Activity-dependent neurotrophic factor (ADNF). An extracellular neuroprotective chaperonin? J Mol Neurosci 1996, 7: 235–244.

22. Hashimoto Y, Niikura T, Ito Y, Sudo H, Hata M, Arakawa E, Abe Y, Kita Y and Nishimoto I. Detailed characterization of neuroprotection by a rescue factor humanin against various Alzheimer's disease-relevant insults. J Neurosci 2001, 21: 9235–9245.

23. Ramirez S H, Sanchez J F, Dimitri C A, Gelbard H A, Dewhurst S and Maggirwar S B. Neurotrophins prevent HIV Tat-induced neuronal apoptosis via a nuclear factor-kappaB (NF-kappaB)-dependent mechanism. J Neurochem 2001, 78: 874–889.

24. Zamostiano R, Pinhasov A, Gelber E, Steingart R A, Seroussi E, Giladi E, Bassan M, Wollman Y, Eyre H J, Mulley J C, Brenneman D E and Gozes I. Cloning and characterization of the human activity-dependent neuroprotective protein. J Biol Chem 2001, 276: 708–714.

25. Sigalov E, Fridkin M, Brenneman D E and Gozes I. VIP-Related protection against Iodoacetate toxicity in pheochromocytoma (PC12) cells: a model for ischemic/hypoxic injury. J Mol Neurosci 2000, 15: 147–154.

26. Offen D, Sherki Y, Melamed E, Fridkin M, Brenneman D E and Gozes I. Vasoactive intestinal peptide (VIP) prevents neurotoxicity in neuronal cultures: relevance to neuroprotection in Parkinson's disease. Brain Res 2000, 854: 257–262.

27. Zemlyak I, Furman S, Brenneman D E and Gozes I. A novel peptide prevents death in enriched neuronal cultures. Reg Peptides 2000, 96: 39–43.

28. Gozes I and Brenneman D E. A new concept in the pharmacology of neuroprotection. J Mol Neurosci 2000, 14: 61–68.

29. Gozes I, Alcalay R, Giladi E, Pinhasov A, Furman S and Brenneman D E. NAP accelerates the performance of normal rats in the water maze. J Mol Neurosci 2002, 19: 167–170.

30. Beni-Adani L, Gozes I, Cohen Y, Assaf Y, Steingart R A, Brenneman D E, Eizenberg O, Trembolver V and Shohami E. A peptide derived from activity-dependent neuroprotective protein (ADNP) ameliorates injury response in closed head injury in mice. J Pharmacol Exp Ther 2001, 296: 57–63.

31. Romano J, Beni-Adani L, Nissenbaum O L, Brenneman D E, Shohami E and Gozes I. A single administration of the peptide NAP induces long-term protective changes against the consequences of head injury: gene Atlas array analysis. J Mol Neurosci 2002, 18: 37–45.

32. Spong C Y, Abebe D T, Gozes I, Brenneman D E and Hill J M. Prevention of fetal demise and growth restriction in a mouse model of fetal alcohol syndrome. J Pharmacol Exp Therap 2001, 297: 774–779.

33. Newton P E, Brenneman D E and Gozes I. 30-day intranasal toxicity studies of NAP in rats and dogs. J Mol Neurosci 2001, 16: 61.

34. Pelsman A, Fernanandez G, Gozes I, Brenneman D E and Busciglio J. *In vitro* degeneration of Down syndrome neurons is prevented by activity-dependent neurotrophic factor-derived peptides. Soc Neurosci Abstracts 1998, 24: 1044.

35. Leker R R, Teichner A, Grigoriadis N, Ovadia H, Brenneman D E, Fridkin M, Giladi E, Romano J and Gozes I. NAP, a femtomolar-acting peptide, protects the brain against ischemic injury by reducing apoptotic death. Stroke 2002, 33: 1085–1092.

36. Smith-Swintosky V L, Gozes I, Brenneman D E and Plata-Salaman C R. Activity dependent neurotrophic factor-9 and NAP promote neurite outgrowth in rat hippocampal and cortical cultures. Soc Neurosci Abstracts 2000, 26: 843.

37. Gozes I, Divinsky I, Pilzer I, Fridkin M, Brenneman D E and Spier A D. From vasoactive intestinal peptide (VIP) through activity-dependent neuroprotective protein (ADNP) to NAP: a view of neuroprotection and cell division. J Mol Neurosci 2003, 20: 315–322.

38. Divinski I, Spier A D and Gozes I. NAP, a peptide derivative of the VIP-regulated gene ADNP, confers neuroprotection through microtubule dynamics. Reg Peptides 2003, 115: 42.

39. Ashur-Fabian O, Giladi E, Furman S, Steingart R A, Wollman Y, Fridkin M, Brenneman D E and Gozes I. Vasoactive intestinal peptide and related molecules induce nitrite accumulation in the extracellular milieu of rat cerebral cortical cultures. Neurosci Lett 2001, 307: 167–170.

40. Steinhoff U, Zugel U, Wand-Wurttenberger A, Hengel H, Rosch R, Munk M E and Kaufmann S H E. Prevention of autoimmune lysis by T cells with specificity for a heat shock protein by antisense oligonucleotide treatment. Proc Natl Acad Sci USA 1994, 91: 5085–5088.

41. Birk O S, Douek D C, Elias D, Takacs K, Dewchand H, Gur S L, Walker M D, van der Zee R, Cohen I R and Altmann D M. A role of hsp60 in autoimmune diabetes: analysis in a transgenic model. Proc Natl Acad Sci USA 1996, 93: 1032–1037.

42. Kurek J B, Bennett T M, Bower J J, Muldoon C M and Austin L. Leukaemia inhibitory factor (LIF) production in a mouse model of spinal trauma. Neurosci Lett 1998, 249: 1–4.

43. Bassan M, Zamostiano R, Giladi E, Davidson A, Wollman Y, Pitman J, Hauser J, Brenneman D E and Gozes I. The identification of secreted heat shock 60-like protein from rat glial cells and a human neuroblastoma cell line. Neurosci Lett 1998, 250: 37–40.

44. Hollander J M, Lin K M, Scott B T and Dillmann W H. Overexpression of PHGPx and HSP60/10 protects against ischemia/reoxygenation injury. Free Radic Biol Med 2003, 35: 742–751.

45. Pinhasov A, Mandel S, Torchinsky A, Giladi E, Pittel Z, Goldsweig A M, Servoss S J, Brenneman D E and Gozes I. Activity-dependent neuroprotective protein: a novel gene essential for brain formation. Brain Res Dev Brain Res 2003, 144: 83–90.

46. Zaltzman R, Beni S M, Giladi E, Pinhasov A, Steingart R A, Romano J, Shohami E and Gozes I. Injections of the neuroprotective peptide NAP to newborn mice attenuate head-injury-related dysfunction in adults. Neuroreport 2003, 14: 481–484.

47. Alcalay R N, Giladi E, Pick C G and Gozes I. Intranasal administration of NAP, a neuroprotective peptide, decreases anxiety-like behavior in aging mice in the elevated plus maze. Neurosci Lett 2004, 361: 128–131.

48. Ashur-Fabian O, Segal-Ruder Y, Skutelsky E, Brenneman D E, Steingart R A, Giladi E and Gozes I. The neuroprotective peptide NAP inhibits the aggregation of the beta-amyloid peptide. Peptides 2003, 24: 1413–1423.

16

Heat Shock Proteins Regulate Inflammation by Both Molecular and Network Cross-Reactivity

Francisco J. Quintana and Irun R. Cohen

16.1. Introduction

Heat shock proteins were initially identified as heterogeneous families of stress-induced proteins characterised by their chaperone activity [1]. Subsequently, they were identified as immunodominant antigens recognised by the host immune system following microbial infection [2] or during the course of autoimmune disease [3–6]. Recently, the role of heat shock proteins as endogenous activators of the innate and adaptive immune system has been unveiled [7]. In this chapter we discuss the relevance of heat shock proteins and their immune activities to the regulation of inflammation and autoimmune disease. We shall see that the regulatory activities of heat shock proteins on inflammation involve two types of cross-reactivity: *molecular* cross-reactivity exists between microbial and self-heat shock proteins and *network* cross-reactivity exists between different self-heat shock proteins.

16.2. Inflammation activates heat shock protein–specific T cells

Although the injection of incomplete Freund's adjuvant (IFA) to BALB/c mice induces local inflammation, Anderton and colleagues demonstrated that the injection of IFA also induces T cells reactive with the mammalian 60-kDa heat shock protein (Hsp60) [8]. These Hsp60-reactive T cells were TCR$\alpha\beta^+$, CD4$^+$ and major histocompatibility complex (MHC) class II-restricted [8]. Notably, Hsp60-specific cells could only be found in the local lymph nodes draining the site of IFA injection, and they were not present in distant lymph nodes. Hsp60-specific T cells are not only induced but also recruited to the site of inflammation [8].

The pro-inflammatory response which drives autoimmune disorders has also been shown to lead to an up-regulation of heat shock protein expression and the recruitment of heat shock protein–specific T cells to the target organ. Mor and colleagues have described that, along with myelin-specific T cells, T cells specific for the mycobacterial 65-kDa (Hsp65) or 71-kDa (Hsp71) heat shock proteins are recruited to the central nervous system (CNS) in rats undergoing experimental autoimmune encephalomyelitis (EAE) [9]. This initial observation was subsequently extended to include self-heat shock proteins and T cells reactive to them, in both EAE and human multiple sclerosis [10–12]. Finally, transplanted organs undergoing rejection show increased levels of expression of endogenous heat shock proteins and are infiltrated by heat shock protein–specific T cells (reviewed in [13]).

In short, heat shock protein–specific T cells are induced by inflammation and are recruited to the sites of inflammation. In this chapter, we will discuss experimental data that support a regulatory role for heat shock proteins and heat shock protein–specific T cells in the control of inflammation.

16.3. Heat shock proteins control inflammation

Adjuvant arthritis (AA) in the Lewis rat [14] and spontaneous autoimmune diabetes in the non-obese diabetic (NOD) mouse [15] are experimental models for two of the most prevalent human autoimmune diseases: rheumatoid arthritis [16] and type 1 diabetes mellitus (T1DM) [17]. Although the clinical signs of the models are naturally different, both experimental diseases are linked by the observation that heat shock proteins can halt the autoimmune attack. We have used these experimental models to study the role of heat shock proteins in the control of autoimmune disease.

16.3.1. Adjuvant arthritis

AA is induced in Lewis rats by a subcutaneous injection of heat-killed *Mycobacterium tuberculosis* in IFA [14]. T cells specific for mycobacterial Hsp65 can both drive and inhibit AA. Although Hsp65-specific CD4+ T cell clones cross-react with cartilage components and transfer AA [18], Hsp65 administered as a protein [19], encoded in a recombinant vaccinia virus [20] or administered as a DNA vaccine [21] can inhibit AA. The administration of Hsp65 can also regulate experimental arthritis triggered by the lipoidal amine CP20961 [22] or by pristane [23].

Inhibition of AA by Hsp65 is thought to involve cross-reactivity with self-Hsp60 [24]. We have studied the specificity of the regulatory immune response that controls AA using DNA vaccines coding for either human Hsp60 (pHsp60) or mycobacterial Hsp65 (pHsp65) [25]. Although both pHsp60 and pHsp65 protect against AA, pHsp60 is significantly more effective [25]. Using DNA vaccines encoding fragments of Hsp60 to identify immunoregulatory regions within Hsp60, the anti-arthritogenic effects of the pHsp60 construct have been shown to reside in the amino acid (aa) 1–260 region of Hsp60 [26]. Using Hsp60-derived overlapping peptides, peptide Hu3 (aa 31–50 of Hsp60) is specifically recognised by T cells of rats protected from AA by DNA vaccination [26]. Vaccination with Hu3, or transfer of splenocytes from Hu3-vaccinated rats, prevents the development of AA, whereas vaccination with the mycobacterial homologue of Hu3 has no effect [26]. Prevention of AA by vaccination with pHsp60, DNA vaccines encoding the N-terminus of Hsp60, or Hu3 was associated with the induction of T cells that secrete IFN-γ, IL-10 and TGF-β1 upon stimulation with Hsp60 [25, 26]. Thus, Hsp60-specific T cells can control the progression of AA. However, what influence do T cells reactive with other heat shock proteins have on such processes?

T cell responses to the mycobacterial 10-kDa heat shock protein (Hsp10) [27] or mycobacterial Hsp71 have also been shown to control the progression of AA [28–30]. We studied whether self-heat shock proteins other than Hsp60 could inhibit AA using DNA vaccines encoding human 70-kDa heat shock protein (Hsp70) or the human 90-kDa heat shock protein (Hsp90). DNA vaccination with Hsp70 or Hsp90 shifted the specific arthritogenic T-cell response from a Th1 to a Th2/3 phenotype and inhibited AA [31]. Thus, Hsp70 and Hsp90 can also modulate arthritogenic T cell responses in AA.

Hsp60-specific responses in patients with rheumatoid arthritis [32, 33] or juvenile chronic arthritis [34] are associated with milder arthritis and a better prognosis. Although no information is yet available on T cell responses to Hsp70 or Hsp90 in human arthritis, these observations suggest that heat shock protein-specific T cells might also have a regulatory role in human autoimmune arthritis. The role of the 70-kDa heat shock protein BiP as a modulator of rheumatoid arthritis is described in detail in Chapter 14.

16.3.2. NOD diabetes

NOD mice spontaneously develop diabetes as a consequence of a T cell–mediated autoimmune process that destroys the insulin-producing β cells of the pancreas [17]. NOD mice have a high frequency of self-reactive T cells [35], which is reflected by a highly self-reactive B-cell repertoire [36]. Several antigens are targeted by diabetogenic T cells, including insulin [37] and glutamic acid decarboxylase (GAD) [38]. Similar to the situation found in AA, T cell reactivity to Hsp65 is a double-edged sword. A peak of Hsp65-specific

T cell reactivity precedes the onset of diabetes [39], and immunisation with Hsp65 can induce a transient hyperglycaemia [39]. However, vaccination with Hsp65 can also inhibit the development of diabetes [39]. These initial reports may be explained by cross-reactivity between mycobacterial Hsp65 and self-Hsp60.

We have shown that self-Hsp60 is targeted by the diabetogenic attack; T cells reactive with the Hsp60 peptide p277 (aa 437–460) can induce diabetes in irradiated NOD recipients [40]. On the other hand, vaccination of NOD mice with peptide p277 has been shown to arrest the development of diabetes [40] and can even induce remission of overt hyperglycaemia [41]. Successful p277 treatment leads to the down-regulation of spontaneous T cell proliferation to p277 and to the induction of a Th1-to-Th2 switch in the immune response to p277 [42]. Other peptides of Hsp60 can also inhibit the development of spontaneous diabetes in NOD mice [43].

NOD mice can also develop a more robust form of diabetes induced by the administration of cyclophosphamide, termed cyclophosphamide-accelerated diabetes (CAD) [44]. Cyclophosphamide is thought to specifically deplete regulatory T cells [44], thereby unleashing a Th1 response which is rich in IFN-γ secreting cells and leads to overt diabetes [45].

We have studied the effect of DNA vaccination with pHsp60 or pHsp65 on CAD. Vaccination with pHsp60, but not with pHsp65, protects NOD mice from CAD [46]. Thus, the efficacy of the pHsp60 DNA vaccine in this situation can be explained by regulatory Hsp60 epitopes that are not shared with Hsp65; indeed well-characterised regulatory epitopes from Hsp60 are not conserved in the sequence of Hsp65 [46]. Vaccination with pHsp60 modulates the T cell responses to Hsp60 and also to GAD and insulin. T cell proliferative responses are significantly reduced, and the cytokine profile induced by stimulation with Hsp60, GAD or insulin revealed an increased secretion of IL-10 and IL-5 and a decreased secretion of IFN-γ, a finding which is compatible with a Th1-to-Th2 shift in the autoimmune response [46].

In conclusion, the administration of Hsp60 peptides, or of whole Hsp60 as a recombinant protein or a DNA vaccine, can halt autoimmune NOD diabetes. Several antigens are targeted during the progression of diabetes [17] and it is therefore remarkable that the immunoregulatory networks triggered by Hsp60 can control diabetogenic T cells that are directed to a range of other antigens, such as insulin and GAD.

B and T cell responses to Hsp70 [47], Hsp60 and p277 [6, 48] have also been described in patients with T1DM. Indeed, a double-blind, phase II clinical trial was designed to study the effects of p277 therapy on newly diagnosed patients

[49]. The administration of p277 after the onset of clinical diabetes preserved the endogenous levels of C-peptide (which fell in the placebo group) and was associated with lower requirements for exogenous insulin, thereby revealing an arrest of β cell destruction [49]. Treatment with p277 led to enhanced Th2 responses to Hsp60 and p277 [49]. Thus, like NOD diabetes, human T1DM appears to be susceptible to immunomodulation by Hsp60 therapy.

Taken together it appears that heat shock proteins can control the progression of inflammation and, in particular, self-heat shock proteins seem to be quite efficient in doing so. However, do we need exogenous heat shock proteins to trigger heat shock protein–based regulatory mechanisms?

16.4. Triggering of heat shock protein–based immunoregulation by innate immune activation

Bacterial DNA stimulates the innate immune system via Toll-like receptor 9 (TLR9) [50] due to the presence of DNA motifs consisting of a central unmethylated CpG dinucleotide flanked by two 5′ purines and two 3′ pyrimidines [51]. Such a sequence is referred to as a CpG motif. We have demonstrated that bacterial CpG motifs can inhibit spontaneous diabetes in NOD mice [52], but not CAD [46]. The prevention of diabetes was characterised by a decreased insulitis [52]. Moreover, we have detected a decrease in the spontaneous proliferative responses of T cells to Hsp60 and its p277 peptide, concomitant with the induction of Th2-like antibodies of the same specificity, thereby revealing a Th1-to-Th2 shift in the autoimmune response of the treated mice [52].

To investigate the mechanisms involved in the regulation of spontaneous NOD diabetes by CpG motifs, we studied the expression of Hsp60 in splenocytes from NOD mice stimulated with a synthetic oligonucleotide containing CpG motifs (CpG). *In vitro* stimulation with CpG led to a dose-dependent upregulation of intracellular Hsp60 levels, as demonstrated by Western blot analysis, and also to the release of Hsp60 into the supernatant. A control oligonucleotide containing an inverted CpG motif (GpC) had no significant effect on the intracellular levels of Hsp60 or on Hsp60 secretion [Quintana and Cohen, manuscript submitted].

CpG also affected the responses of T cell clones specific for the Hsp60 peptides p12 (aa 166–185) or p277 (aa 437–460). In the presence of irradiated antigen-presenting cells (APCs), CpG triggered the dose-dependent proliferation of both Hsp60-specific T cell clones, but not of an anti-ovalbumin T cell line [Quintana and Cohen, manuscript submitted]. All the T cells were activated by their target antigen, but not by lipopolysaccharide (LPS), thereby ruling out the possibility that some of the observed proliferation was due to the presence of contaminating B cells [Quintana and Cohen, manuscript submitted]. The analysis of cytokine secretion revealed that CpG stimulation triggered the secretion of higher amounts of IL-10 and lower amounts of IFN-γ than did activation with the target Hsp60 peptides (p12 or p277) [Quintana and Cohen, manuscript submitted]. The Hsp60-specific T cell lines were not activated by CpG in the absence of APCs, and CpG-induced proliferation was inhibited by anti-MHC class II antibodies [Quintana and Cohen, manuscript submitted]. Thus, CpG activates Hsp60-specific T cells by stimulating the presentation of peptides derived from endogenous Hsp60 in the MHC class II molecules of the APC. Because IL-10 is known to have suppressor effects on immune responses [53], the relative increase in IL-10 secretion by Hsp60-specific T cells might explain the protective effect of CpG on NOD diabetes. The reader should refer to Chapters 13 and 14 for a discussion of chaperones that selectively induce IL-10 over IL-1/tumour necrosis factor synthesis.

Figure 16.1 depicts our model for the action of CpG on spontaneous NOD diabetes. The activation of APCs or of other cell types via TLR9 leads to the up-regulation of intracellular levels of Hsp60 and eventually to its secretion. Hsp60 is then presented on the surface of the APC via MHC class II molecules. Hsp60-specific regulatory T cells are therefore activated, halting the progression of NOD diabetes.

A paper by Kumaraguru and colleagues reports that CpG triggers the up-regulation and release of Hsp70 from macrophages; however, the effects of CpG on Hsp70-specific T cell lines were not studied [54]. Based on the APC function of macrophages, it is likely that Hsp70 peptides presented in the MHC molecules of CpG-treated macrophages can modulate Hsp70-specific immunity.

TLR9-mediated activation has been shown to control several experimental models of autoimmune disease including EAE [55], colitis [56] and arthritis [57, 58]. Ligands for other TLRs, such as poly I:C [59] or LPS [60–62], have also been reported to inhibit experimental autoimmunity. Whether the activation of heat shock protein–based immunoregulatory mechanisms is a feature shared by several TLR-dependent signalling cascades remains to be seen. Nevertheless, our results suggest that regulatory Hsp60-specific T cell responses can be triggered by the activation of innate networks that lead to the release of endogenous heat shock proteins leading, in turn, to the activation of the specific T cell populations. Could we use these innate networks to diversify the heat shock protein–specific immune response? In other words, could we administer a particular heat shock protein and induce T cell responses directed to a different heat shock protein?

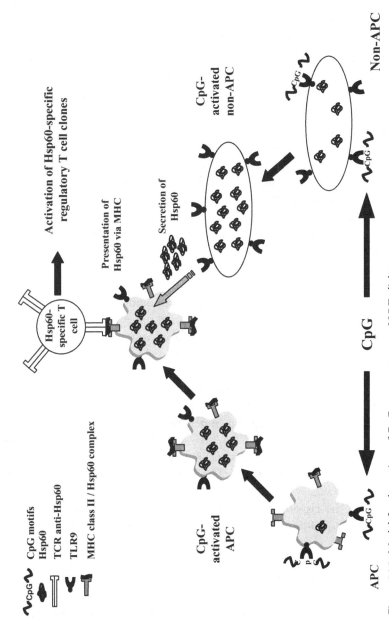

Figure 16.1. Model for the action of CpG on spontaneous NOD diabetes.

Activation of Hsp60-specific
regulatory T cell clones

Presentation of
Hsp60 via MHC

Secretion of
Hsp60

CpG-
activated
non-APC

Non-APC

Hsp60-
specific T
cell

CpG

CpG-
activated
APC

APC

⌇CpG⌇   CpG motifs
         Hsp60
   Y   TCR anti-Hsp60
   T   TLR9
         MHC class II / Hsp60 complex

269

16.5. Connectivity between different heat shock protein–specific immune responses

We have demonstrated that DNA vaccines coding for Hsp60 (pHsp60), Hsp70 (pHsp70) or Hsp90 (pHsp90) can inhibit AA [25, 26, 30]. Moreover, DNA vaccines coding for Hsp70 or Hsp90 can modulate the Hsp65-specific T cell response which drives AA, in a similar manner to that previously demonstrated for Hsp60 [25, 26, 30]. Hsp60, Hsp70 and Hsp90 bear no significant sequence homology or immune cross-reactivity. However, might immunisation with an exogenous heat shock protein trigger the presentation of a different endogenous heat shock protein, leading to the diversification of the immune response induced by vaccination with a particular heat shock protein?

DNA vaccination with pHsp70 or pHsp90 induces antigen-specific proliferative responses: pHsp70-vaccinated rats manifest T cell responses to Hsp70, and pHsp90-vaccinated rats manifest T cell responses to Hsp90 [31]. However, DNA vaccination with pHsp70 or pHsp90 could also induce T cells that proliferated and secreted IFN-γ, TGF-β1 and IL-10 upon stimulation with Hsp60 [31]. Thus different heat shock protein molecules are linked immunologically.

To characterise this connection, we compared the epitope specificity of the Hsp60-specific T cell response induced by pHsp60 with that induced by pHsp70 using a panel of overlapping peptides derived from the human Hsp60 sequence [31]. We had previously found that pHsp60 DNA-vaccination-induced regulatory T cells were reactive with a single Hsp60 peptide epitope, Hu3 (aa 31–50) [26]. However, lymph node cells (LNCs) from pHsp70-vaccinated rats responded to several other Hsp60 peptides: Hu19 (aa 271–290), Hu24 (aa 346–365), Hu25 (aa 361–380), Hu27 (aa 391–410), Hu28 (aa 406–425), Hu30 (aa 436–455), Hu32 (aa 466–485), Hu33 (aa 481–500) and Hu34 (aa 271–290) [31]. Thus, although both pHsp60 and pHsp70 can induce Hsp60-specific T cells, the fine specificities of the T cell responses induced are different. The cross-talk between the Hsp60- and the Hsp70-specific T cell responses is reciprocal, in that pHsp60-vaccinated rats showed significant T cell responses upon stimulation with Hsp70 [31]. These findings are schematically represented in Figure 16.2.

Hsp60, Hsp70 and Hsp90 share no sequence homology and are not immunologically cross-reactive. One possible explanation for the induction of Hsp60-specific T cell responses by pHsp70 or pHsp90 is self-vaccination with endogenous self-Hsp60 which is induced and/or released as a result of the DNA vaccinations. Indeed, we could detect increased levels of circulating Hsp60 in pHsp70-vaccinated rats [Quintana et al., manuscript submitted]. The upregulation of Hsp60 levels in the circulation was dependent on the presence

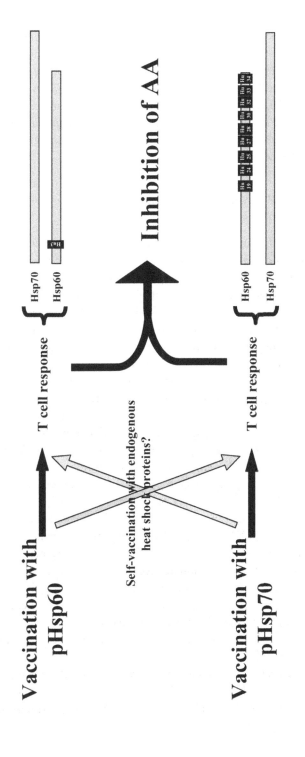

Figure 16.2. Connectivity between Hsp60- and Hsp70-specific T cell responses.

of the *hsp70* gene on the pHsp70 DNA construct; as a control, empty plasmid had no effect on circulating Hsp60 levels [Quintana et al., manuscript submitted]. Reciprocally, vaccination with pHsp60 induced a T cell response to Hsp70; however, we have not yet been able to measure the levels of Hsp70 in the blood of pHsp60-vaccinated rats. Although the molecular mechanisms require further study, the present findings demonstrate that heat shock protein–specific responses are inter-regulated and highlight the multiple immune signalling activities of these molecules.

16.6. Immunoregulatory mechanisms triggered by heat shock proteins

Earlier, we have seen that inflammation induces heat shock protein–specific T cells and that, despite a lack of immunological cross-reactivity between the molecules, T cell responses to different heat shock proteins are connected. We have shown that heat shock proteins can control autoimmune disease, and we have referred to experimental data suggesting that heat shock protein–based regulatory networks can be triggered in the absence of exogenous immunisation with heat shock proteins. We are therefore left with the question of how heat shock protein molecules might control autoimmunity.

The autoimmune attack which leads to overt autoimmune disease is a process that simultaneously engages several cell types and molecular mechanisms that are not restricted to the immune system. Faced by such a multi-front attack, it is not surprising that heat shock proteins bear several features that might prove helpful for the control of the autoimmune response. Heat shock proteins, as described in Chapter 1, are intracellular chaperones which facilitate the correct folding of newly synthesised proteins [1]. Moreover, they also improve antigen processing and presentation by APCs [63]. Heat shock proteins facilitate the induction of T cell responses to free peptide epitopes which are bound by circulating heat shock proteins and taken up by APCs through heat shock protein–specific receptors [63, 64]. Circulating heat shock proteins, not loaded with any peptide, can directly activate several cell types via innate receptors [7]. Heat shock proteins can activate immune system cells, such as dendritic cells [65–67], and also non-immune cells, such as endothelial cells [68]. Finally, heat shock proteins bear regulatory T cell epitopes [25, 69]. However, as we have seen, heat shock proteins can also be targeted by the pathogenic T cells that characterise autoimmune diseases such as T1DM [6].

The sites at which heat shock proteins are expressed can influence their immunoregulatory functions. The intracellular levels of heat shock proteins are increased upon cellular stress. Viral or bacterial infections up-regulate heat shock protein expression [70–73], and necrotic cells release heat shock proteins [74]. Inflammation is a source of cellular stress, and heat shock proteins are over-expressed at the sites of inflammation, such as in the synovium in arthritis [75]. Strikingly, heat shock proteins are also up-regulated in activated macrophages [76] and T cells [77]. Thus, heat shock proteins simultaneously mark the cells targeted by the autoimmune attack and the pathogenic immune cells that carry out the attack.

Based on the intra- and extra-cellular functions of heat shock protein and their localisation, several mutually non-exclusive mechanisms, involving adaptive and non-adaptive immunity, can contribute to the immunoregulatory properties of heat shock proteins.

16.6.1. Adaptive immunity

16.6.1.1. Environmental regulation of heat shock protein–specific immunity

Heat shock proteins are immunodominant bacterial antigens [78]. Because mucosal immunisation is known to induce antigen-specific regulatory responses [79], exposure to bacterial heat shock proteins from the intestinal flora might be a source of heat shock protein–specific regulatory T cells. Indeed, Moudgil and colleagues have demonstrated that environmental microbes can induce Hsp65-specific T cells directed to regulatory epitopes that are cross-reactive with self-Hsp60 [69, 80]. Vaccination with heat shock proteins or their peptides might simply amplify this naturally acquired regulation. However, based on this mechanism, any cross-reactive protein conserved through evolution from bacteria to mammals should be immunoregulatory. This is not always the case, as recently reported by Prakken and colleagues [81].

16.6.1.2. Boost of regulatory T cell responses

Heat shock proteins can bind free peptides and induce peptide-specific immune responses, even in the presence of low amounts of the target peptide [82]. Thus, heat shock protein molecules could be loaded *in vivo* with regulatory self-peptides and subsequently boost or amplify specific regulatory T cell responses. Indeed, Chandawarkar and colleagues have reported that gp96 can both induce and down-regulate tumour-specific immune responses [83]. Furthermore, heat shock proteins purified from the inflamed CNS of EAE rats (and not from naïve rats) can vaccinate naïve rats against EAE [84]. Thus, Hsp70–peptide complexes synthesised at the sites of active inflammation can trigger tissue-specific anti-inflammatory T cell responses [84]. Nevertheless, this mechanism does not explain the immunomodulatory effects of heat shock protein–derived fragments or peptides. See Chapters 17 and 18 for more information on chaperones and peptide-specific immune responses.

16.6.1.3. Cytokine-mediated bystander inhibition

Inflammation leads to the local up-regulation of heat shock proteins. Heat shock protein–specific T cells might therefore be recruited to sites of inflammation, where they could control pathogenic T cell clones by the secretion of regulatory cytokines. Heat shock protein–specific T cells induced by vaccination with immunoregulatory DNA vaccines or peptides secrete regulatory cytokines (IL-10 and TGF-β1) [25, 29, 30, 46].

16.6.1.4. Anti-ergotypic regulation

T cells reactive to activated T cells (but not to resting T cells) can control experimental autoimmune disease [85–87]. The T cell receptor expressed by these regulatory T cells recognises peptides derived from activation markers (ergotopes), such as the α-chain of the IL-2 receptor [86, 87] or the TNF-α receptor [87]. These cells are termed anti-ergotypic [85]. Now it has been reported that mRNAs encoding for heat shock proteins are up-regulated upon T cell activation [77]. Thus, Hsp60 too might serve as an ergotope. We studied whether vaccination with DNA vaccines encoding Hsp60, or with the regulatory peptide Hu3, might induce anti-ergotypic responses. To serve as an ergotope, Hsp60 would have to fulfil two requirements. Firstly, Hsp60 must be up-regulated in activated T cells. Secondly, activated T cells must present Hsp60-derived peptides to Hsp60-specific T cells.

The activation of T cells by the mitogen Concanavalin A, or by specific antigen, up-regulates intracellular levels of Hsp60 [Quintana et al., manuscript submitted]. Thus, the first condition is fulfilled: T cell activation triggers Hsp60 expression. Moreover, activated T cells can present Hsp60. Hsp60-specific T cells proliferate to activated T cells and secrete both IFN-γ and TGF-β1 [Quintana et al., manuscript submitted]. The activation of Hsp60-specific T cells was MHC class II (RT1.B) restricted, since it could be inhibited with the OX6 monoclonal antibody [Quintana et al., manuscript submitted]. Thus, Hsp60 can function as an ergotope *in vitro*; however, can functional Hsp60-specific anti-ergotypic responses be induced *in vivo*?

DNA vaccination with pHsp60 has been found to induce anti-ergotypic T cell responses that are MHC class II (RT1.B) and MHC class I restricted [Quintana et al., manuscript submitted]. In contrast, vaccination with Hu3 induced only an MHC class II restricted (RT1.B) anti-ergotypic T cell response [Quintana et al., manuscript submitted]. Thus, Hsp60-specific CD4$^+$ and CD8$^+$ anti-ergotypic T cells can be induced *in vivo*.

LNCs from rats with AA stimulated with the immunodominant 180–88 T cell epitope of Hsp65 (mt180) secrete high levels of IFN-γ [25]. Since T cells specific for this epitope have been shown to transfer AA [18, 88], the reactivity of LNCs of AA to mt180 is thought to reflect the behaviour of the arthritogenic T cells. LNCs of AA rats stimulated with mt180 in the presence of Hsp60-specific anti-ergotypic T cells (but not with a control anti-myelin bask protein (MBP) line) secrete significantly less IFN-γ [Quintana et al., manuscript submitted]. Thus, anti-ergotypic responses can control the arthritogenic response *in vitro*. Our model for the role of Hsp60-specific T cells in anti-ergotypic response is depicted in Figure 16.3. However, the contribution of the anti-ergotypic response to the regulatory functions of heat shock protein–specific T cells in AA and other autoimmune disorders *in vivo* is still unknown.

16.6.2. Innate immunity

16.6.2.1. Innate activation of regulatory T cells

Heat shock proteins are endogenous ligands for innate receptors. Hsp60 and Hsp70 activate TLR4 and TLR2 [89]; Hsp70 and Hsp90 have also been reported to signal via CD40 and CD91 [90, 91]. See Chapters 7, 8 and 10 for more details of the receptors for chaperones. Caramalho and colleagues have reported that regulatory CD25$^+$ T cells are activated via TLR4 [92]. Thus, it is possible that self-heat shock proteins directly activate regulatory cells via innate receptors. This hypothesis is partially supported by the findings made by Dr. Gabriel Nussbaum in our laboratory, who has generated NOD mice lacking a functional TLR4. NOD mice carrying a non-functional *tlr4* allele show an early onset and an increase in the incidence of spontaneous diabetes. Interestingly, the sensitivity of those NOD mice to CAD remains unchanged (Dr. Gabriel Nussbaum, personal communication). Cyclophosphamide is thought to deplete regulatory cells [44]; thus, these findings suggest that TLR4-mediated signals triggered by self-ligands do activate regulatory cells involved in the control of autoimmune diabetes.

16.6.3. Hsp60 triggers anti-inflammatory activities in T cells via TLR2

Hsp60 and p277 can directly inhibit chemotaxis and activate anti-inflammatory activities in human T cells, via TLR2 [93]. Human T cells activated by mitogen in the presence of Hsp60 or p277 also show a decreased secretion of IFN-γ and an increased secretion of IL-10 (unpublished observations). Thus, soluble Hsp60

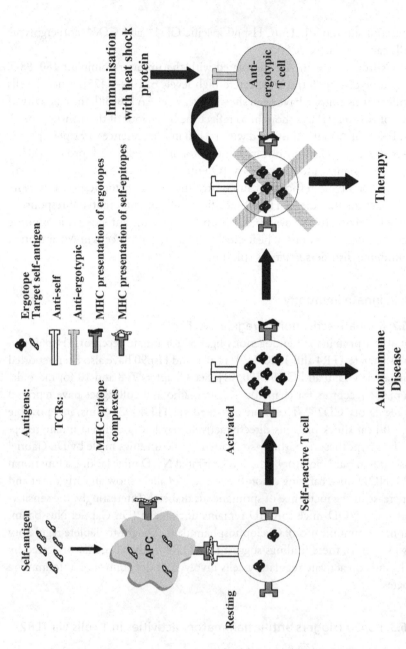

Figure 16.3. Anti-ergotypic response mediated by Hsp60-specific T cells.

Antigens:
🦴 Ergotope
ℰ Target self-antigen

TCRs:
⊤ Anti-self
⊤ Anti-ergotypic

MHC-epitope complexes:
⊤ MHC presentation of ergotopes
⊤ MHC presentation of self-epitopes

Self-antigen → APC

Resting → Activated → Self-reactive T cell

Autoimmune Disease

Immunisation with heat shock protein → Anti-ergotypic T cell

Therapy

or p277, acting via TLR2, can modulate T cells involved in the progression of inflammation.

16.7. Heat shock proteins: physiological modulators of inflammation

Inflammation is physiological [94]; it plays a role in processes ranging from wound healing [95] to neuroprotection [96]. However, uncontrolled inflammation can lead to disease and, as a consequence, precise mechanisms have been selected through evolution for the tight control of inflammation.

Inflammation induces heat shock proteins and heat shock protein–specific immune responses. However, heat shock proteins and the immune responses directed against them can both promote and inhibit inflammation. Heat shock proteins are central nodes in physiological networks that control inflammation; they integrate the intra-cellular response to stress with the inter-cellular signals that spread a cascade of pro- or anti-inflammatory responses (Figure 16.4A). The variety of the anti-inflammatory responses co-ordinated by heat shock proteins is as diverse as the biological activities of heat shock proteins. In the short term, heat shock proteins can activate regulatory mechanisms via innate receptors. In the long term, heat shock proteins can also trigger adaptive immunoregulatory T cell responses directed against heat shock proteins or other self-proteins. Heat shock proteins bridge the innate and adaptive immune responses involved in the physiological control of inflammation.

Regulatory networks centred on heat shock proteins can be boosted by several methods to treat autoimmune disease (Figure 16.4B). Indeed, these therapies could operate by simply mimicking the effects that the environment has on the immune system. The rise in the standard of living achieved during the past century in the developed world seems to have diminished the microbial stimulation of the immunoregulatory functions that keep immune balance. This reduction in immune stimulation might contribute to the increased incidence of autoimmune diseases observed in developed countries, as we have discussed elsewhere [97].

16.7.1. Network cross-reactivity

It is striking that immunisation with defined heat shock proteins leads to the induction of T cell responses directed to other structurally unrelated heat shock protein molecules [31]. The definition of immunological cross-reactivity, usually found in immunology textbooks, would not account for this unexpected finding. Herein, we would like to propose a new definition for cross-reactivity.

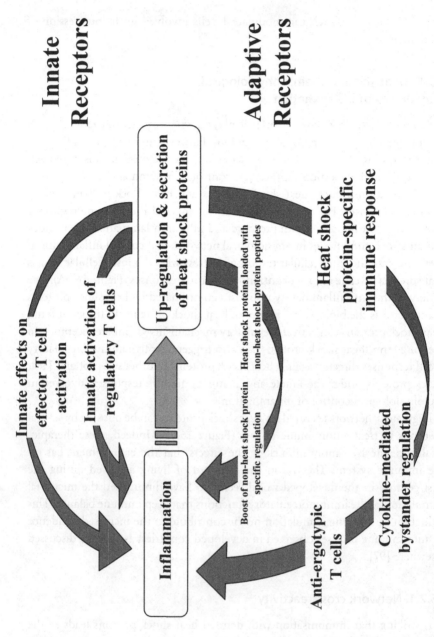

Innate Receptors

Adaptive Receptors

Innate effects on effector T cell activation

Innate activation of regulatory T cells

Up-regulation & secretion of heat shock proteins

Heat shock protein-specific immune response

Inflammation

Heat shock proteins loaded with non-heat shock protein peptides

Boost of non-heat shock protein specific regulation

Anti-ergotypic T cells

Cytokine-mediated bystander regulation

Figure 16.4. Heat shock proteins as regulators of inflammation: (A) physiological regulation of inflammation by heat shock proteins and (B) therapeutic/environmental regulation of inflammation by heat shock proteins.

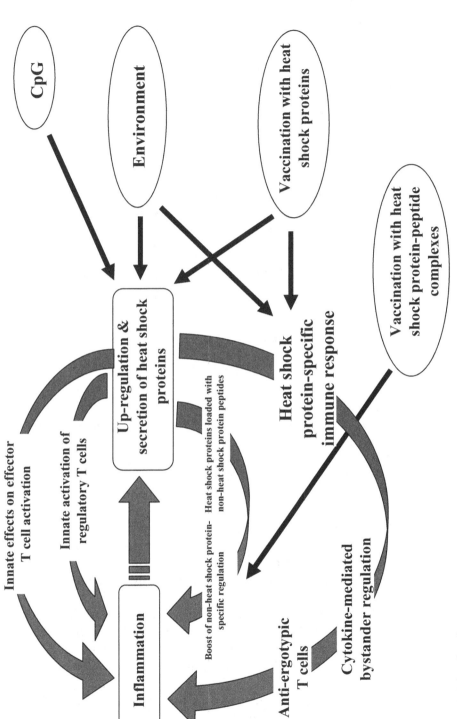

Figure 16.4. (*continued*)

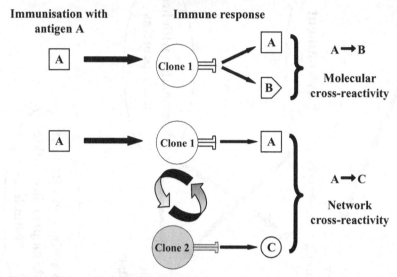

Figure 16.5. Immunological cross-reactivity.

We define *molecular* cross-reactivity as the classical cross-reactivity that exists between antigens that share sequence homology, leading to their recognition by the same T or B cell clones (Figure 16.5). We define *network* cross-reactivity as the immune connection existing between molecules that bear no sequence homology, like Hsp60 and Hsp70, but whose specific immune responses are somehow interconnected (by self-vaccination or another mechanism). Thus, the organisation of the immune network is such that immunisation with antigen 1 can induce an immune response that targets not only antigen 1, but also antigen 2, in the absence of any single T or B cell clone that recognises both antigens (Figure 16.5). The regulatory properties of heat shock proteins might result from the molecular cross-reactivity existing between self and microbial heat shock proteins and the network cross-reactivity that exists between different endogenous heat shock proteins.

The study of the pro- and anti-inflammatory mechanisms mediated by heat shock proteins could lead to the design of novel therapies for autoimmunity: therapies aimed at reinforcing the built-in mechanisms that are based on the physiological control of the immune function by heat shock proteins [94, 98]. The initial success of the Hsp60 peptide p277 in treating human T1DM shows the feasibility of this approach [49]. A deeper understanding of the multiple activities of heat shock proteins on the immune system and body homeostasis might allow us to extend these findings to other autoimmune disorders.

Acknowledgements

Professor Irun R. Cohen is the incumbent of the Mauerberger Chair in Immunology. We thank Ms. Danielle Sabah-Israel for excellent secretarial assistance.

REFERENCES

1. Hartl F U and Hayer-Hartl M. Molecular chaperones in the cytosol: from nascent chain to folded protein. Science 2002, 295: 1852–1858.
2. Young D B, Ivanyi J, Cox J H and Lamb J R. The 65kDa antigen of mycobacteria – a common bacterial protein? Immunol Today 1987, 8: 215–219.
3. Gaston J S. Heat shock proteins and arthritis – new readers start here. Autoimmunity 1997, 26: 33–42.
4. Dhillon V, Latchman D and Isenberg D. Heat shock proteins and systemic lupus erythematosus. Lupus 1991, 1: 3–8.
5. Lehner T. The role of heat shock protein, microbial and autoimmune agents in the aetiology of Behcet's disease. Int Rev Immunol 1997, 14: 21–32.
6. Abulafia-Lapid R, Elias D, Raz I, Keren-Zur Y, Atlan H and Cohen I R. T cell proliferative responses to type 1 diabetes patients and healthy individuals to human Hsp60 and its peptides. J Autoimmunity 1999, 12: 121–129.
7. Beg A A. Endogenous ligands of Toll-like receptors: implications for regulating inflammatory and immune responses. Trends Immunol 2002, 23: 509–512.
8. Anderton S M, van der Zee R and Goodacre J A. Inflammation activates self hsp60-specific T cells. Eur J Immunol 1993, 23: 33–38.
9. Mor F and Cohen I R. T cells in the lesion of experimental autoimmune encephalomyelitis. Enrichment for reactivities to myelin basic protein and to heat shock proteins. J Clin Invest 1992, 90: 2447–2455.
10. Birnbaum G and Kotilinek L. Heat shock or stress proteins and their role as autoantigens in multiple sclerosis. Ann NY Acad Sci 1997, 835: 157–167.
11. Birnbaum G. Stress proteins: their role in the normal central nervous system and in disease states, especially multiple sclerosis. Springer Semin Immunopathol 1995, 17: 107–118.
12. Gao Y L, Brosnan C F and Raine C S. Experimental autoimmune encephalomyelitis. Qualitative and semiquantitative differences in heat shock protein 60 expression in the central nervous system. J Immunol 1995, 154: 3548–3556.
13. Pockley A G. Heat shock proteins, heat shock protein reactivity and allograft rejection. Transplantation 2001, 71: 1503–1507.
14. Wauben M H M, Wagenaar-Hilbers J P A and van Eden W. Adjuvant arthritis. In Cohen, I. R. and Miller, A. (Eds.) *Autoimmune Disease Models.* Academic Press, Inc., New York 1994, pp 201–216.
15. Elias D. The NOD mouse: a model for autoimmune insulin-dependent diabetes. In Cohen, I. R. and Miller, A. (Eds.) *Autoimmune Disease Models.* Academic Press, Inc., New York 1994, pp 147–161.
16. 16. Feldmann M, Brennan F M and Maini R N. Rheumatoid arthritis. Cell 1996, 85: 307–310.
17. Tisch R and McDevitt H. Insulin-dependent diabetes mellitus. Cell Biol Int 1996, 85: 291–297.

18. van Eden W, Holoshitz J, Nevo Z, Frenkel A, Klajman A and Cohen I R. Arthritis induced by a T-lymphocyte clone that responds to *Mycobacterium tuberculosis* and to cartilage proteoglycans. Proc Natl Acad Sci USA 1985, 82: 5117–5120.
19. Billingham M E J, Carney S, Butler R and Colston M J. A mycobacterial 65-kd heat shock protein induces antigen-specific suppression of adjuvant arthritis, but is not itself arthritogenic. J Exp Med 1990, 171: 339–344.
20. Hogervorst E J, Schouls L, Wagenaar J P, Boog C J, Spaan W J, van Embden J D and van Eden W. Modulation of experimental autoimmunity: treatment of adjuvant arthritis by immunization with a recombinant vaccinia virus. Infect Immun 1991, 59: 2029–2035.
21. Ragno S, Colston M J, Lowrie D B, Winrow V R, Blake D R and Tascon R. Protection of rats from adjuvant arthritis by immunization with naked DNA encoding for mycobacterial heat shock protein 65. Arthritis Rheum 1997, 40: 277–283.
22. Anderton S M, van der Zee R, Prakken B, Noordzij A and van Eden W. Activation of T cells recognizing self 60-kD heat shock protein can protect against experimental arthritis. J Exp Med 1995, 181: 943–952.
23. Thompson S J, Francis J N, Siew L K, Webb G R, Jenner P J, Colston M J and Elson C J. An immunodominant epitope from mycobacterial 65-kDa heat shock protein protects against pristane-induced arthritis. J Immunol 1998, 160: 4628–4634.
24. van Eden W, van der Zee R, Paul A G A, Prakken B J, Wendling U, Anderton S M and Wauben M H M. Do heat shock proteins control the balance of T-cell regulation in inflammatory diseases? Immunol Today 1998, 19: 303–307.
25. Quintana F J, Carmi P, Mor F and Cohen I R. Inhibition of adjuvant arthritis by a DNA vaccine encoding human heat shock protein 60. J Immunol 2002, 169: 3422–3428.
26. Quintana F J, Carmi P, Mor F and Cohen I R. DNA fragments of the human 60-kDa heat shock protein (HSP60) vaccinate against adjuvant arthritis: identification of a regulatory HSP60 peptide. J Immunol 2003, 171: 3533–3541.
27. Ragno S, Winrow V R, Mascagni P, Lucietto P, Di Pierro F, Morris C J and Blake D R. A synthetic 10-kD heat shock protein (hsp10) from *Mycobacterium tuberculosis* modulates adjuvant arthritis. Clin Exp Immunol 1996, 103: 384–390.
28. Kingston A E, Hicks C A, Colston M J and Billingham M E J. A 71-kD heat shock protein (hsp) from *Mycobacterium tuberculosis* has modulatory effects on experimental rat arthritis. Clin Exp Immunol 1996, 103: 77–82.
29. Wendling U, Paul L, van der Zee R, Prakken B, Singh M and van Eden W. A conserved mycobacterial heat shock protein (Hsp) 70 sequence prevents adjuvant arthritis upon nasal administration and induces IL-10-producing T cells that cross-react with the mammalian self-hsp70 homologue. J Immunol 2000, 164: 2711–2717.
30. Tanaka S, Kimura Y, Mitani A, Yamamoto G, Nishimura H, Spallek R, Singh M, Noguchi T and Yoshikai Y. Activation of T cells recognizing an epitope of heat-shock protein 70 can protect against rat adjuvant arthritis. J Immunol 1999, 163: 5560–5565.
31. Quintana F J, Carmi P, Mor F and Cohen I R. Inhibition of adjuvant-unduced as arthritis by DNA vaccination with the 70-kd or the 90-kd human heat-shock protein: immune cross-regulation with the 60-kd heat-shock protein. Arth Rheum 2004, 50: 3712–3720.

32. Macht L M, Elson C J, Kirwan J R, Gaston J S H, Lamont A G, Thompson J M and Thompson S J. Relationship between disease severity and responses by blood mononuclear cells from patients with rheumatoid arthritis to human heat-shock protein 60. Immunology 2000, 99: 208–214.

33. van Roon J A G, van Eden W, van Roy J L A M, Lafeber F J P G and Bijlsma J W J. Stimulation of suppressive T cell responses by human but not bacterial 60-kD heat shock protein in synovial fluid of patients with rheumatoid arthritis. J Clin Invest 1997, 100: 459–463.

34. Prakken A B, van Eden W, Rijkers G T, Kuis W, Toebes E A, de Graeff-Meeder E R, van der Zee R and Zegers B J. Autoreactivity to human Hsp60 predicts disease remission in oligoarticular juvenile rheumatoid arthritis. Arthritis Rheum 1996, 39: 1826–1832.

35. Kanagawa O, Martin S M, Vaupel B A, Carrasco-Marin E and Unanue E R. Autoreactivity of T cells from nonobese diabetic mice: an I-Ag7-dependent reaction. Proc Natl Acad Sci USA 1998, 95: 1721–1724.

36. Quintana F J and Cohen I R. Autoantibody patterns in diabetes-prone NOD mice and in standard C57BL/6 mice. J Autoimmunity 2001, 17: 191–197.

37. Wegmann D R, Norbury-Glaser M and Daniel D. Insulin-specific T cells are a predominant component of islet infiltrates in pre-diabetic NOD mice. Eur J Immunol 1994, 24: 1853–1857.

38. Tisch R, Yang X D, Singer S M, Liblau R S, Fugger L and McDevitt H O. Immune response to glutamic acid decarboxylase correlates with insulitis in non-obese diabetic mice. Nature 1993, 366: 72–75.

39. Elias D, Markovits D, Reshef T, van der Zee R and Cohen I R. Induction and therapy of autoimmune diabetes in the non-obese diabetic mouse by a 65-kDa heat shock protein. Proc Natl Acad Sci USA 1990, 87: 1576–1580.

40. Elias D, Reshef T, Birk O S, van der Zee R, Walker M D and Cohen I R. Vaccination against autoimmune mouse diabetes with a T cell epitope of the human 65-kDa heat shock protein. Proc Natl Acad Sci USA 1991, 88: 3088–3091.

41. Elias D and Cohen I R. Peptide therapy for diabetes in NOD mice. Lancet 1994, 343: 704–706.

42. Elias D, Meilin A, Ablamunits V, Birk O S, Carmi P, Konen-Waisman S and Cohen I R. Hsp60 peptide therapy of NOD mouse diabetes induces a Th2 cytokine burst and downregulates autoimmunity to various β-cell antigens. Diabetes 1997, 46: 758–764.

43. Bockova J, Elias D and Cohen I R. Treatment of NOD diabetes with a novel peptide of the Hsp60 molecule induces Th2-type antibodies. J Autoimmunity 1997, 10: 323–329.

44. Yasunami R and Bach J F. Anti-suppressor effect of cyclophosphamide on the development of spontaneous diabetes in NOD mice. Eur J Immunol 1988, 18: 481–484.

45. Ablamunits V, Quintana F, Reshef T, Elias D and Cohen I R. Acceleration of autoimmune diabetes by cyclophosphamide is associated with an enhanced IFN-γ secretion pathway. J Autoimmunity 1999, 13: 383–392.

46. Quintana F J, Carmi P and Cohen I R. DNA vaccination with heat shock protein 60 inhibits cyclophosphamide-accelerated diabetes. J Immunol 2002, 169: 6030–6035.

284   Francisco J. Quintana and Irun R. Cohen

47. Abulafia-Lapid R, Gillis D, Yosef O, Atlan H and Cohen I R. T cells and autoantibodies to human HSP70 in Type 1 diabetes in children. J Autoimmunity 2003: 313–321.
48. Horváth L, Cervenak L, Oroszlán M, Proháska Z, Uray K, Hudecz F, Baranyi É, Madácsy L, Singh M, Romics L, Füst G and Pánczél P. Antibodies against different epitopes of heat shock protein 60 in children with type 1 diabetes mellitus. Immunol Lett 2002, 80: 155–162.
49. Raz I, Elias D, Avron A, Tamir M, Metzger M and Cohen I R. Beta-cell function in new-onset type 1 diabetes and immunomodulation with a heat-shock protein peptide (DiaPep277): a randomised, double-blind, phase II trial. Lancet 2001, 358: 1749–1753.
50. Hemmi H, Takeuchi O, Kawai T, Kaisho T, Sato S, Sanjo H, Matsumoto M, Hoshino K, Wagner H, Takeda K and Akira S. A Toll-like receptor recognizes bacterial DNA. Nature 2000, 408: 740–745.
51. Krieg A M. CpG motifs in bacterial DNA and their immune effects. Ann Rev Immunol 2002, 20: 709–760.
52. Quintana F J, Rotem A, Carmi P and Cohen I R. Vaccination with empty plasmid DNA or CpG oligonucleotide inhibits diabetes in nonobese diabetic mice: modulation of spontaneous 60-kDa heat shock protein autoimmunity. J Immunol 2000, 165: 6148–6155.
53. Akdis C A and Blaser K. Mechanisms of interleukin-10-mediated immune suppression. Immunology 2001, 103: 131–136.
54. Kumaraguru U, Pack C D and Rouse B T. Toll-like receptor ligand links innate and adaptive immune responses by the production of heat-shock proteins. J Leuk Biol 2003, 73: 574–583.
55. Boccaccio G L, Mor F and Steinman L. Non-coding plasmid DNA induces IFN-γ in vivo and suppresses autoimmune encephalomyelitis. Int Immunol 1999, 11: 289–296.
56. Rachmilewitz D, Karmeli F, Takabayashi K, Hayashi T, Leider-Trejo L, Lee J, Leoni L M and Raz E. Immunostimulatory DNA ameliorates experimental and spontaneous murine colitis. Gastroenterology 2002, 122: 1428–1441.
57. Zeuner R A, Ishii K J, Lizak M J, Gursel I, Yamada H, Klinman D M and Verthelyi D. Reduction of CpG-induced arthritis by suppressive oligodeoxynucleotides. Arthritis Rheum 2002, 46: 2219–2224.
58. Zeuner R A, Verthelyi D, Gursel M, Ishii K J and Klinman D M. Influence of stimulatory and suppressive DNA motifs on host susceptibility to inflammatory arthritis. Arthritis Rheum 2003, 48: 1701–1707.
59. Serreze D V, Hamaguchi K and Leiter E H. Immunostimulation circumvents diabetes in NOD/Lt mice. J Autoimmunity 1989, 2: 759–776.
60. Tian J, Zekzer D, Hanssen L, Lu Y, Olcott A and Kaufman D L. Lipopolysaccharide-activated B cells down-regulate Th1 immunity and prevent autoimmune diabetes in nonobese diabetic mice. J Immunol 2001, 167: 1081–1089.
61. Sai P and Rivereau A S. Prevention of diabetes in the nonobese diabetic mouse by oral immunological treatments. Comparative efficiency of human insulin and two bacterial antigens, lipopolysacharide from Escherichia coli and glycoprotein extract from Klebsiella pneumoniae. Diabetes Metab 1996, 22: 341–348.

62. Iguchi M, Inagawa H, Nishizawa T, Okutomi T, Morikawa A, Soma G I and Mizuno D. Homeostasis as regulated by activated macrophage. V. Suppression of diabetes mellitus in non-obese diabetic mice by LPSw (a lipopolysaccharide from wheat flour). Chem Pharm Bull (Tokyo) 1992, 40: 1004–1006.

63. Li Z, Ménoret A and Srivastava P. Roles of heat-shock proteins in antigen presentation and cross-presentation. Cur Opin Immunol 2002, 14: 45–51.

64. Srivastava P. Roles of heat-shock proteins in innate and adaptive immunity. Nat Rev Immunol 2002, 2: 185–194.

65. Vabulas R M, Braedel S, Hilf N, Singh-Jasuja H, Herter S, Ahmad-Nejad P, Kirschning C J, Da Costa C, Rammensee H G, Wagner H and Schild H. The endoplasmic reticulum-resident heat shock protein Gp96 activates dendritic cells via the Toll-like receptor 2/4 pathway. J Biol Chem 2002, 277: 20847–20853.

66. Bethke K, Staib F, Distler M, Schmitt U, Jonuleit H, Enk A H, Galle P R and Heike M. Different efficiency of heat shock proteins to activate human mono-cytes and dendritic cells: Superiority of HSP60. J Immunol 2002, 169: 6141–6148.

67. Flohé S B, Bruggemann J, Lendemans S, Nikulina M, Meierhoff G, Flohé S and Kolb H. Human heat shock protein 60 induces maturation of dendritic cells versus a Th1-promoting phenotype. J Immunol 2003, 170: 2340–2348.

68. Bulut Y, Faure E, Thomas L, Karahashi H, Michelsen K S, Equils O, Morrison S G, Morrison R P and Arditi M. Chlamydial heat shock protein 60 activates macrophages and endothelial cells through Toll-like receptor 4 and MD2 in a MyD88-dependent pathway. J Immunol 2002, 168: 1435–1440.

69. Moudgil K D, Chang T T, Eradat H, Chen A M, Gupta R S, Brahn E and Sercarz E E. Diversification of T cell responses to carboxy-terminal determinants within the 65-kD heat-shock protein is involved in regulation of autoimmune arthritis. J Exp Med 1997, 185: 1307–1316.

70. Hirono S, Dibrov E, Hurtado C, Kostenuk A, Ducas R and Pierce G N. *Chlamydia pneumoniae* stimulates proliferation of vascular smooth muscle cells through induction of endogenous heat shock protein 60. Circ Res 2003, 93: 710–716.

71. Beimnet K, Soderstrom K, Jindal S, Gronberg A, Frommel D and Kiessling R R. Induction of heat shock protein 60 expression in human monocytic cell lines infected with *Mycobacterium leprae*. Infect Immun 1996, 64: 4356–4358.

72. Wainberg Z, Oliveira M, Lerner S, Tao Y and Brenner B G. Modulation of stress protein (hsp27 and hsp70) expression in CD4$^+$ lymphocytic cells following acute infection with human immunodeficiency virus type-1. Virology. 1997, 233: 364–373.

73. Saito K, Katsuragi H, Mikami M, Kato C, Miyamaru M and Nagaso K. Increase of heat-shock protein and induction of γ/δ T cells in peritoneal exudate of mice after injection of live *Fusobacterium nucleatum*. Immunology 1997, 90: 229–235.

74. Basu S, Binder R J, Suto R, Anderson K M and Srivastava P K. Necrotic but not apoptotic cell death releases heat shock proteins, which deliver a partial maturation signal to dendritic cells and activates the NF-κB pathway. Int Immunol 2000, 12: 1539–1546.

75. Boog C J P, de Graeff-Meeder E R, Lucassen M A, van der Zee R, Voorhorst Ogink M M, van Kooten P J S, Geuze H J and van Eden W. Two monoclonal antibodies

generated against human hsp60 show reactivity with synovial membranes of patients with juvenile arthritis. J Exp Med 1992, 175: 1805–1810.

76. Teshima S, Rokutan K, Takahashi M, Nikawa T and Kishi K. Induction of heat shock proteins and their possible roles in macrophages during activation by macrophage colony-stimulating factor. Biochem J 1996, 315: 497–504.

77. Ferris D K, HarelBellan A, Morimoto R I, Welch W J and Farrar W L. Mitogen and lymphokine stimulation of heat shock proteins in T lymphocytes. Proc Natl Acad Sci USA 1988, 85: 3850–3854.

78. Zugel U and Kaufmann S H. Immune response against heat shock proteins in infectious diseases. Immunobiology 1999, 201: 22–35.

79. Chen Y, Kuchroo V K, Inobe J, Hafler D A and Weiner H L. Regulatory T cell clones induced by oral tolerance: suppression of autoimmune encephalomyelitis. Science 1994, 265: 1237–1240.

80. Moudgil K D, Kim E, Yun O J, Chi H H, Brahn E and Sercarz E E. Environmental modulation of autoimmune arthritis involves the spontaneous microbial induction of T cell responses to regulatory determinants within heat shock protein 65. J Immunol 2001, 166: 4237–4243.

81. Prakken B J, Wendling U, van der Zee R, Rutten V P M, Kuis W and van Eden W. Induction of IL-10 and inhibition of experimental arthritis are specific features of microbial heat shock proteins that are absent for other evolutionarily conserved immunodominant proteins. J Immunol 2001, 167: 4147–4153.

82. Blachere N E, Li Z L, Chandawarkar R Y, Suto R, Jaikaria N S, Basu S, Udono H and Srivastava P K. Heat shock protein-peptide complexes, reconstituted in vitro, elicit peptide-specific cytotoxic T lymphocyte response and tumor immunity. J Exp Med 1997, 186: 1315–1322.

83. Chandawarkar R Y, Wagh M S and Srivastava P K. The dual nature of specific immunological activity of tumour-derived gp96 preparations. J Exp Med 1999, 189: 1437–1442.

84. Galazka G, Walczak A, Berkowicz T and Selmaj K. Effect of Hsp70-peptide complexes generated in vivo on modulation EAE. Adv Exp Med Biol 2001, 495: 227–230.

85. Lohse A W, Mor F, Karin N and Cohen I R. Control of experimental autoimmune encephalomyelitis by T cells responding to activated T cells. Science 1989, 244: 820–822.

86. Mimran A, Mor F, Carmi P, Quintana F J, Rotter V and Cohen I R. DNA vaccination with CD25 protects rats from adjuvant arthritis and induces an antiergotypic response. J Clin Invest 2004, 113: 924–932.

87. Mor F, Reizis B, Cohen I R and Steinman L. IL-2 and TNF receptors as targets of regulatory T-T interactions: isolation and characterization of cytokine receptor-reactive T cell lines in the Lewis rat. J Immunol. 1996, 157: 4855–4861.

88. van Eden W, Thole J E R, van der Zee R, Noordzij A, van Embden J D A, Hensen E J and Cohen I R. Cloning of the mycobacterial epitope recognized by T lymphocytes in adjuvant arthritis. Nature 1988, 331: 171–173.

89. Vabulas R M, Ahmad-Nejad P, da Costa C, Miethke T, Kirschning C J, Hacker H and Wagner H. Endocytosed HSP60s use Toll-like receptor 2 (TLR2) and TLR4 to activate the Toll/interleukin-1 receptor signaling pathway in innate immune cells. J Biol Chem 2001, 276: 31332–31339.

90. Becker T, Hartl F U and Wieland F. CD40, an extracellular receptor for binding and uptake of Hsp70-peptide complexes. J Cell Biol 2002, 158: 1277–1285.
91. Basu S, Binder R J, Ramalingam T and Srivastava P K. CD91 is a common receptor for heat shock proteins gp96, hsp90, hsp70 and calreticulin. Immunity 2001, 14: 303–313.
92. Caramalho I, Lopes-Carvalho T, Ostler D, Zelenay S, Haury M and Demengeot J. Regulatory T cells selectively express Toll-like receptors and are activated by lipopolysaccharide. J Exp Med 2003, 197: 403–411.
93. Zanin-Zhorov A, Nussbaum G, Franitza S, Cohen I R and Lider O. T cells respond to heat shock protein 60 via TLR2: activation of adhesion and inhibition of chemokine receptors. FASEB J 2003, 17: 1567–1569.
94. Cohen I R, Quintana F J, Nussbaum G, Cohen M, Zanin A and Lider O. HSP60 and the regulation of inflammation: physiological and pathological. In van Eden, W. (Ed.) *Heat Shock Proteins and Inflammation*. Birkhauser Verlag A G, Basel 2004, pp 1–13.
95. Werner S and Grose R. Regulation of wound healing by growth factors and cytokines. Physiol Rev 2003, 83: 835–870.
96. Cohen I R and Schwartz M. Autoimmune maintenance and neuroprotection of the central nervous system. J Neuroimmunol 1999, 100: 111–114.
97. Quintana F J and Cohen I R. Type I diabetes mellitus, infection and Toll-like receptors. In Shoenfeld, Y. and Rose, N. (Eds.) *Infection and Autoimmunity*. Elsevier, Amsterdam 2004.
98. Cohen I R. *Tending Adam's Garden: Evolving the Cognitive Immune Self*. Academic Press, London: 2000.

17

Heat Shock Protein Fusions: A Platform for the Induction of Antigen-Specific Immunity

Lee Mizzen and John Neefe

17.1. Introduction

The unusual immunogenicity of heat shock proteins (also known as stress proteins) was discovered in studies of the immune response to microbial infection, in which a large proportion of the humoral and cellular immune response to diverse microbial pathogens was found to be specific for pathogen-derived heat shock protein [1]. These studies demonstrated that immune recognition of pathogen-derived heat shock proteins occurs in natural and experimental settings in animals and man. This is discussed in detail in Chapter 16. Immune responses elicited to mycobacterial heat shock proteins have been particularly well studied. In man, recognition of mycobacterial heat shock protein by CD4$^+$ T cells occurs in the context of numerous human leukocyte antigen (HLA) alleles, and epitopes have been identified that are presented by multiple HLA molecules [2]. The promiscuous recognition of mycobacterial heat shock proteins supports their utility as 'universal' immunogens for the genetically diverse human population. The immunogenic properties of microbial heat shock proteins have accordingly led to their application in a variety of immunisation formats as prophylactic and therapeutic agents in models of infectious disease and cancer [3]. In these studies, heat shock proteins have been delivered as subunit vaccines, carrier proteins in chemical conjugates, recombinant fusion proteins and DNA expression vectors for induction of humoral and cellular immunity.

To explain the disproportionate focus of the immune response on a small subset of pathogen antigens, heat shock proteins were proposed to act as 'red flags' – alerting the immune system to the presence of a foreign invader [4]. Given their ubiquity in microbial pathogens and their over-production by microbes in response to 'stressful' conditions experienced within the infected host [5], pathogen-derived heat shock proteins represent highly effective targets for sensing infection. However, evidence is accumulating that the immune system may also respond to endogenous (i.e., self) heat shock proteins during various pathophysiological states, such as inflammation or necrotic cell death. Indeed, in this context, heat shock proteins are postulated to function as universal 'danger signals', alerting the immune system to the presence of stressed, infected or diseased tissue [6].

Given the central role of professional antigen-presenting cells (APCs) in initiating immune responses, the induction of antigen-specific immunity by heat shock proteins may be explained, in part, by the identification of candidate heat shock protein receptors on dendritic cells (DCs) and macrophages. On murine and/or human monocytes, receptors for Hsp90, Hsp70 and Hsp60 homologues include CD14, members of the Toll-like receptor (TLR) family, TLR4 and TLR2, scavenger receptors (SRs) CD91 and LOX-1, and the co-stimulatory molecule CD40 [7–14]. In addition, CD94 is implicated as a receptor for Hsp70 on human NK cells [15]. The receptors for chaperones are discussed in detail in Chapters 7, 8 and 10. A common function of TLRs and SRs is the binding of exogenous or endogenous 'pattern recognition' ligands, respectively, which triggers innate immune responses [16, 17]. Similarly, stimulation of natural killer (NK) cells via CD94 can activate their innate tumour-killing function, as observed with Hsp70 [18]. Finally, co-stimulation of DCs through CD40 is a critical activation signal for initiating adaptive T cell responses [19]. Hence, receptor molecules implicated in heat shock protein recognition share the property of transmitting activation signals from the innate to the adaptive immune system.

Specific heat shock protein recognition and uptake by APCs is also supported by *in vitro* studies demonstrating activation and maturation of DCs exposed to exogenous heat shock proteins [20, 21] as well as the 'cross-priming' of CD8$^+$ T cell responses to antigens associated with heat shock proteins [22, 23]. At present, direct biochemical confirmation of receptor binding by heat shock proteins is lacking; the more prevalent view is that identified molecules are part of signalling pathways in APCs that may involve as yet unidentified co-receptors.

17.2. Heat shock protein fusions: a new approach to immunisation

Given the ability of heat shock proteins to target APCs and to enhance immune responses to associated antigens, a practical approach for the design and manufacture of new antigen-specific vaccine formulations is the construction of heat shock protein fusions by recombinant DNA technology. Here, DNA sequences for heat shock proteins and target antigens are spliced together in-frame ('fused') using standard techniques, and the resultant chimaeric genes or encoded proteins are employed as immunogens. As reviewed next, immunisation with heat shock protein fusions in protein, RNA, DNA or viral vector formats has yielded promising results in animal models of infectious disease and cancer. Significantly, this approach has progressed to the first human testing of a heat shock protein fusion protein as a viral immunotherapeutic.

17.2.1. Heat shock protein fusion proteins

In the first published example of this concept, immunisation with an HIV p24-*Mycobacterium tuberculosis* Hsp70 fusion protein was shown to induce p24-specific immune responses in mice, as characterised by the production of Th1 and Th2 cytokines and the induction of a significant IgG titre that persisted for over a year [24]. This work alerted researchers to the potential of heat shock protein fusions for the induction of immune responses and highlighted some important features that would be confirmed in subsequent research: lack of dependence on adjuvants, induction of antigen-specific B and T cell responses and the significant enhancement of antigen-specific immune responses by covalent linkage (fusion) of heat shock proteins to the antigen.

The ability of heat shock protein fusion proteins to enhance antibody responses was extended by studies employing *Leishmania infantum* Hsp70 or Hsp83 fusions with *Escherichia coli* maltose-binding protein (MBP) [25–27]. A consistent finding in these studies was the induction of a Th1 immune response to MBP, as measured by cytokine secretion and IgG isotype profile. Notably, immunisation of athymic *nu/nu* mice with the fusion protein generated anti-MBP IgG$_{2a}$ antibodies, which is consistent with a T cell–dependent response [26]. As noted in another study, the induction of IgG antibodies by heat shock proteins can possess unusual features [28].

Although heat shock protein fusions can elicit markedly enhanced antigen-specific humoral responses, the DC targeting property of heat shock proteins has been exploited by a majority of researchers seeking to augment cellular immunity to fused antigens. In the first example of cross-priming by a heat shock protein fusion protein, immunisation of mice with an ovalbumin–*M. tuberculosis* Hsp70 fusion induced ovalbumin-specific CD8$^+$ cytotoxic T lymphocytes (CTLs) [29]. These CTLs recognised the H-2^b-restricted ovalbumin SIINFEKL peptide (amino acids (aa) 257–264), indicating that heat shock protein fusion immunisation led to 'natural' antigen processing *in vivo*. Mice immunised with the fusion were protected against challenge with an ovalbumin-expressing tumour cell line, implicating ovalbumin-specific CTLs in tumour rejection.

In another study concerning *in vivo* antigen processing, *Trypanosoma cruzi* Hsp70 was chosen as a vehicle for the discovery of potential human CTL epitopes in a target antigen from the same parasite. In this study, immunisation of HLA-A2/K^b transgenic mice with a kinetoplastid membrane protein-11

(KMP11)–Hsp70 fusion permitted the identification of two HLA-A2–restricted CTL epitopes in KMP11 [30].

In the first example of CTL induction by a fusion protein containing an Hsp60 homologue, peptide-specific CD8$^+$ CTLs were elicited in mice of H-2^b and H-2^d haplotypes immunised with *Mycobacterium bovis* BCG Hsp65-influenza nucleo-protein fusions [31]. In this study, significant CTL activity appeared following a single immunisation. This persisted for a minimum of four months and was boosted by a second immunisation, thereby suggesting the presence of memory T cells. This study also demonstrated that, as observed for Hsp70 fusions, Hsp65 fusion immunisation elicits CTLs specific for defined immunodominant epitopes present within fused antigens.

With these features in mind, a tumour 'vaccine' using *M. bovis* BCG Hsp65 was created and tested in a mouse model of human cervical cancer [32]. Here, prophylactic and therapeutic immunisation with an Hsp65–human papillomavirus (HPV)-type 16 E7 fusion (HspE7) led to the rejection of an HPV16 E7-expressing tumour, TC-1. HspE7 immunisation was associated with induction of a Th1-like cell-mediated immune response, based on the cytokine secretion profile, and the presence of cytolytic activity against TC-1 cells. Tumour rejection following therapeutic immunisation with HspE7 was dependent on the presence of CD8$^+$ T cells during priming and was associated with long-term survival (over 253 days) in 80 percent of treated animals.

A number of studies have provided insight into the mechanisms by which heat shock protein fusion proteins elicit CTLs. A particularly revealing observation is that CD8$^+$ CTLs are induced by mycobacterial Hsp70 and Hsp65 fusions in CD4$^+$ T cell-deficient mice [32–36]. Induction of CTLs without CD4$^+$ T cell 'help' can occur by direct activation of APCs, as shown by CD40 ligation or acute viral infection [37, 38]. Consistent with mechanisms by which APCs 'licence' CD8$^+$ T cell responses, mycobacterial Hsp65 fusion proteins have been shown to directly activate murine bone-marrow-derived DCs *in vitro* whereas, *in vivo*, myeloid DCs recovered from fusion immunised mice displayed an activated phenotype [33]. To obtain more quantitative measurements on the potency of CTL induction by heat shock protein fusions, an ovalbumin–*M. tuberculosis* Hsp70 fusion protein was used to immunise OT-1 mice transgenic for a T cell receptor recognising the ovalbumin SIINFEKL peptide [36]. By comparison with animals receiving the SIINFEKL peptide in complete Freund's adjuvant (CFA), immunisation of mice with the fusion protein induced stronger and more durable proliferation of adoptively transferred OT-1 cells *in vivo*. On a molar basis, delivery of the SIINFEKL peptide within an ovalbumin protein fragment fused to Hsp70 was several hundred-fold more effective in eliciting CTL responses than peptide in CFA. These observations illustrate, in this transgenic model, the quantitative and qualitative superiority of a heat shock protein fusion over the gold-standard adjuvant CFA as a vehicle for CTL induction.

To map the region of Hsp70 involved in CTL induction to fused antigen, mice were immunised with heat shock protein fusions composed of an ovalbumin fragment joined to different portions of the *M. tuberculosis* Hsp70 protein [34]. This study revealed that the ability to elicit SIINFEKL-specific CTLs resided within a region of Hsp70 located in the ATPase domain (aa 161–370). It is noteworthy that the amino acid sequences in Hsp70 associated with induction of antibody or CTL responses to fused antigens have been mapped to the conserved ATPase domain [25, 34, 35]. The region of human Hsp70 responsible for CD40 binding has also been mapped to the ATPase domain [13], in contrast to findings with mycobacterial Hsp70 [12]. However, this discrepancy might be related to the conformation of Hsp70 induced by nucleotide and peptide binding [13]. The structure–function relationship of Hsp70 with respect to binding to CD40 is described in detail in Chapter 10.

The high degree of sequence and functional conservation among heat shock proteins in evolution and predictions from the 'danger hypothesis' [6] suggest that the immunogenic properties of heat shock proteins will extend from microbes to higher eukaryotes, including mammals. In the context of a heat shock protein fusion, evidence for this has been provided in studies in mice that have demonstrated CTL induction to antigens fused to murine (i.e., self-) Hsp70 proteins [34, 35]. In one study, CTLs have been elicited to five different major histocompatibility complex (MHC)-restricted peptide sequences fused individually to murine Hsc70 as minimal epitopes without flanking sequences [35].

17.2.2. Heat shock protein gene fusions

As noted earlier, current research indicates that the induction of antigen-specific CTLs by heat shock protein fusion proteins derives largely from their ability to act upon myeloid APCs in at least two ways: firstly, to efficiently deliver fused antigen sequences into the MHC class I pathway for presentation to $CD8^+$ T cells, and secondly, to deliver activation signals that upregulate co-stimulatory functions and the secretion of cytokines that promote cellular immunity. At present, the intracellular pathways traversed by heat shock protein fusion proteins in APCs are largely unknown. Because endogenous expression of proteins by DNA vaccines can enhance the induction of CTLs [39], it is instructive to compare immune responses induced by heat shock protein fusions encountered by APCs as exogenous proteins versus endogenously expressed proteins.

To enhance cellular immune responses to HPV16 E7, prototype vaccines have been tested in mice based on gene fusions between HPV16 E7 sequences and *M. tuberculosis* Hsp70, delivered in plasmid DNA [40, 41], adeno-associated virus

[42], Sindbis virus RNA replicon [43, 44] and vaccinia virus vectors [41]. In all of these studies, E7-specific CD8$^+$ T cell responses were induced, and when tested these were superior to those that were induced by the E7 gene alone or other vector controls. Moreover, these studies demonstrated the prophylactic or therapeutic activity of the various E7–Hsp70 fusion constructs against TC-1 tumour cells, including, in one instance, a TC-1 variant with reduced class I expression, a phenotype common to many human cancers [41]. These studies contained another important observation: that immunisation with E7–Hsp70 fusions in plasmid DNA or RNA replicon formats induces CD8$^+$ T cells/CTLs and/or anti-tumour responses that are wholly or predominantly independent of CD4$^+$ T cells [40, 41, 43]. In addition to CTL effectors, a subset of these studies also demonstrated the essential contribution of NK cells and IFN-γ in tumour rejection following the administration of RNA or DNA vectors expressing E7–Hsp70 [41, 43].

In a direct extension of previous work using a corresponding heat shock protein fusion protein, immunisation of mice with plasmid DNA encoding a *T. cruzi* KMP11-Hsp70 fusion elicited production of anti-KMP11 IgG$_{2a}$ antibodies and CD8$^+$ CTL recognising predicted HLA-A2 epitopes in KMP11 and conferred prophylactic immunity against challenge with *T. cruzi* [45]. Extending the utility of a DNA vaccine approach to a mammalian heat shock protein fusion partner, a rabbit calreticulin–E7 plasmid DNA vaccine has been shown to induce high levels of E7-specific antibodies and CD8$^+$ T cells in mice and provides prophylactic and therapeutic immunity against TC-1 tumour cells [46]. Notably, CTL induction with the CRT–E7 DNA vaccine does not require CD4$^+$ T cells during the priming phase of immunisation. Therefore, one feature of CTL induction shared by many heat shock protein fusions delivered as protein or nucleic acid–based immunogens is the relative lack of dependence on CD4$^+$ T cells. The induction of both CTL and antibody responses by heat shock protein fusions, as surrogates for delivery of antigen into the MHC class I and class II pathways, respectively, presumably reflects the multiple pathways of heat shock protein trafficking within APCs [47, 48].

17.3. Clinical testing of a heat shock protein fusion protein, HspE7

Based on immunotherapeutic activity in a pre-clinical model of HPV-associated cancer, an *M. bovis* BCG Hsp65–HPV16 E7 fusion protein, HspE7 [32], has progressed into clinical testing in humans. HspE7 (formal designation SGN-00101) is being developed by Stressgen Biotechnologies Inc. (San Diego, CA, U.S.A.) for the treatment of diseases caused by HPV, including genital warts (GW), recurrent respiratory (or laryngeal) papillomatosis (RRP or warts of the respiratory tract), the cancer precursors known as cervical and anal intraepithelial neoplasia

(CIN and AIN respectively) and cervical cancer. HPV is ubiquitous in humans and is estimated to infect over 70 percent of the sexually active population [49]. Based on nucleotide sequence information, there are over one hundred geno-types of HPV [50]. Infection with a subset of 'high-risk' oncogenic HPV types, such as type 16, is associated with CIN, AIN and cancer, whereas infection with 'low-risk' HPV types 6 and 11 is commonly associated with GW and RRP [51].

In the clinical trials, HspE7 is being administered to patients by subcutaneous injection in a buffered saline vehicle. HspE7 treatment is well tolerated, with the most common adverse experience being an injection site reaction, typical of vac-cines, which resolves without treatment. An active treatment regimen for HspE7 has been identified: 500 µg given three times at monthly intervals. To date, the trial results in the HPV indications suggest that therapeutic immunisation with HspE7 is active in AIN, GW [52] and RRP. Clinical activity, measured as reduc-tion or complete resolution of indicated HPV lesions, is observed in a majority of patients receiving HspE7 immunisation and, when followed, is durable in a majority of patients for follow-up periods ranging from several months to two years. For patients with high-grade AIN and RRP, the response to HspE7 ther-apy means a potential reduction or avoidance of invasive surgical procedures. For GW patients, the durability of response to HspE7 therapy is potentially superior to that typically observed with use of caustic/ablative techniques or topical immunodulators.

The activity of HspE7, which contains HPV16 E7, against GW and RRP, which are caused primarily by HPV types 6 and 11, suggests induction of cross-reactive immunity to HPV E7 sequences present in non-HPV16 genotypes. This evidence supports the use of HspE7 as a broad-spectrum immunotherapeutic for diseases caused by multiple HPV genotypes. Other notable clinical findings include the activity of HspE7 in the HPV-infected adult (GW, AIN) and paediatric (RRP) populations, and activity in treating HPV-associated lesions of the anogenital skin and mucosa, and mucosa of the upper respiratory tract. Future clinical trials plan to test HspE7 in HIV$^+$ patients with anogenital HPV disease. The rationale for testing HspE7 in this population is directly supported by the observed activity of HspE7 in CD4-deficient mice [32].

17.4. Conclusions

Heat shock protein fusions, delivered as exogenous protein immunogens or as endogenously expressing nucleic acid and viral-based vectors, engender po-tent humoral and cellular immune responses. The literature demonstrates that immune responses can be induced with heat shock protein fusions containing antigen sequences that are 1) of varying lengths and character, 2) contained within natural flanking residues or as minimal epitopes and 3) fused to the N- or C-termini of heat shock protein sequences. In addition, antigen-specific immune responses have been demonstrated with heat shock protein fusion partners derived from 1) different heat shock protein families (e.g., Hsp90, Hsp70, Hsp60) and 2) prokaryotic and eukaryotic species, including the same species as the immunised recipient (i.e., 'self'). These observations speak to the breadth and reproducibility of this approach as a robust immunisation platform.

The induction of immune responses and, in particular, $CD8^+$ CTLs by heat shock protein fusions is an important phenomenon from both theoretical and practical standpoints. Firstly, it demonstrates efficient CTL priming to antigens by mechanisms apparently independent of the 'natural' chaperone function hypothesised to underlie CTL induction by non-covalent heat shock protein complexes [53]. Secondly, CTL responses, which are paramount for eradication of intracellular pathogens and transformed cells, are routinely achieved by heat shock protein fusions without adjuvant. This represents a new avenue of inquiry for developers of vaccines and immunotherapies, to whom decades of experience have indicated that CTL induction to soluble protein antigens requires adjuvants that pose safety and toxicity risks [54, 55]. Based on the pre-clinical and clinical experience described in this review, heat shock protein fusions offer considerable promise as a platform for future development of therapeutic vaccines to treat chronic viral infections and cancer.

Acknowledgements

We thank Lori Kernaghan for expert assistance in the preparation of this manuscript.

REFERENCES

1. Suzue K and Young R A. Heat shock proteins as immunological carriers and vaccines. In Fiege, U. (Ed.) *Stress-Inducible Cellular Responses*. Birkhauser Verlag, Basel 1996, pp 451–465.
2. Mustafa A S. HLA-restricted immune response to mycobacterial antigens: relevance to vaccine design. Human Immunol 2000, 61: 166–171.
3. Mizzen L. Immune responses to stress proteins: applications to infectious disease and cancer. Biotherapy 1998, 10: 173–189.
4. Murray P and Young R A. Stress and immunological recognition in host-pathogen interactions. J Bacteriol 1992, 174: 4193–4196.
5. Garbe T R. Heat shock proteins and infection: interactions of pathogen and host. Experientia 1992, 48: 635–639.
6. Gallucci S and Matzinger P. Danger signals: SOS to the immune system. Cur Opin Immunol 2001, 13: 114–119.

7. Asea A, Kraeft S-K, Kurt-Jones E A, Stevenson M A, Chen L B, Finberg R W, Koo G C and Calderwood S K. Hsp70 stimulates cytokine production through a CD14-dependent pathway, demonstrating its dual role as a chaperone and cytokine. Nat Med 2000, 6: 435–442.

8. Kol A, Lichtman A H, Finberg R W, Libby P and Kurt-Jones E A. Heat shock protein (HSP) 60 activates the innate immune response: CD14 is an essential receptor for HSP60 activation of mononuclear cells. J Immunol 2000, 164: 13–17.

9. Ohashi K, Burkart V, Flohé S and Kolb H. Heat shock protein 60 is a putative endogenous ligand of the Toll-like receptor-4 complex. J Immunol 2000, 164: 558–561.

10. Basu S, Binder R J, Ramalingam T and Srivastava P K. CD91 is a common receptor for heat shock proteins gp96, hsp90, hsp70 and calreticulin. Immunity 2001, 14: 303–313.

11. Vabulas R M, Ahmad-Nejad P, da Costa C, Miethke T, Kirschning C J, Hacker H and Wagner H. Endocytosed HSP60s use Toll-like receptor 2 (TLR2) and TLR4 to activate the toll/interleukin-1 receptor signaling pathway in innate immune cells. J Biol Chem 2001, 276: 31332–31339.

12. Wang Y, Kelly C G, Karttunen T, Whittall T, Lehner P J, Duncan L, MacAry P, Younson J S, Singh M, Oehlmann W, Cheng G, Bergmeier L and Lehner T. CD40 is a cellular receptor mediating mycobacterial heat shock protein 70 stimulation of CC-chemokines. Immunity 2001, 15: 971–983.

13. Becker T, Hartl F U and Wieland F. CD40, an extracellular receptor for binding and uptake of Hsp70-peptide complexes. J Cell Biol 2002, 158: 1277–1285.

14. Delneste Y, Magistrelli G, Gauchat J, Haeuw J, Aubry J, Nakamura K, Kawakami-Honda N, Goetsch L, Sawamura T, Bonnefoy J and Jeannin P. Involvement of LOX-1 in dendritic cell-mediated antigen cross-presentation. Immunity 2002, 17: 353–362.

15. Gross C, Hansch D, Gastpar R and Multhoff G. Interaction of heat shock protein 70 peptide with NK cells involves the NK receptor CD94. Biol Chem 2003, 384: 267–279.

16. Krieger M. The other side of scavenger receptors: pattern recognition for host defense. Cur Opin Lipidol 1997, 8: 275–280.

17. Janeway C A J and Medzhitov R. Innate immune recognition. Ann Rev Immunol 2002, 20: 197–216.

18. Multhoff G, Mizzen L, Winchester C C, Milner C M, Wenk S, Eissner G, Kampinga H H, Laumbacher B and Johnson J. Heat shock protein 70 (Hsp70) stimulates proliferation and cytolytic activity of natural killer cells. Exp Hematol 1999, 27: 1627–1636.

19. Yang Y and Wilson J M. CD40 ligand-dependent T cell activation: requirement of B7-CD28 signalling through CD40. Science 1996, 273: 1862–1864.

20. Kuppner M C, Gastpar R, Gelwer S, Nossner E, Ochmann O, Scharner A and Issels R D. The role of heat shock protein (hsp70) in dendritic cell maturation: hsp70 induces the maturation of immature dendritic cells but reduces DC differentiation from monocyte precursors. Eur J Immunol 2001, 31: 1602–1609.

21. Bethke K, Staib F, Distler M, Schmitt U, Jonuleit H, Enk A H, Galle P R and Heike M. Different efficiency of heat shock proteins to activate human monocytes and dendritic cells: Superiority of HSP60. J Immunol 2002, 169: 6141–6148.

22. Arnold D, Faath S, Rammensee H-G and Schild H. Cross-priming of minor histo-compatibility antigen-specific cytotoxic T cells upon immunization with the heat shock protein gp96. J Exp Med 1995, 182: 885–889.
23. Suto R and Srivastava P K. A mechanism for the specific immunogenicity of heat shock protein-chaperoned peptides. Science 1995, 269: 1585–1588.
24. Suzue K and Young R A. Adjuvant-free hsp70 fusion protein system elicits humoral and cellular immune responses to HIV-1 p24. J Immunol 1996, 156: 873–879.
25. Rico A I, Angel S O, Alonso C and Requena J M. Immunostimulatory properties of the *Leishmania infantum* heat shock proteins hsp70 and hsp83. Mol Immunol 1999, 36: 1131–1139.
26. Rico A I, Del Real G, Soto M, Quijada L, Martinez-A C, Alonso C and Requena J M. Characterization of the immunostimulatory properties of *Leishmania infantum* Hsp70 by fusion to the *Escherichia coli* maltose-binding protein in normal and nu/nu BALB/c mice. Infect Immun 1998, 66: 347–352.
27. Echeverria P, Dran G, Pereda G, Rico A I, Requena J M, Alonso C, Guarnera E and Angel S O. Analysis of the adjuvant effect of recombinant *Leishmania infantum* Hsp83 protein as a tool for vaccination. Immunol Lett 2001, 76: 107–110.
28. Bonorino C, Nardi N B, Zhang X and Wysocki L J. Characteristics of the strong antibody response to mycobacterial hsp70: a primary, T cell-dependent IgG response with no evidence of natural priming or γδ T cell involvement. J Immunol 1998, 161: 5210–5216.
29. Suzue K, Zhou X, Eisen H N and Young R A. Heat shock fusion proteins as vehicles for antigen delivery into the major histocompatibility complex class I presentation pathway. Proc Nat Acad Sci USA 1997, 94: 13146–13151.
30. Maranon C, Thomas M C, Planelles L and Lopez M C. The immunization of A2/K^b transgenic mice with the kmp11-hsp70 fusion protein induces CTL response against human cells expressing the T. cruzi kmp11 antigen: identification of A2-restricted epitopes. Mol Immunol 2001, 38: 279–287.
31. Anthony L S D, Wu H, Sweet H, Turnnir C, Boux L and Mizzen L A. Priming of CD8$^+$ CTL effector cells in mice by immunization with a stress protein – influenza virus nucleoprotein fusion molecule. Vaccine 1999, 17: 373–383.
32. Chu N R, Wu H B, Boux L J, Siegel M I and Mizzen L A. Immunotherapy of a human papillomavirus (HPV) type 16 E7-expressing tumour by administration of fusion protein comprising *Mycobacterium bovis* bacille Calmette-Guerin (BCG) hsp65 and HPV16 E7. Clin Exp Immunol 2000, 121: 216–225.
33. Cho B K, Palliser D, Guillen E, Wisniewski J, Young R A, Chen J and Eisen H N. A proposed mechanism for the induction of cytotoxic T lymphocyte production by heat shock fusion proteins. Immunity 2000, 12: 263–272.
34. Huang Q, Richmond J F, Suzue K, Eisen H N and Young R A. *In vivo* cytotoxic T lymphocyte elicitation by mycobacterial heat shock protein 70 fusion proteins maps to a discrete domain and is CD4$^+$ T cell independent. J Exp Med 2000, 191: 403–408.
35. Udono H, Yamano T, Kawabata Y, Ueda M and Yui K. Generation of cytotoxic T lymphocytes by MHC class I ligands fused to heat shock cognate protein 70. Int Immunol 2001, 13: 1233–1242.
36. Harmala L A E, Ingulli E G, Curtsinger J M, Lucido M M, Schmidt C S, Weigel B J, Blazar B R, Mescher M F and Pennell C A. The adjuvant effects of *Mycobacterium*

tuberculosis heat shock protein 70 result from the rapid and prolonged activation of antigen-specific CD8$^+$ T cells *in vivo*. J Immunol 2002, 169: 5622–5629.

37. Rahemtulla A, Fung-Leung W P, Schillham M W, Kundig T M, Sambhara S R, Narendran A, Arabian A, Wakeham A, Paige C J, Zinkernagel R M, Miller R G and Mak T W. Normal development and function of CD8$^+$ cells but markedly decreased helper cell activity in mice lacking CD4. Nature 1991, 353: 180–184.

38. Ridge J P, Di Rosa F and Matzinger P. A conditioned dendritic cell can be a temporal bridge between a CD4$^+$ T-helper and a T-killer cell. Nature 1998, 393: 474–478.

39. Gurunathan S, Klinman D M and Seder R A. DNA vaccines: immunology, application, and optimization. Ann Rev Immunol 2000, 18: 927–974.

40. Chen C-H, Wang T-L, Hung C-F, Yang Y, Young R A, Pardoll D M and Wu T-C. Enhancement of DNA vaccine potency by linkage of antigen gene to an hsp70 gene. Cancer Res 2000, 60: 1035–1042.

41. Cheng W F, Hung C F, Lin K Y, Juang J, He L, Lin C T and Wu T-C. CD8$^+$ T cells, NK cells and IFN-γ are important for control of tumor with downregulated MHC class I expression by DNA vaccination. Gene Therapy 2003, 10: 1311–1320.

42. Liu D-W, Tsao Y-P, Kung J T, Ding Y-A, Sytwu H-K, Xiao X and Chen S-L. Recombinant adeno-associated virus expressing human papillomavirus type 16 E7 peptide DNA fused with heat shock protein DNA as a potential vaccine for cervical cancer. J Virol 2000, 74: 2888–2894.

43. Cheng W-F, Hung C-F, Chai C-Y, Hsu K-F, He L, Rice C M, Ling M and Wu T-C. Enhancement of Sindbis virus self-replicating RNA vaccine potency by linkage of *Mycobacterium tuberculosis* heat shock protein 70 gene to an antigen gene. J Immunol 2001, 166: 6218–6226.

44. Hsu K-F, Hung C-F, Cheng W-F, He L, Slater L A, Ling M and Wu T-C. Enhancement of suicidal DNA vaccine potency by linking *Mycobacterium tuberculosis* heat shock protein 70 to an antigen. Gene Therapy 2001, 8: 376–383.

45. Plannelles L, Thomas M C, Alonso C and Lopez M C. DNA immunization with *Trypanosoma cruzi* Hsp70 fused to the kmp11 protein elicits a cytotoxic and humoral immune response against the antigen and leads to protection. Infect Immun 2001, 69: 6558–6563.

46. Cheng W F, Hung C F, Chai C Y, Hsu K F, He L, Ling M and Wu T C. Tumor-specific immunity and antiangiogenesis generated by a DNA vaccine encoding calreticulin linked to a tumor antigen. J Clin Invest 2001, 108: 669–678.

47. Wassenberg J J, Dezfulian C and Nicchitta C V. Receptor mediated and fluid phase pathways for internalization of the ER hsp90 chaperone grp94 in murine macrophages. J Cell Sci 1999, 112: 2167–2175.

48. Castellino F, Boucher P E, Eichelberg K, Mayhew M, Rothman J E, Houghton A N and Germain R N. Receptor-mediated uptake of antigen/heat shock protein complexes results in major histocompatibility complex class I antigen presentation via two distinct pathways. J Exp Med 2000, 191: 1957–1964.

49. Koutsky L. Epidemiology of genital human papillomavirus infection. Am J Med 1997, 102 (suppl 5A): 3–8.

50. Gissmann L, Osen W, Muller M and Jochmus I. Therapeutic vaccines for human papillomaviruses. Intervirology 2001, 44: 167–175.

51. Richart R M, Masood S, Syrjanen K J, Vassilakos P, Kaufman R H, Meisels A, Olszewski W T, Sakamoto A, Stoler M H, Vooijs G P and Wilbur D C. Human papillomavirus. IAC task force summary. Acta Cytologica 1998, 42: 50–58.
52. Goldstone S E, Palefsky J M, Winnett M T and Neefe J R. Activity of HspE7, a novel immunotherapy, in patients with anogenital warts. Dis Colon Rectum 2002, 45: 502–507.
53. Srivastava P K, Udono H, Blachere N E and Li Z. Heat shock proteins transfer peptides during antigen processing and CTL priming. Immunogenetics 1994, 39: 93–98.
54. Raychaudhuri S and Morrow W J W. Can soluble antigens induce CD8$^+$ cytotoxic T-cell responses? A paradox revisited. Immunol Today 1993, 14: 344–348.
55. Gupta R K and Siber G R. Adjuvants for human vaccines – current status, problems and future prospects. Vaccine 1995, 13: 1263–1276.

18

Molecular Chaperones as Inducers of Tumour Immunity

Pinaki P. Banerjee and Zihai Li

18.1. Introduction

Tumour antigens can be broadly classified into four categories: (i) those that are expressed in larger quantities in tumours than their normal counterparts (e.g., tumour-associated carbohydrate antigens) [1], (ii) onco-fetal antigens (e.g., carcinoembryonic antigen) [2], (iii) differentiation antigens (e.g., melanoma differentiation antigen) [3, 4] and (iv) tumour-specific antigens. Tumour antigens in the first three categories could serve as useful markers for diagnostic and prognostic purposes. Although some of these antigens are being used in immunotherapy, none can be called tumour-specific in a true sense. Only the last group includes antigens that are truly specific for tumour cells, in that they contain tumour-specific mutations that are unique for individual tumours such as the tumour-specific point mutation that is found in cyclin-dependent kinase-4. Such a mutation gives rise to a novel antigenic epitope which can be recognised by cytotoxic T lymphocytes (CTLs) [5]. However, for these antigens to be of any value as therapeutic agents, they must be detected in and epitopes isolated from a large range of cancers, and this makes the general use of these antigens difficult.

In the past two decades, evidence has accumulated to support the concept that molecular chaperones or heat shock proteins can be used as a potent source of cancer vaccines [6, 7]. Molecular chaperones, particularly those derived from the Hsp70 and Hsp90 families, are now being tested in the clinical arena for therapeutic efficacy against a range of cancers (Table 18.1). The concept of vaccinating against cancer using molecular chaperones developed from an observation which was made half a century ago, namely that mice immunised with a particular type of irradiated tumour cells were protected against challenge with live tumour cells from the same, but not a different kind, of tumour [8] (Figure 18.1). It was evident that, although tumour A and tumour B were induced by

Table 18.1. Gp96 in clinical trials for different types of cancer

Heat shock protein vaccine	Type of cancer	Status
gp96	Renal cell carcinoma	Phase III
	Melanoma	Phase III
	Colorectal cancer	Phase II
	Gastric cancer	Phase I and II
	Pancreatic cancer	Phase II
	Sarcoma	Phase II
	Ovarian cancer	Phase I

the same carcinogen, were derived from the same histological type or were even present in the same host, the antigenic determinants of each of these tumours appeared to be distinct (see review by Li et al. [9]).

It was postulated that each tumour bears unique tumour-specific transplantation antigens (TSTAs) that could be used to induce protective responses against a particular type of tumour, but not others [10]. A pioneering experiment in which the immunogenic TSTA fraction from tumour lysates was purified led to the identification of gp96 and other heat shock proteins as the protective agents against autologous tumours [11]. However, gp96 is not immunogenic *per se*; rather the immunogenicity of the preparation is defined by novel tumour-derived peptides that are carried by the gp96 and are co-purified with it [12, 13]. Immunisation with gp96 elicits a potent CTL response against the tumour from which gp96 is isolated, the specificity of which is defined by the gp96-associated peptides [14–17]. It has since been shown that gp96 devoid of peptides can activate $CD8^+$ T cells in a dose-dependent manner [18] and can provide co-stimulation to $CD4^+$ T cells which results in the generation of a type 2 helper T cell (Th2) immune response [19]. This chapter will review features and mechanisms underlying the capacity of autologous tumour-derived gp96 and other heat shock proteins to elicit tumour-specific immunity and act as anti-cancer vaccines.

18.2. Application of extracellular heat shock protein gp96 in cancer immunity

18.2.1. Autologous, soluble gp96 as a cancer vaccine

Gp96 was first discovered as the glucose-regulated 'stress' protein [20], the primary function of which was to serve the role of a molecular chaperone [21].

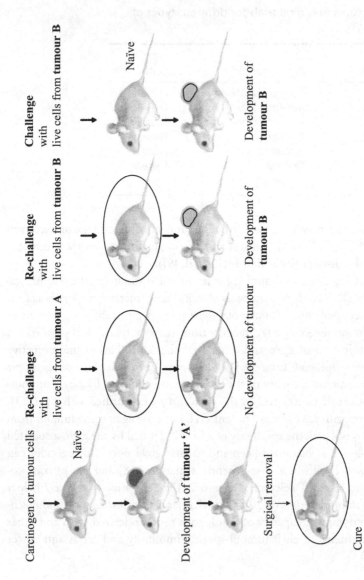

Figure 18.1. When cured surgically, mice bearing tumour A (encircled) became resistant to the challenge from the same tumour cells, but much like the naïve animals remained susceptible to tumour B. This experiment symbolises the existence of tumour antigens and also the generation of immune response against these antigens.

Because this protein is found to be regulated by glucose levels it is, therefore, also known as the 94-kDa glucose-regulated protein (grp94) [22]. This protein is a major glycoprotein resident in the endoplasmic reticulum (ER) and is thus named endoplasmin [23] or 99-kDa endoplasmic reticulum protein (Erp99) [24]. However, under stressful conditions, gp96 tends to redistribute to the Golgi apparatus [25], is found to be enriched to some extent in the nucleus [20] and can also be expressed on the outer surface of the plasma membrane [26]. It has been suggested that the glycosylation pattern of gp96 changes after cellular stress, as denoted by an increased resistance to endoglycosidase H digestion, which depends on the cell types [25]. However, this phenomenon has also been observed in several disease states such as cancer [27], thereby suggesting that cells might be in a state of stress which also results in partial translocation of ER chaperones to the Golgi apparatus.

The protective immunity elicited by gp96 vaccination is exquisitely specific. However, questions have been raised about the polymorphism of the molecule and also there has been doubt about the existence of the mutated form of gp96 in cancer cells, either of which could be a tumour antigen. Molecular cloning and sequencing approaches have rejected these two hypotheses and categorically mapped only one true gene locus, named tra-1 in humans [28]. Sequencing of gp96 from normal and tumour tissues has demonstrated absolute sequence homology between gp96 from the two sources, thereby confirming gp96 as being solely a carrier of immunogenic peptide from the tumour tissues.

Much of the recent studies on the immunologic properties of gp96 have focussed on understanding the interaction between purified gp96 and a variety of immune effector cells including T cells and professional antigen-presenting cells (APCs). APCs such as dendritic cells (DCs) and macrophages carry the receptors of gp96 on the cell surface, one of which is CD91, or α-2 macroglobulin (α-2M) receptor [29]. Binding of tumour-derived gp96 to CD91 leads to a receptor-mediated endocytosis of gp96 complete with its associated tumour-derived peptides (gp96-tumour peptide), and this forms the basis of the initial critical step leading to the induction of tumour peptide-specific immunity [30]. The receptor–ligand interaction activates, matures and leads to the cross-presentation of tumour peptides via major histocompatibility complex (MHC) class I molecules by the DCs. Once activated, DCs from the peripheral tissues migrate to the lymph node in which CD8[+] T cell–DC interactions result in the priming and activation of peptide-specific T cells (Figure 18.2). The induction of tumour immunity by gp96 could be a representation of 'danger theory' itself [31]. To clarify this proposal, stressful conditions such as glucose deprivation, hypoxia and acidosis might induce the release of gp96-peptide complexes from the tumour cells, which acts as a danger signal to the immune system. This

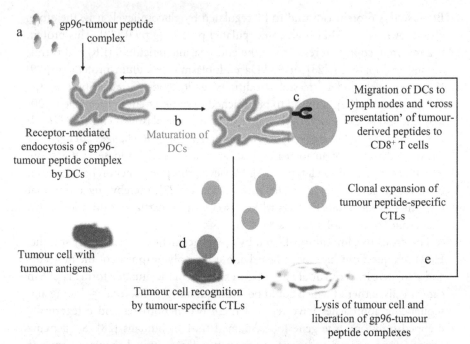

Figure 18.2. Generation of CTL responses by gp96 vaccination. (a) Gp96–tumour peptide complexes can (b) mature DCs and these DCs (c) generate a potent CTL response by cross-presentation of the peptide. Effector CTLs (d) migrate to the site of the tumour to elicit tumour-specific cytolysis. The gp96–tumour peptide from lysed cancer cells (e) can further interact with DCs to prime the anti-tumour response.

extracellular gp96 might activate macrophages and DCs to initiate a cascade of downstream events that result in the induction of an effective anti-tumour immune response.

Intradermal injection of as little as 1 μg tumour-derived gp96 can protect against tumour challenge. The 'mathematical derivation' behind this phenomenon of gp96 is indeed very interesting to note [32]. In brief, such an immunisation incites 10,000 activated DCs to migrate to the lymph node and results in the presentation of an estimated 3,000 MHC tumour-specific peptide complexes to CD8+ cells. As commented by Srivastava, 'one tenth as many DCs presenting one tenth as many antigenic peptides are powerful enough to elicit a potent T cell response' [32]. Other routes of gp96 immunisation such as subcutaneous and intraperitoneal approaches are also effective, however only when 10- to 50-fold higher doses of gp96 are used, respectively [33]. It has been interesting to note that, irrespective of the route of administration, doses of soluble gp96 above or below the optimal do not elicit tumour-protective responses. Higher doses of gp96 can lead to the generation of transferable suppressive T cell subsets that bear CD4$^+$ molecules on the cell surface. A detailed insight into the mechanisms underlying this particular phenomenon has yet to be offered. However, these findings highlight the need for more insight into the immunologic properties of gp96.

18.2.2. Surface expression of gp96 and its role in CTL response

Gp96 contains the C-terminal KDEL (Lys-Asp-Glu-Leu) motif, which is known as the ER retention signal [34]. However, it has been estimated that about 3% of the total amount of gp96 synthesised could be expressed on the surface of the Meth-A tumour cells [35]. Cell surface expression of gp96 has also been observed in a variety of other cell types (see the review by Li et al. [9]). In each case, surface expression was not due to cell death, because surface expression was clearly dependent on the active transport from ER to Golgi. It has been hypothesised that the export of gp96 from the ER to the cell surface is a critical signal to the immune system. To test this concept, we have targeted gp96 to the surfaces of tumour cells by transfection with an engineered gp96 expression vector [36]. The engineered gp96, named 96tm, does not possess the KDEL motif and to the C-terminus of gp96 was attached a transmembrane domain of a platelet-derived growth factor receptor.

Surface expression of 96tm does not lead to changes in the level of endogenous gp96, nor does it have any effect on the folding or transport of MHC class I molecules. 96tm was targeted onto cell surfaces of many tumour types in a type I orientation. A number of important observations were made with these reagents [36, 37]: (i) the immunogenicity of Meth-A fibrosarcomas and CT-26 colon cancers expressing cell surface gp96 is increased; (ii) co-culturing of bone marrow (BM)-derived immature DCs with tumour cells expressing gp96 on their cell surface induces phenotypic maturation of DCs as evidenced by upregulation of CD80, CD86, CD40, MHC class I and MHC class II expression; (iii) DCs activated by cell surface gp96 produce a large amount of pro-inflammatory cytokines such as TNF-α, IL-1β, IL-12 and the chemokine MCP-1; (iv) DC maturation induced by 96tm-expressing tumour cells requires direct cell-to-cell contact; (v) 96tm-expressing tumour cells recruit both CD4$^+$ and CD8$^+$ cells to the site of tumours *in vivo*; (vi) tumour cells expressing cell surface gp96 can cross-prime antigen-specific CD8$^+$ T cells more efficiently and (vii) immunisation with tumour cell expressing cell surface gp96 leads to the development of long-lasting protective CD4$^+$ and CD8$^+$ T cell memory. Thus, over-expression of surface gp96 can also be an effective and alternative strategy for inducing effective cell-mediated anti-tumour immunity against less immunogenic tumours.

Using a similar strategy, the immunological consequence of constitutive expression of gp96 on cell surfaces was examined by generating a gp96tm transgenic mouse model [38]. Interestingly, these animals spontaneously develop a DC-mediated autoimmune disease, which highlights the intrinsic pro-inflammatory properties of gp96. This transgenic murine model might prove invaluable for further mechanistic study of gp96 in immune responses.

18.3. A comparison of the functional mechanism between soluble gp96 and gp96tm

In general, a protective CTL response can be sequentially described by three consecutive events: initiation, maintenance and the induction of immunological memory. Soluble gp96 and membrane-bound gp96 might elicit CTL responses in a subtle, but different mechanism [9, 39]. Important functional differences in immune responses induced by gp96-tm and naturally occurring cell surface gp96 might also exist, and these might be revealed following whole-cell immunisation. Invariably, the chemistry underlying different types of immune responses relies on the nature of the interaction between cells and the gp96 molecule and its associated peptides/proteins in the vaccine preparations. Thus, it would be pertinent to address and also to compare the involvement of various APCs and T cells in the different phases of the anti-tumour immune response which is induced by different forms of gp96-based vaccines.

In the priming phase, the requirement of $CD8^+$ T cells, but not the $CD4^+$ T cells, is important, especially when animals are immunised with the soluble gp96-peptide complex [39]. This phenomenon is found to be largely dependent on the presence of carrageenan-susceptible cells, which could be macrophages and other, as yet to be identified APCs. In contrast, the priming of CTLs following immunisation with irradiated cancer cell requires the presence of $CD4^+$ T cells. During the effector phase, both T cell subsets are important for effective immunisation in response to tumour-derived gp96 or irradiated tumour cells. Much like the priming phase, macrophages play a pivotal role in the CTL effector response induced by purified gp96.

An interesting observation is that although immunisation with soluble protein generally initiates the $CD4^+$ T cell response, the depletion of $CD4^+$ T cells during gp96 administration does not affect the generation of tumour immunity in the host. This might be due to the fact that gp96-tumour peptide is targeted directly to macrophages and DCs [30], which cross-present the 'antigen' to $CD8^+$ cells. Thus, depletion of macrophages from the immune milieu compromises the antigenic response. Both soluble gp96 [40] and membrane-bound 96tm [37] have been shown to elicit memory anti-tumour immune responses.

In the case of 96tm immunisation, both CD4$^+$ and CD8$^+$ T cells are required for tumour rejection in the priming as well as memory phase. The requirement for CD4$^+$ T cells in the generation of CD8$^+$ T cell memory to soluble gp96 immunisation, in vivo, has yet to be carefully studied. Suffice to say, both the membrane-bound and the soluble gp96 can elicit a potent anti-tumour immunity. However, the mechanisms under which immune responses are generated are differentially regulated and the mechanistic differences have yet to be elucidated.

18.4. Immunotherapy with other forms of gp96

In addition to the soluble and membrane-bound forms of gp96, gp96-Ig, the secreted form of gp96 from tumours, has also attracted considerable interest for its potency in cancer therapy [41, 42]. A gp96-Ig fusion protein has been made by replacing the KDEL sequence with CH2 and CH3 domains of the murine IgG$_1$ molecule [43]. This construct was transfected to the ovalbumin-expressing E.G7 lymphoma cells, and this tumour model was used as a source of vaccine. Gp96-Ig secreting E.G7 cells, but not gp96-Ig negative control cells, were able to prime adoptively transferred T cell receptor transgenic CD8$^+$ T cells specific for the MHC class I–restricted SIINFEKL peptide sequence of ovalbumin. To determine whether the N terminus or the C terminus are crucial for gp96-mediated peptide presentation, the C terminus of gp96 has been fused with a CTL epitope, and this had a limited effects on the capacity to prime CTL responses [44]. Covalent fusion of gp96 with peptides might inhibit the presentation of these peptides via MHC class I molecules, because heat shock protein-tumour peptides require enzymatic processing before being presented via proteosome-dependent or -independent pathways [45, 46]. Some believe that the N-terminal domain of gp96 does not have peptide binding capability, but that it might induce cross-protection in various tumours [47], whereas others believe that the N-terminal domain of gp96 can indeed carry tumour peptide [13].

Gp96-Ig can efficiently elicit anti-tumour responses in the absence of CD4$^+$ T cells and macrophages [43], and gp96-Ig and gp96-tm are more effective than soluble gp96 when mice are challenged with 5×10^5 live Meth A tumour cells, a dose which is five-fold greater than the maximum number of cells up to which immunity mediated by soluble gp96 can be retained. Gp96-tm has been shown to be slightly more effective than gp96-Ig in a tumour rejection experiment [37].

To increase the efficiency of gp96 in tumour rejection, it was thought that it was necessary to enhance the macrophage/DC population in vivo at the time of gp96 vaccine administration. Thus, the administration of Lewis lung cancer (LLC) cells transduced with granulocyte macrophage colony stimulating factor (GM-CSF) in combination with 1 μg tumour-derived gp96 was found to be more effective than either one administered separately [48]. This strategy of combination vaccination can mature, activate and increase the number of the DCs in the draining lymph node within 36 hours after vaccination. Further, this LLC-GM-gp96 therapy also depends on the optimum amount of gp96. DC-mediated activation of CD8$^+$ T lymphocytes is also CD4$^+$ T cell– and natural killer (NK) cell–dependent. Markers of DC maturation (MHC class II, B7.2) are up-regulated by gp96 with no significant change in CD40 expression level. However, the expression of gp96 receptor (CD91) is regarded as being constitutively expressed and unrelated to the presence or absence of gp96 immunisation [48].

18.5. Role of CD91 and gp96 in innate and adaptive immunity

From the preceding discussion, the question might be asked how gp96-mediated anti-tumour immunity is regulated, especially given that its receptor is constitutively expressed on DCs. Attempts to elucidate this fact revealed that cell lysates prepared from necrotic cells, but not apoptotic cells, can deliver a maturation signal to DCs via the NF-κB pathway, and that it is through this route that gp96 delivers it signals to the DCs [49]. Apoptotic cells are known to bear apoptotic markers such as phosphatidylserine (PS) on their cell surface and for which DCs bear the receptor (PSR). Engagement of PS with PSR down-regulates the expression of co-stimulatory molecules by DCs and concomitantly increases the secretion of anti-inflammatory cytokine transforming growth factor (TGF)-β [50]. Other receptors such as CD14 and CD36 recognise apoptotic cell-associated ligands and also transmit similar tolerogenic signals to macrophages and DCs. However, during necrosis the binding of gp96 to its receptor on DCs elicits the secretion of IL-12 and TNF-α, and also up-regulates the expression of co-stimulatory molecules. In the peripheral circulation, any potential effects of heat shock protein–peptide complexes can be neutralised by the presence of α-2M, another ligand for CD91. However, in the tissues, in which α-2M is not present, CD91 becomes accessible to gp96 released from damaged or stressed cells [51].

It is important to point out that CD91 is not the only receptor which is proposed to bind to heat shock proteins. A CD91-independent pathway has been reported to mediate cross-presentation of gp96-chaperoned peptides [52]. Toll-like receptors (TLRs) [53] and scavenger receptor-A [54] are such candidate receptors for gp96. The receptors for heat shock proteins are discussed in detail in Chapters 7, 8 and 10. Furthermore, the observation that gp96 devoid of peptides can activate DCs strongly indicates the co-existence of innate immunity in tumour rejection [55]. Binding of gp96 to macrophages and DCs can elicit the secretion of IL-12, TNF-α, type-1 interferon, IL-1β and GM-CSF. IL-12 activates NK cells which lyse tumour cells directly. On tumour cell lysis, gp96 and its associated peptides are released, and this binds to CD91 on neighbouring APCs, thereby generating an adaptive immune response. Thus, as is the case with other heat shock proteins, gp96 bridges innate and adaptive immune responses, and this capacity probably developed as early as the emergence of vertebrates [7, 41, 56–58]. To exemplify, autologous tumour-derived gp96- and Hsp70-based immunisation induces significant protection against a transplantable tumour in *Xenopus*, the earliest vertebrate to possess an adaptive immune system [59].

18.6. The role of other heat shock protein family members in anti-cancer immunity and a comparison with gp96

Soon after the establishment of gp96 as a tumour 'antigen', attention focussed on the possibility that other molecular chaperones exhibited similar properties. Two of the Hsp90 family members, named 84- and 86-kDa TSTA (i.e., p84 and p86), are also immunogenic for cancer [60]. Unlike gp96, these molecules are devoid of sugar moieties and do not bind to lectins; however, much like gp96 these molecules do not bear tumour-specific polymorphic DNA sequences and are found to protect only against autologous tumours. The overall homology between p84/86 and gp96 is 49%. Interestingly, at the lowest effective dose the immunogenicity of these molecules is dependent on the purification process. Thus, fractions purified using Mono-Q columns offer comparatively better protection against a lethal challenge with Meth A cells than those purified using hydroxyapatite. However, in contrast to gp96, both the purified fractions of p84/86 offered a similar, almost 80%, protection when used in higher dose (i.e., 10–40 μg). Much like p84/86, the immunogenicity of Hsp70 also depends on its preparation protocol. Thus, the replacement of ADP-agarose from the ATP-agarose in the chromatography column completely restores the ability of the Hsp70(-peptide) to elicit anti-tumour immunity [61, 62]. ATP binding and hydrolysis have been shown to release Hsp70-associated peptides [63] and the loss of immunogenicity after ATP treatment provides strong evidence that the peptides chaperoned by Hsp70 are the immunogens [62] as, for example, Hsp70 chaperoned tyrosinase peptide [64].

18.7. Mechanisms underlying Hsp70-mediated tumour regression

To determine which portion of the Hsp70 is essential for the generation of the most potent CTL response against the peptide linked to it, five different class I restricted CTL epitopes have been covalently linked to either the N or C terminus of Hsc70 (the constitutive member of the Hsp70 family) and these have been immunised to mice via different routes [44]. In contrast to gp96-tumour peptide immunisation, intravenous rather than intradermal injection of Hsp70-tumour peptide is the most effective route of immunisation. When compared to a dose of 1 μg, the intravenous administration of 10 μg chaperone–peptide complex generates a much more vigorous CTL response, as determined by the number of IFN-γ–producing cells. The N-terminal region and the C-terminal region flanking peptides are equally effective as far as intravenous immunisation is concerned. However, in an overall 'score' in the CTL assay, and depending on different routes of immunisation, the C-terminal-bound flanking peptides are more efficient than those bound at the N-terminal end.

Studies attempting to elucidate the mechanism resulting in Hsp70-tumour peptide-mediated CTL responses have shown that, much like gp96-peptide, carragenan-susceptible cells but not CD4$^+$ T cells are essential for the cytolytic activity and have demonstrated that Hsp70 binds to newly processed peptides within as little as 1 hour at 25 °C [62]. An elegant study to identify the CTL epitopes of HLA-B46 positive cancer patients resulted in the discovery of two oligopeptides, nine amino acids long, derived from a truncated form of Hsc70 (81% homologous). The sequence of these epitopes varied to a small extent from the wild-type form and was mapped to a position outside of the peptide binding domain [65]. Further, the recognition of surface Hsp72 (a family member of Hsp70) by a distinct population of NK cells has also been documented in the literature [66–69]. Taken together, these findings confirm the probable role of Hsp70 in eliciting the innate and adaptive immunity in tumour regression.

18.8. The recipient of TAP-transported peptides as another group of chaperone vaccines

From the data discussed so far, it is evident that any of the three heat shock proteins (gp96, Hsp90 or Hsp70) could be used as a source of cancer vaccine against autologous tumours. However, to choose the best, a comparison between these three heat shock proteins has been undertaken using administration by the subcutaneous route [70]. This demonstrated that Hsp70 and gp96 are of equal efficiency, whereas Hsp90 is only 10% as effective as its counterparts. Thus, whereas 9 μg gp96 or Hsp70–peptide complex protects mice from tumour challenge, the dose of Hsp90 required for similar protection is 90 μg. Interestingly all three of these heat shock proteins bind equal amounts of peptide. The molecular and cellular bases for the differential immunogenicity of these proteins are unclear.

It is proposed that Hsp70 and Hsp90 act as peptide transporters, in that they carry their peptide cargo to the transporter associated with antigen processing

(TAP), whereas gp96 and calreticulin facilitate the binding of the MHC class I-$\beta 2$ microglobulin peptide complex in the ER. This reasoning suggests that, much like gp96, other recipients of the TAP-transported peptides such as calreticulin (CRT) and the 170-kDa glucose regulated protein grp170 might also act as tumour vaccines. Grp170 has been tested as a vaccine in a metastatic tumour model, in which it was found to be effective in terms of reducing the number of metastatic colonies and growth of further tumour challenge [71].

CRT has also been found to bind tumour peptides and upon subcutaneous immunisation CRT and gp96 are comparable in their protective efficiency against a tumour challenge [72]. As CRT-tumour peptides can be cross-presented by DCs to generate a CTL response, it is therefore suggested to be the key mechanism behind CRT-induced anti-tumour immunity. Interestingly, DNA vaccination of CRT, or CRT-E7 (CRT DNA linked to HPV-16 DNA) equally reduces pulmonary tumour nodules due to an inhibition of bFGF-induced angiogenesis. This 'T cell–independent tumour regression' has been further proven in a nude mice model in which a slight advantage of CRT-E7 DNA vaccination over the CRT vaccinated mice has been observed [73].

CRT, Hsp70, Hsp90 and gp96 bind to the same receptor on DCs, CD91. CRT-tumour peptides exhibit the greatest capacity to induce IFN-γ secretion from antigen-specific CTLs *in vitro*, and this is almost twice that of gp96 [74]. Taken together, these data clarify the involvement of innate and adaptive immunity in CRT-mediated tumour rejection. The capacity of CRT in the cross-presentation experiment is comparable to Hsp70, whereas the efficiency of Hsp90 lies in between that of gp96 and Hsp70 [70]. Further, a comparative analysis of heat shock protein binding to CD11b$^+$ cells has also shown that, even at the very high concentration (200 μg/ml), Hsp70 binding to CD11b$^+$ cells is not saturated, whereas Hsp90 and gp96 were saturated at this concentration [75]. This observation suggests either a high affinity of Hsp70 for CD91 or the existence of another receptor for Hsp70 on CD11b$^+$ cells. Indeed, a new receptor for Hsp70 (LOX-1) from the group of the scavenger receptor family has now been documented in the literature [76]. The existence of other receptors for Hsp70 argue against the observations that an anti-CD91 antibody can completely abrogate the re-presentation of Hsp70–peptide complex [74] and that Toll-like receptor (TLR-2 and TLR-4) pathways are involved in gp96-mediated DC activation which is preceded by endocytosis [30, 53, 77].

The *in vivo* data provide insight into the relative merits of different heat shock proteins as anti-tumour vaccines and it would appear that gp96 has a number of advantages given the low doses required to induce CTL responses and tumour rejection. Indeed, gp96 is currently being evaluated in a number of clinical trials (Table 18.1). Hsp70 is also being evaluated in a phase II clinical trial. Studies into the therapeutic potential of other heat shock proteins such as Hsp60, Hsp40 and Hsp27 have also been initiated; however, the intense focus on gp96 has to date overshadowed their contribution [78].

18.9. Conclusion

It is evident that gp96 is not a tumour antigen, nor 'probably' are any of the heat shock proteins; rather the nature of the immune response induced by them is dictated by the peptides with which they are associated. In general, heat shock protein–tumour peptide complexes elicit CTL responses against a wide variety of cancers [15, 79, 80]. However, the mechanisms of action of each heat shock protein and their various forms (membrane bound, soluble or secreted) have yet to be elucidated in any great detail. For example, the mechanistic basis for the distinct $CD4^+$ T cell requirement by soluble and membrane-bound gp96 is unclear. Does each heat shock protein engage with same or different sets of receptors? Do heat shock proteins differ in their abilities to activate innate and adaptive immunity? Despite these uncertainties, these chaperone–peptide complexes exhibit unique advantages over other vaccines as follows:

- They represent the entire repertoire of peptides generated in a tumour cell, thereby abolishing the need to identify and isolate immunogenic tumour epitopes.
- Their use is not restricted to a particular type of cancer, or to patients with a particular MHC haplotype.
- The risks that are associated with the use of transforming DNA, attenuated organisms or immunosuppressive factors such as TGF-β are eliminated.

Finally, with little or no treatment-related toxicity, tumour-derived heat shock proteins likely fulfil all of the immunological criteria to justify their clinical evaluation as cancer vaccines.

REFERENCES

1. Livingston P. Ganglioside vaccines with emphasis on GM2. Semin Oncol 1998, 25: 636–645.
2. Greiner J W, Zeytin H, Anver M R and Schlom J. Vaccine-based therapy directed against carcinoembryonic antigen demonstrates antitumor activity on spontaneous intestinal tumors in the absence of autoimmunity. Cancer Res 2002, 62: 6944–6951.
3. Boon T, Cerottini J C, van den Eynde B, van der Bruggen P and van Pel A. Tumor antigens recognized by T lymphocytes. Ann Rev Immunol 1994, 12: 337–365.

4. Denkberg G, Lev A, Eisenbach L, Benhar I and Reiter Y. Selective targeting of melanoma and APCs using a recombinant antibody with TCR-like specificity directed toward a melanoma differentiation antigen. J Immunol 2003, 171: 2197–2207.

5. Wolfel T, Hauer M, Schneider J, Serrano M, Wolfel C, Klehmann-Hieb E, De Plaen E, Hankeln T, Meyer zum Buschenfelde K H and Beach D. A p16INK4a-insensitive CDK4 mutant targeted by cytolytic T lymphocytes in a human melanoma. Science 1995, 269: 1281–1284.

6. Srivastava P K. Immunotherapy of human cancer: lessons from mice. Nat Immunol 2000, 1: 363–366.

7. Srivastava P. Roles of heat-shock proteins in innate and adaptive immunity. Nat Rev Immunol 2002, 2: 185–194.

8. Foley E J. Antigenic properties of methylcholanthrene-induced tumors in mice of the strain of origin. Cancer Res 1953, 13: 835–837.

9. Li Z, Dai J, Zheng H, Liu B and Caudill M. An integrated view of the roles and mechanisms of heat shock protein gp96-peptide complex in immune response. Frontiers Biosci 2002, 7: 731–751.

10. Prehn R T and Main J M. Immunity to methylcholanthrene-induced sarcomas. J Nat Cancer Inst 1957, 18: 769–778.

11. Srivastava P K, DeLeo A B and Old L J. Tumor rejection antigens of chemically induced sarcomas of inbred mice. Proc Natl Acad Sci USA 1986, 83: 3407–3411.

12. Li Z and Srivastava P K. Tumor rejection antigen gp96/grp94 is an ATPse: implications for protein folding and antigen presentation. EMBO J 1993, 12: 3143–3151.

13. Linderoth N A, Popowicz A and Sastry S. Identification of the peptide-binding site in the heat shock chaperone/tumor rejection antigen gp96 (Grp94). J Biol Chem 2000, 275: 5472–5477.

14. Arnold D, Faath S, Rammensee H-G and Schild H. Cross-priming of minor histocompatibility antigen-specific cytotoxic T cells upon immunization with the heat shock protein gp96. J Exp Med 1995, 182: 885–889.

15. Blachere N E, Li Z L, Chandawarkar R Y, Suto R, Jaikaria N S, Basu S, Udono H and Srivastava P K. Heat shock protein-peptide complexes, reconstituted in vitro, elicit peptide-specific cytotoxic T lymphocyte response and tumor immunity. J Exp Med 1997, 186: 1315–1322.

16. Blachere N E and Srivastava P K. Heat shock protein-based cancer vaccines and related thoughts on immunogenicity of human tumors. Semin Cancer Biol 1995, 6: 349–355.

17. Nieland T J F, Tan M C A A, Monee-van Muijen M, Koning F, Kruisbeek A M and van Bleek G M. Isolation of an immunodominant viral peptide that is endogenously bound to the stress protein GP96/GRP94. Proc Natl Acad Sci USA 1996, 93: 6135–6139.

18. Breloer M, Fleischer B and von Bonin A. In vivo and in vitro activation of T cells after administration of Ag-negative heat shock proteins. J Immunol 1999, 162: 3141–3147.

19. Banerjee P P, Vinay D S, Mathew A, Raje M, Parekh V, Prasad D V, Kumar A, Mitra D and Mishra G C. Evidence that glycoprotein 96 (B2), a stress protein, functions as a Th2-specific costimulatory molecule. J Immunol 2002, 169: 3507–3518.

20. Welch W J, Garrels J I, Thomas G P, Lin J J and Feramisco J R. Biochemical characterization of the mammalian stress proteins and identification of two stress proteins as glucose- and Ca^{2+}-ionophore-regulated proteins. J Biol Chem 1983, 258: 7102–7111.

21. Csermely P, Schnaider T, C. S, Prohászka Z and Nardai G. The 90-kDa molecular chaperone family: structure, function, and clinical applications. A comprehensive review. Pharmacol Ther 1998, 79: 129–168.

22. Lee A S. The accumulation of three specific proteins related to glucose-regulated proteins in a temperature-sensitive hamster mutant cell line K12. J Cell Physiol 1981, 106: 119–125.

23. Koch G, Smith M, Macer D, Webster P and Mortara R. Endoplasmic reticulum contains a common, abundant calcium-binding glycoprotein, endoplasmin. J Cell Sci 1986, 86: 217–222.

24. Lewis M J, Mazzarella R A and Green M. Structure and assembly of the endoplasmic reticulum. The synthesis of three major endoplasmic reticulum proteins during lipopolysaccharide-induced differentiation of murine lymphocytes. J Biol Chem 1985, 260: 3050–3057.

25. Booth C and Koch G L. Perturbation of cellular calcium induces secretion of luminal ER proteins. Cell Biol Int 1989, 59: 729–737.

26. Teriukova N P, Tiuriaeva I I, Grandilevskaia A B and Ivanov V A. The detection of membrane tumor-associated antigens of Zajdela's hepatoma on the surface of cultured rat cells. Tsitologiia 1997, 39: 577–581.

27. Feldweg A M and Srivastava P K. Molecular heterogeneity of tumor rejection antigen/heat shock protein GP96. Int J Cancer 1995, 63: 310–314.

28. Maki R G, Eddy R L J, Byers M, Shows T B and Srivastava P K. Mapping of the genes for human endoplasmic reticular heat shock protein gp96/grp94. Somat Cell Mol Genetics 1993, 19: 73–81.

29. Binder R J, Han D K and Srivastava P K. CD91: a receptor for heat shock protein gp96. Nat Immunol 2000, 1: 151–155.

30. Singh-Jasuja H, Toes R E M, Spee P, Münz C, Hilf N, Schoenberger S P, Ricciardi-Castagnoli P, Neefjes J, Rammensee H-G, Arnold-Schild D and Schild H. Cross-presentation of glycoprotein 96-associated antigens on major histocompatibility complex molecules requires receptor-mediated endocytosis. J Exp Med 2000, 191: 1965–1974.

31. Matzinger P. Tolerance, danger, and the extended family. Ann Rev Immunol 1994, 12: 991–1045.

32. Srivastava P. Interaction of heat shock proteins with peptides and antigen presenting cells: chaperoning of the innate and adaptive immune responses. Ann Rev Immunol 2002, 20: 395–425.

33. Chandawarkar R Y, Wagh M S and Srivastava P K. The dual nature of specific immunological activity of tumour-derived gp96 preparations. J Exp Med 1999, 189: 1437–1442.

34. Munro S and Pelham H R. A C-terminal signal prevents secretion of luminal ER proteins. Cell Biol Int 1987, 48: 899–907.

35. Altmeyer A, Maki R G, Feldweg A M, Heike M, Protopopov V P, Masur S K and Srivastava P K. Tumor-specific cell surface expression of the KDEL containing, endoplasmic reticular heat shock protein gp96. Int J Cancer 1996, 69: 340–349.

36. Zheng H, Dai J, Stoilova D and Li Z. Cell surface targeting of heat shock protein gp96 induces dendritic cell maturation and antitumor immunity. J Immunol 2001, 167: 6731–6735.

37. Dai J, Liu B, Caudill M, Zheng H, Qiao Y, Podack E R and Li Z. Cell surface expression of heat shock protein gp96 enhances cross-presentation of cellular antigens and the generation of tumor-specific T cell memory. Cancer Immun 2003, 3: 1–5.

38. Liu B, Dai J, Zheng H, Stoilova D, Sun S and Li Z. Cell surface expression of an endoplasmic reticulum resident heat shock protein gp96 triggers MyD88-dependent systemic autoimmune diseases. Proc Nat Acad Sci USA 2003, 100: 15824–15829.

39. Udono H, Levey D L and Srivastava P K. Cellular requirements for tumor-specific immunity elicited by heat shock proteins: tumor rejection antigen gp96 primes CD8+ T cells in vivo. Proc Natl Acad Sci USA 1994, 91: 3077–3081.

40. Janetzki S, Blachere N E and Srivastava P K. Generation of tumor-specific cytotoxic T lymphocytes and memory T cells by immunization with tumor-derived heat shock protein gp96. J Immunotherapy 1998, 21: 269–276.

41. Strbo N, Oizumi S, Sotosek-Tokmadzic V and Podack E R. Perforin is required for innate and adaptive immunity induced by heat shock protein gp96. Immunity 2003, 18: 381–390.

42. Strbo N, Yamazaki K, Lee K, Rukavina D and Podack E R. Heat shock fusion protein gp96-Ig mediates strong CD8 CTL expansion in vivo. Am J Reprod Immunol 2002, 48: 220–225.

43. Yamazaki K, Nguyen T and Podack E R. Tumour secreted heat shock-fusion protein elicits CD8 cells for rejection. J Immunol 1999, 163: 5178–5182.

44. Udono H, Yamano T, Kawabata Y, Ueda M and Yui K. Generation of cytotoxic T lymphocytes by MHC class I ligands fused to heat shock cognate protein 70. Int Immunol 2001, 13: 1233–1242.

45. Binder R J, Blachere N E and Srivastava P K. Heat shock protein-chaperoned peptides but not free peptides introduced into the cytosol are presented efficiently by major histocompatibility complex I molecules. J Biol Chem 2001, 276: 17163–17171.

46. Castellino F, Boucher P E, Eichelberg K, Mayhew M, Rothman J E, Houghton A N and Germain R N. Receptor-mediated uptake of antigen/heat shock protein complexes results in major histocompatibility complex class I antigen presentation via two distinct pathways. J Exp Med 2000, 191: 1957–1964.

47. Baker-LePain J C, Sarzotti M, Fields T A, Li C Y and Nicchitta C V. GRP94 (gp96) and GRP94 N-terminal geldanamycin binding domain elicit tissue nonrestricted tumor suppression. J Exp Med 2002, 196: 1447–1459.

48. Kojima T, Yamazaki K, Tamura Y, Ogura S, Tani K, Konishi J, Shinagawa N, Kinoshita I, Hizawa N, Yamaguchi E, Dosaka-Akita H and Nishimura M. Granulocyte-macrophage colony-stimulating factor gene-transduced tumor cells combined with tumor-derived gp96 inhibit tumor growth in mice. Human Gene Therapy 2003, 14: 715–728.

49. Basu S, Binder R J, Suto R, Anderson K M and Srivastava P K. Necrotic but not apoptotic cell death releases heat shock proteins, which deliver a partial maturation signal to dendritic cells and activates the NF-κB pathway. Int Immunol 2000, 12: 1539–1546.

50. Huynh M L, Fadok V A and Henson P M. Phosphatidylserine-dependent ingestion of apoptotic cells promotes TGF-β1 secretion and the resolution of inflammation. J Clin Invest 2002, 109: 41–50.

51. Schild H and Rammensee H-G. gp96 – the immune system's Swiss army knife. Nat Immunology 2000, 1: 100–101.

52. Berwin B, Hart J P, Pizzo S V and Nicchitta C V. CD91-independent cross-presentation of grp94(gp96)-associated peptides. J Immunol 2002, 168: 4282–4286.

53. Vabulas R M, Braedel S, Hilf N, Singh-Jasuja H, Herter S, Ahmad-Nejad P, Kirschning C J, Da Costa C, Rammensee H G, Wagner H and Schild H. The endoplasmic reticulum-resident heat shock protein Gp96 activates dendritic cells via the Toll-like receptor 2/4 pathway. J Biol Chem 2002, 277: 20847–20853.

54. Berwin B, Hart J P, Rice S, Gass C, Pizzo S V, Post S R and Nicchitta C V. Scavenger receptor-A mediates gp96/GRP94 and calreticulin internalization by antigen-presenting cells. EMBO J 2003, 22: 6127–6136.

55. Nicchitta C V. Re-evaluating the role of heat-shock protein-peptide interactions in tumour immunity. Nat Rev Immunol 2003, 3: 427–432.

56. Bausinger H, Lipsker D and Hanau D. Heat-shock proteins as activators of the innate immune system. Trends Immunol 2002, 23: 342–343.

57. 57. Gaston J S H. Heat shock proteins and innate immunity. Clin Exp Immunol 2002, 127: 1–3.

58. Wallin R P A, Lundqvist A, Moré S H, von Bonin A, Kiessling R and Ljunggren H-G. Heat-shock proteins as activators of the innate immune system. Trends Immunol 2002, 23: 130–135.

59. Robert J, Gantress J, Rau L, Bell A and Cohen N. Minor histocompatibility antigen-specific MHC-restricted CD8 T cell responses elicited by heat shock proteins. J Immunol 2002, 168: 1697–1703.

60. Ullrich S J, Robinson E A, Law L W, Willingham M and Appella E. A mouse tumor-specific transplantation antigen is a heat shock-related protein. Proc Natl Acad Sci USA 1986, 83: 3121–3125.

61. Peng P, Ménoret A and Srivastava P K. Purification of immunogenic heat shock protein 70-peptide complexes by ADP-affinity chromatography. J Immunol Methods 1997, 204: 13–21.

62. Udono H and Srivastava P K. Heat shock protein 70-associated peptides elicit specific cancer immunity. J Exp Med 1993, 178: 1391–1396.

63. Flynn G C, Chappell T G and Rothman J E. Peptide binding and release by proteins implicated as catalysts of protein assembly. Science 1989, 245: 385–390.

64. Noessner E, Gastpar R, Milani V, Brandl A, Hutzler P J, Kuppner M C, Roos M, Kremmer E, Asea A, Calderwood S K and Issels R D. Tumor-derived heat shock protein 70 peptide complexes are cross-presented by human dendritic cells. J Immunol 2002, 169: 5424–5432.

65. Azuma K, Shichijo S, Takedatsu H, Komatsu N, Sawamizu H and Itoh K. Heat shock cognate protein 70 encodes antigenic epitopes recognised by HLA-B4601-restricted cytotoxic T lymphocytes from cancer patients. Brit J Cancer 2003, 89: 1079–1085.

66. Gehrmann M, Schmetzer H, Eissner G, Haferlach T, Hiddemann W and Multhoff G. Membrane-bound heat shock protein 70 (Hsp70) in acute myeloid leukemia: a tumor specific recognition structure for the cytolytic activity of autologous NK cells. Haematologica 2003, 88: 474–476.

67. Gross C, Koelch W, DeMaio A, Arispe N and Multhoff G. Cell surface-bound heat shock protein 70 (Hsp70) mediates perforin-independent apoptosis by specific binding and uptake of granzyme B. J Biol Chem 2003, 278: 41173–41181.

68. Moser C, Schmidbauer C, Gurtler U, Gross C, Gehrmann M, Thonigs G, Pfister K and Multhoff G. Inhibition of tumor growth in mice with severe combined

immunodeficiency is mediated by heat shock protein 70 (Hsp70)-peptide-activated, CD94 positive natural killer cells. Cell Stress Chaperon 2002, 7: 365–373.

69. Multhoff G. Activation of natural killer cells by heat shock protein 70. Int J Hyperthermia 2002, 18: 576–585.

70. Udono H and Srivastava P K. Comparison of tumor-specific immunogenicities of stress-induced proteins gp96, hsp90 and hsp70. J Immunol 1994, 152: 5398–5403.

71. Wang X Y, Kazim L, Repasky E A and Subjeck J R. Immunization with tumor-derived ER chaperone grp170 elicits tumor-specific CD8⁺ T-cell responses and reduces pulmonary metastatic disease. Int J Cancer 2003, 105: 226–231.

72. Basu S and Srivastava P. Calreticulin, a peptide-binding chaperone of the endoplasmic reticulum, elicits tumor- and peptide-specific immunity. J Exp Med 1999, 189: 797–802.

73. Cheng W F, Hung C F, Chai C Y, Hsu K F, He L, Ling M and Wu T C. Tumor-specific immunity and antiangiogenesis generated by a DNA vaccine encoding calreticulin linked to a tumor antigen. J Clin Invest 2001, 108: 669–678.

74. Basu S, Binder R J, Ramalingam T and Srivastava P K. CD91 is a common receptor for heat shock proteins gp96, hsp90, hsp70 and calreticulin. Immunity 2001, 14: 303–313.

75. Binder R J, Harris M L, Ménoret A and Srivastava P K. Saturation, competition, and specificity in interaction of heat shock proteins (hsp) gp96, hsp90, and hsp70 with CD11b⁺ cells. J Immunol 2000, 165: 2582–2587.

76. Delneste Y, Magistrelli G, Gauchat J, Haeuw J, Aubry J, Nakamura K, Kawakami-Honda N, Goetsch L, Sawamura T, Bonnefoy J and Jeannin P. Involvement of LOX-1 in dendritic cell-mediated antigen cross-presentation. Immunity 2002, 17: 353–362.

77. Vabulas R M, Wagner H and Schild H. Heat shock proteins as ligands of toll-like receptors. Cur Topics Microbiol Immunol 2002, 270: 169–184.

78. Ménoret A and Bell G. Purification of multiple heat shock proteins from a single tumor sample. J Immunol Methods 2000, 237: 119–130.

79. Suto R and Srivastava P K. A mechanism for the specific immunogenicity of heat shock protein-chaperoned peptides. Science 1995, 269: 1585–1588.

80. Tamura Y, Peng P, Liu K, Daou M and Srivastava P K. Immunotherapy of tumors with autologous tumor-derived heat shock protein preparations. Science 1997, 278: 117–120.

Extracellular Biology of Molecular Chaperones: What Does the Future Hold?

19

Gazing into the Crystal Ball: The Unfolding Future of Molecular Chaperones

Lawrence E. Hightower

Predicting the future of the exobiology of molecular chaperones is bound to be risky business: after all, unravelling the intracellular lives of the chaperones has become legendary for its unexpected twists and turns. Can we expect differently for their extracellular capers? I cannot claim the clearest crystal, but I do have a unique perspective on the field from my perch as Editor-in-Chief of the major specialty journal in the field, *Cell Stress & Chaperones*. I will refer to papers in recent issues that will lead interested readers to other papers in key areas that I believe provide insights into the future as well. Perhaps we can begin to illuminate the crystal ball by listing major unsolved problems and by identifying the disciplines of the investigators that these problems are now attracting into the field.

One of the exciting and renewing aspects of the heat shock field, as it was known historically, has been the succession of colleagues from different disciplines that have entered and moved the field forward. The chance initial finding of the heat shock response in *Drosophila* by Ritossa in 1962 [1] was pursued by a small group of *Drosophila* biologists until about 1978 when the response was discovered in a variety of other organisms. Molecular geneticists were attracted to the heat shock genes as models of inducible eukaryotic gene expression, and the field took on a more global interest. The 1982 Cold Spring Harbor Heat Shock Meeting: From Bacteria to Man was dominated by molecular geneticists describing gene organisation and transcription, chromatin structure and regulation in several systems besides *Drosophila* [2]. There were a few talks from biochemists, cell biologists and a smattering of physiologists, the vanguard of many more to come who would work out the function of the heat shock proteins, create the names 'molecular chaperones', 'chaperonins' and 'co-chaperones', and establish the physiological setting of cellular stress responses, as the heat shock response came to be known. Progress by the biochemists in the purification and characterisation of the substantial amounts of molecular chaperones attracted structural biologists, and we got our first three-dimensional look at chaperones at the atomic level (see Chapter 1).

Joining this rich interdisciplinary mix of investigators who continue to build on a now solid foundation of basic science, we now find investigators interested in translational research including medical scientists focused on specific diseases along with a few clinicians and M.D./Ph.D. colleagues who are beginning to take the field from the laboratory bench to the patient bedside. Cellular immunologists in particular are having a major impact. So, here is the first light in our crystal ball: the future will bring more research on molecular chaperones in human biology and disease, the effects of environmental stress on human populations, and the use of animal model systems, particularly those with sequenced genomes, to study the complex biology of stress response physiology and cytoprotection, the morphed sibling of thermotolerance and *Drosophila* phenocopy protection.

Our crystal brightens more, because colleagues with translational interests in our field work in biotechnology companies as well as in academia, and my second prediction is that they will succeed in producing vaccines against cancers and dysplasias of viral origin that are both therapeutic and preventative. Some of these vaccines will be directed against the tumour cells of individual patients; that is, our field will spawn one of the first really dramatic successes in personalised medicine (see Chapters 17 and 18). In addition, the process of stress-conditioning to induce tissue cytoprotection will become part of pre-operative patient care, at least for elective surgeries, and new drugs will be marketed to stimulate in some cases and turn off in others the stress response in humans. Perhaps some of the new drugs needed to accomplish these tasks will be based on the cytokine activities of molecular chaperones, knowledge of their extracellular receptors, and the signal transduction pathways to which they are linked. My confidence in these predictions is high and we have opened for 2004 a new section in *Cell Stress & Chaperones* entitled 'Stress Response Translational Research'.

The path to translational research on molecular chaperones is not quite the linear one just outlined. Two of the talks on thermotolerance and heat shock proteins at the 1982 meeting were given by colleagues crossing over from the discipline of radiation oncology. The clinical use of hyperthermia in cancer treatment to kill cancer cells is a line of translational research which has paralleled and sometimes crossed paths with the cellular stress response field. Researchers on both paths have come to view thermotolerance or cytoprotection as a key cellular state that needs to be understood and controlled, and both have come to appreciate the central roles of heat shock proteins in this altered state of cellular physiology. There is renewed interest in the clinical application of mild hyperthermia and fever-like responses, and both groups share a fascination with apoptosis. What will the future hold? I note that the Cell Stress Society International and the North American Hyperthermia Society held a joint meeting in Québec in 2003, The First International Congress on Stress Responses in Biology and Medicine. My third prediction is that more than a few productive collaborations on translational research in our field will trace back to this seminal international meeting.

Perhaps the major unanswered questions are: How do stress proteins egress from cells, and how are they taken up by other cells? Path-breaking work in the 1980s initially identified heat shock proteins among the set known as glia-axon transfer proteins, demonstrating that at least a couple of the heat shock proteins can be transferred cell-to-cell and showed that these same heat shock proteins are released from cultured mammalian cells by a non-ER–Golgi mechanism [3, 4]. Hsp70, Hsc70 and Hsp110 were released from heat-stressed newborn rat cell cultures, and Hsc70 was released from non-heat-stressed cultures, possibly stimulated by medium changes during the experimental protocol [4]. Actin was also released into the medium and transferred cell-to-cell, raising the possibility of a microfilament-aided release or transfer mechanism. The release of actin by a non-ER–Golgi pathway was reported independently by Rubenstein and colleagues [5].

There appear to be at least three general ways in which a nucleocytoplasmic protein like Hsp/Hsc70 might become extracellular. *In vivo*, it might be released locally by tissue damage during wounding (i.e., tissue trauma). If the injury is large enough, sufficient amounts could be released or taken into blood vessels to create a system distribution. Secondly, there appears to be a release mechanism via a non-ER–Golgi pathway which might be very sensitive to triggering by breaks in tissue homeostasis. And finally, there appears to be a cell-to-cell transfer mechanism which might be particularly important in nervous tissue in which glial cells capable of producing heat shock proteins might transfer them to neuronal cells that do not during physiological stress. Tytell and colleagues have pursued a possible therapeutic application for the transfer of protection by introducing purified Hsp70 into neuronal cells as a model for treating spinal cord injury [6].

A portion of Hsp70 is associated with plasma membranes, as early localisation studies in *Drosophila* showed. In fact, the propensity of Hsp/Hsc70 to bind fatty acids and to interact with lipid bi-layers in interesting ways, to produce channels for example, expands to possibilities for egress and uptake mechanisms. It will no doubt take a while longer to sort out these mechanisms, even using model cells such as yeast, in which non-ER–Golgi release has also been observed (See Chapter 3). What is very obvious now is that once heat shock proteins are released from cells, they find receptors on other cells, particularly cells of the immune system and tumour cells, on which they have biological effects. In addition to nucleocytoplasmic heat shock proteins, their relatives in the endoplasmic reticulum, members of the glucose-regulated protein set, can escape, presumably by an ER–Golgi secretory pathway, to contribute to these biological effects.

Another approach to illuminating our crystal is to try to predict some of the keywords that might come to dominate our field. For homeothermic vertebrates, I would include inflammation, vascular endothelium and innate or natural immunity among the key search terms of the future. Whereas external temperature cues appear to dominate the thermal induction of the heat shock response of poikilotherms, it is more likely that the heat generated during localised tissue inflammation, along with a host of other inducers, acts in concert and probably synergistically to activate the heat shock or cellular stress response most frequently in homeotherms. We can also include systematic responses such as fever and the effects of elevated blood levels of pro-inflammatory cytokines on the vascular endothelium. It has been clear since some of the earliest work on the induction of heat shock proteins in the tissues of heat-shocked rats by F. P. White [7] that cells associated with blood vessels are among the most responsive in a broad range of tissues. Innate immunity is considered to be the most ancient of the arms of the immune response, and it appears that a very ancient and evolutionarily well-conserved set of proteins, the heat shock proteins, has become intertwined in this rather non-specific first line of defence against microbes.

The components of innate immunity include physical barriers such as skin, mucosal epithelia and the anti-microbial molecules that they produce such as defensins, the complement system, macrophages, neutrophils and natural killer cells. Some potentially interesting overlaps in our keywords are now apparent. Components of innate immunity contribute to inflammatory responses, and the recruitment of neutrophils into areas of tissue damage requires a regulated response from the vascular endothelium in the vicinity of the wounded or inflamed tissue. In these venues, extracellular molecular chaperones have ample opportunity to exercise their activities as chemokines to regulate tissue-level inflammatory processes. This will be a rich hunting ground for additional responsive cells, more receptors and signalling pathways regulated by extracellular chaperones.

Coincidentally, the issue of *Cell Stress & Chaperones* which is most current to the writing of this chapter contains three articles in succession that would have appeared in our futuristic keyword search. The search term 'inflammation' would have brought up an article by Barton and colleagues providing immuno-histochemical evidence of Hsp32 (hemoxygenase 1) in normal and inflamed human stomach and colon [8]. Hsp32 was detected in inflammatory cells and gastric epithelial cells of normal human gastric and colonic mucosa. It was expressed at higher levels in inflamed gastric mucosa, independent of *Helicobacter pylori* infection, and was particularly high in inflamed colon samples from patients with active ulcerative colitis [8]. In rats, increased expression of Hsp32 has been correlated with decreased inflammation in chemically induced colitis [9]. Whether or not Hsp32 plays a similar regulatory role in the types of chronic human inflammatory diseases described in this study remains to be determined.

The search term 'vascular endothelium' would identify the next paper in this issue, a study by Kabakov and colleagues on the cytoprotective effects of over-expressing either Hsp70 or Hsp27 in human endothelial cells [10]. The authors used an *in vitro* model of ischaemia-reperfusion injury, as may occur during myocardial infarctions and strokes, to show that the expression of either of these heat shock proteins within the first six hours of post-hypoxic re-oxygenation results in significant reductions in endothelial cell apoptosis [10]. This is a more realistic test of the potential for therapeutic intervention than previous animal models in which stress conditioning has been applied prior to the experimental induction of a myocardial infarct or stroke, since these are not predictable events that lead to the initial ischaemia. I fully agree with the authors' contention that protection of the vascular endothelium from oxidative damage and apoptosis is likely one of the most important considerations in facilitating recovery with minimal tissue damage.

The third paper in this sequence describes the finding that Hsp70 reactivity independent of major histocompatibility complex (MHC) class I was associated with increased densities of the C-type lectin receptor CD94 and the neuronal adhesion molecule CD56 on the surface of human primary natural killer (NK) cells following stimulation by the peptide terminal localised Hsp70 sequence 'TKDnnllgrfelsg' (TKD, 99 450–463) [11]. This paper by Multhoff and colleagues provides a glimpse into the exobiology of the human molecular chaperone Hsp70. The cell surface receptors CD94 and CD56 in primary NK cells were selectively upregulated following treatment with the Hsp70 peptide TKD. Elevated densities of these two receptors were correlated with an increased cytolytic response of these activated NK cells against target tumour cells displaying Hsp70 on their surface membranes. An important point to note about all three of these papers is that they employ human cells to address possible links between heat shock proteins and human diseases – they represent translational research contributions. This trend in the molecular chaperone field toward translational research will likely expand in the near future.

REFERENCES

1. Ritossa F A. A new puffing pattern induced by temperature shock and DNP in *Drosophila*. Experientia 1962, 18: 571–573.
2. Schlesinger M J, Tissières A and Ashburner M (Eds.). *Heat Shock, from Bacteria to Man*. Cold Spring Harbor Laboratory Press 1982.
3. Tytell M, Greenberg S G and Lasek R J. Heat shock-like protein is transferred from glia to axon. Brain Res 1986, 363: 161–164.
4. Hightower L E and Guidon P T. Selective release from cultured mammalian cells of heat-shock (stress) proteins that resemble glia-axon transfer proteins. J Cell Physiol 1989, 138: 257–266.
5. Rubenstein P, Ruppert T and Sandra A. Selective isoactin release from cultured embryonic skeletal muscle cells. J. Cell Biol 1982, 92: 164–169.
6. Tidwell J L, Houenou L J and Tytell M. Administration of Hsp70 *in vivo* inhibits motor and sensory neuron degeneration. Cell Stress Chaperon 2004, 9: 88–98.
7. Currie R W and White F P. Characterization of the synthesis and accumulation of a 71-kilodalton protein induced in rat tissues after hyperthermia. Can J Biochem Cell Biol 1983, 61: 438–446.
8. Barton S G R G, Rampton D S, Winrow V R, Domizio P and Feakins R M. Expression of heat shock protein 32 (hemoxygenase-1) in the normal and inflamed human stomach and colon: an immunohistochemical study. Cell Stress Chaperon 2003, 8: 329–334.
9. Wang W P, Guo X, Koo M W, Wong B C, Lam S K, Ye Y N and Cho C H. Protective role of heme oxygenase-1 on trinitrobenzene sulfonic acid-induced colitis in rats. Am J Physiol Gastrointest Liver Physiol 2001, 281: G586–594.
10. Kabakov A E, Budagova K R, Bryantsev A L and Latchman D S. Heat shock protein 70 or heat shock protein 27 overexpressed in human endothelial cells during posthypoxic reoxygenation can protect from delayed apoptosis. Cell Stress Chaperon 2003, 8: 335–347.
11. Gross C, Schmidt-Wolf I G H, Nagaraj S, Gastpar R, Ellwart J, Kunz-Schughart L A and Multhoff G. Heat shock protein 70-reactivity is associated with increased cell surface density of CD94/CD56 on primary natural killer cells. Cell Stress Chaperon 2003, 8: 348–360.

Index